GRADE 8

STP MATHEMATICS
for Jamaica

F W Ali   S Chandler   A Shepherd

L Bostock   C E Layne   E Smith

Nelson Thornes

Contents

Contents

Introduction

This book attempts to satisfy your needs as you begin your study of mathematics in the secondary school. We are very conscious of the need for success together with the enjoyment everyone finds in getting things right. With this in mind we have divided most of the exercises into three types of question:

> The first type, identified by plain numbers, e.g. **12**, helps you to see if you understand the work. These questions are considered necessary for every chapter you attempt.

> The second type, identified by a single underline, e.g. <u>12</u>, are extra, but not harder, questions for quicker workers, for extra practice or for later revision.

> The third type, identified by a double underline, e.g. <u><u>12</u></u>, are for those of you who manage Type 1 questions fairly easily and therefore need to attempt questions that are a little harder.

Most chapters end with 'mixed exercises'. These will help you revise what you have done, either when you have finished the chapter or at a later date.

All of you should be able to use a calculator accurately by the time you leave school. It is wise, in your first and second years, to use it mainly to check your answers. Whether you use the calculator or do the working yourself, always estimate your answer and always ask yourself the question, 'Is my answer a reasonable one?'

Preface

To the teacher

The general aims of the series are:

(1) to help students to
- attain solid mathematical skills
- connect mathematics to their everyday lives and understand its role in the development of our contemporary society
- see the importance of thinking skills in everyday problems
- discover the fun of doing mathematics and reinforce their positive attitudes to it.

(2) to encourage teachers to include historical information about mathematics in their program.

In writing this three-book series the authors attempted to present topics in such a way that students will understand the connections in mathematics, and be encouraged to see and use mathematics as a means to help make sense in the real world.

Topics from the history of mathematics have been incorporated to ensure that mathematics is not dissociated from its past. This should lead to an increase in the levels of enthusiasm, interest and fascination for mathematics, and should also enrich the teaching of it.

Careful grading of exercises makes the books approachable.

Some suggestions:

(1) Before each lesson give a brief outline of the topic to be covered in the lesson. As examples are given, refer back to the outline to show how the example fits into it.

(2) List terms on the chalkboard that you consider new to the students. Solicit additional words from the class and encourage students to read from the text and make their own vocabulary.
Remember that mathematics is a foreign language. The ability to communicate mathematically must involve the careful use of the correct terminology.

(3) When possible have students construct alternative ways to phrase questions. This ties in with seeing mathematics as a language. Students tend to concentrate on the numerical or 'maths' part of the question and pay little attention to the instructions, which give information that is required to solve the problem.

(4) When solving problems have students identify their own problem-solving strategies and listen to others. This practice should create an atmosphere of discussion in the class centred around different approaches to the same problem.

As the students try to solve problems on their own they will make mistakes. This is healthy, as this was the experience of the inventors of mathematics: they tried, guessed, made many mistakes and worked for hours, days and sometimes years before reaching a solution.

There are enough problems in the exercises to allow the students to try and try again. The excitement, disappointment and struggle with a problem until a solution is found provide a healthy classroom atmosphere.

To the student

These books are written for you. As you study:

Try to break up the material in a chapter into manageable bits.

Always have paper and pencil when you study mathematics.

When you meet a new word write it down together with its meaning.

Read your questions carefully and rephrase them in your own words.

The information that you need to solve your problem is given in the wording of the problem, not the number part only.

Your success in mathematics may be achieved through practice.

You are therefore advised to try to solve as many problems as you can.

Always try more problems than those set by your teacher for homework.

Remember that the greatest cricketer or netball player became great by practising for many hours.

We have provided enough problems in the books to allow you to practise.

Above all do not be afraid to make mistakes as you are learning. The greatest mathematicians all made many mistakes as they tried to solve problems.

You are now on your way to success in mathematics – GOOD LUCK!

Picture credits

Pascal machine, page 119, Poyet/La Nature 1904/MEPL

1 Directed numbers

At the end of this chapter you should be able to...

1 Plot points with positive and negative coordinates.

2 Use positive or negative numbers to describe displacements on one side or the other of a given point on a line.

3 Apply positive and negative numbers, where appropriate, in a physical situation.

4 Perform operations of addition, subtraction, multiplication or division on positive and negative numbers.

5 Multiply expressions with directed numbers.

6 Solve linear equations involving directed numbers.

Did you know?

David Blackwell is, to mathematicians, the most famous, perhaps the greatest, African American mathematician. A biannual prize has been inaugurated in his honour: the Blackwell-Tapia Prize. The first recipient was Dr Arlie O. Petters. Dr Petters emigrated from Belize to the United States in 1979 and became a US citizen in 1990.

You need to know...

✔ how to add, subtract, multiply and divide simple numbers mentally

✔ the basic properties of the square, rectangle and parallelogram

Key words

diagonal, directed number, negative, negative coordinate, number line, parallelogram, positive, positive coordinate, quadrilateral, rectangle, square

1

Negative coordinates

If A(2, 0), B(4, 2) and C(6, 0) are three corners of a square ABCD, we can see
that the fourth corner, D, is two squares below the *x*-axis.

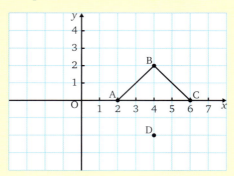

To describe the position of D we need to extend the scale on the *y*-axis below zero.
To do this we use the negative numbers

$$-1, -2, -3, -4, \ldots$$

In the same way we can use the negative numbers −1, −2, −3, … to extend
the scale on the *x*-axis to the left of zero.

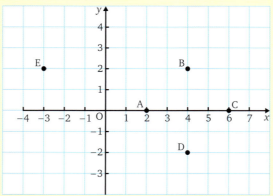

The *y*-coordinate of the point D is written −2 and is called 'negative 2'.

The *x*-coordinate of the point E is written −3 and is called 'negative 3'.

The numbers 1, 2, 3, 4, … are called positive numbers. They could be written
as +1, +2, +3, +4, … but we do not usually put the + sign in.

Now D is   4 squares to the right of O so its *x*-coordinate is 4
and   2 squares below the *x*-axis so its *y*-coordinate is −2,

D is the point (4, −2)

E is   3 squares to the left of O so its *x*-coordinate is −3
and   2 squares up from O so its *y*-coordinate is 2,

E is the point (−3, 2)

Exercise 1a

Use this diagram for questions 1 and 2.

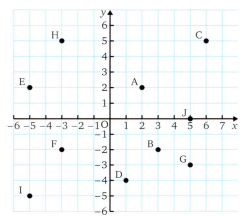

1   Write down the *x*-coordinate of each of the points A, B, C, D, E, F, G, H, I, J and O (the origin).

2   Write down the *y*-coordinate of each of the points A, B, C, D, E, H, I and J.

How many squares above or below the *x*-axis is each of the following points?

3   P: the *y*-coordinate is −5

4   L: the *y*-coordinate is +3

5   M: the *y*-coordinate is −1

6   B: the *y*-coordinate is 10

7   A: the *y*-coordinate is 0

8   D: the *y*-coordinate is −4

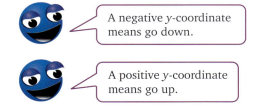

A negative *y*-coordinate means go down.

A positive *y*-coordinate means go up.

How many squares to the left or to the right of the *y*-axis is each of the following points?

9   Q: the *x*-coordinate is 3

10 R: the *x*-coordinate is −5

11 T: the *x*-coordinate is +2

12 S: the *x*-coordinate is −7

13 V: the *x*-coordinate is 0

14 G: the *x*-coordinate is −9

15 Write down the coordinates of the points A, B, C, D, E, F, G, H, I and J.

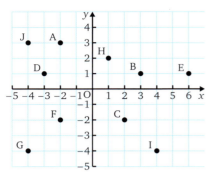

The *x*-coordinate comes first, followed by the *y*-coordinate. Each coordinate may be positive or negative.

In questions **16** to **21** draw your own set of axes and scale each one from −5 to 5:

16 Mark the points A(−3, 4),   B(−1, 4),   C(1, 3),   D(1, 2), E(−1, 1),   F(1, 0),   G(1, −1),   H(−1, −2),   I(−3, −2). Join the points in alphabetical order and join I to A.

17 Mark the points A(4, −1),   B(4, 2),   C(3, 3),   D(2, 3), E(2, 4),   F(1, 4),   G(1, 3),   H(−2, 3),   I(−3, 2), J(−3, −1). Join the points in alphabetical order and join J to A.

18 Mark the points A(2, 1),   B(−1, 3),   C(−3, 0),   D(0, −2). Join the points to make the figure ABCD. What is the name of the figure?

19 Mark the points A(1, 3),   B(−1, −1),   C(3, −1). Join the points to make the figure ABC and describe ABC.

20 Mark the points A(−2, −1),   B(5, −1),   C(5, 2),   D(−2, 2). Join the points to make the figure ABCD and describe ABCD.

21 Mark the points A(−3, 0),   B(1, 3),   C(0, −4). What kind of triangle is ABC?

Exercise 1b

Draw your own set of axes for each question in this exercise. Mark a scale on each axis from −10 to +10.

In questions **1** to **10** mark the points A and B and then find the length of the line AB:

1 A(2, 2) B(−4, 2)   **4** A(1, −6) B(1, −8)

2 A(−2, −1) B(6, −1)   **5** A(3, 2) B(5, 2)

3 A(−4, −4) B(−4, 2)   **6** A(5, −1) B(5, 6)

7 A(−2, 4)   B(−7, 4)   **9** A(−3, 5)   B(−3, −6)

8 A(−1, −2)   B(−8, −2)   **10** A(−2, −4)   B(−2, 7)

In questions **11** to **20**, the points A, B and C are three corners of a square ABCD. Mark the points and find the point D. Give the coordinates of D:

11 A(1, 1)   B(1, −1)   C(−1, −1)   **16** A(−3, −1)   B(−3, 2)   C(0, 2)

12 A(1, 3)   B(6, 3)   C(6, −2)   **17** A(0, 4)   B(−2, 1)   C(1, −1)

13 A(3, 3)   B(3, −1)   C(−1, −1)   **18** A(1, 0)   B(3, 2)   C(1, 4)

14 A(−2, −1)   B(−2, 3)   C(−6, 3)   **19** A(−2, −1)   B(2, −2)   C(3, 2)

15 A(−5, −3)   B(−1, −3)   C(−1, 1)   **20** A(−3, −2)   B(−5, 2)   C(−1, 4)

In questions **21** to **30**, mark the points A and B and the point C, the midpoint of the line AB. Give the coordinates of C:

21 A(2, 2)   B(6, 2)   **26** A(2, 1)   B(6, 2)

22 A(2, 3)   B(2, −5)   **27** A(2, 1)   B(−4, 5)

23 A(−1, 3)   B(−6, 3)   **28** A(−7, −3)   B(5, 3)

24 A(−3, 5)   B(−3, −7)   **29** A(−3, 3)   B(3, −3)

25 A(−1, −2)   B(−9, −2)   **30** A(−7, −3)   B(5, 3)

Straight lines

Exercise 1c

1

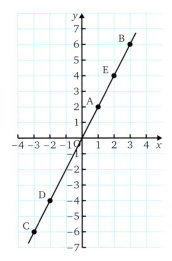

The points A, B, C, D and E are all on the same straight line.

a   Write down the coordinates of the points A, B, C, D and E.

b   F is another point on the same line. The *x*-coordinate of F is 5.
   Write down the *y*-coordinate of F.

c G, H, I, J, K, L and M are also points on this line. Fill in the missing coordinates: G(8, □)   H(10, □)   I(−4, □)   J(□, 12)   K(□, 18) L(□, −10)   M(a, □).

The points A, B, C, D, E, F and G are all on the same straight line.

a Write down the coordinates of the points A, B, C, D, E, F and G.

b How is the y-coordinate of each point related to its x-coordinate?

c H is another point on this line. Its x-coordinate is 8; what is its y-coordinate?

d I, J, K, L, M, N are further points on this line. Fill in the missing coordinates: I(12, □),   J(20, □),   K(30, □),   L(−12, □), M(□, 9),   N(a, □).

The points A, B, C, D, E, F and G are all on the same straight line.

a Write down the coordinates of the points A, B, C, D, E, F and G.

b H, I, J, K. L, M, N, P and Q are further points on the same line. Fill in the missing coordinates: H(7, □), I(10, □), J(12, □), K(20, □), L(−7, □), M(−9, □), N(□, 10), P(□, −8), Q(□, 12).

Exercise 1d

In the following questions we are going to investigate the properties of the diagonals of the special quadrilaterals. You will need your own set of axes for each question. Mark a scale on each axis from −5 to +5. Mark the points A, B, C and D and join them to form the quadrilateral ABCD.

1   A(5, −2),   B(2, 4),   C(−3, 4),   D(0, −2)

     a   What type of quadrilateral is ABCD?

     b   Join A to C and B to D. These are the diagonals of the quadrilateral. Mark with an E the point where the diagonals cross.

     c   Measure the diagonals. Are they the same length?

     d   Is E the midpoint of either, or both, of the diagonals?

     e   Measure the four angles at E. Do the diagonals cross at right angles?

Now repeat question **1** for the following points:

2   A(2, −2),   B(2, 4),   C(−4, 4),   D(−4, −2)

3   A(2, −2),   B(5, 4),   C(−3, 4),   D(−1, −2)

4   A(2, 0),   B(0, 4),   C(−2, 0),   D(0, −4)

5   A(1, −4),   B(1, −1),   C(−5, −1),   D(−5, −4)

6   Name the quadrilaterals in which the two diagonals are of equal length.

7   Name the quadrilaterals in which the diagonals cut at right angles.

8   Name the quadrilaterals in which the diagonals cut each other in half.

! Investigation

Draw your own set of *x*- and *y*-axes and scale each of them from −6 to +8.

Plot the points A(−1, 3), B(3, −1) and C(−1, −5).

a   Can you write down

     i   the coordinates of a point D such that ABCD is a square

     ii   the coordinates of a point E such that ACBE is a parallelogram

     iii   the coordinates of a point F such that CDEF is a rectangle?

b   Can you give the name of the special quadrilateral EDBF?

Use of positive and negative numbers

Positive and negative numbers are collectively known as directed numbers.

Directed numbers can be used to describe any quantity that can be measured above or below a natural zero. For example, a distance of 50 m above sea level and a distance of 50 m below sea level could be written as +50 m and −50 m respectively.

They can also be used to describe time before and after a particular event. For example, 5 seconds before the start of a race and 5 seconds after the start of a race could be written as −5 s and +5 s respectively.

Directed numbers can also be used to describe quantities that involve one of two possible directions. For example, if a car is travelling north at 70 km/h and another car is travelling south at 70 km/h they can be described as going at +70 km/h and −70 km/h respectively.

A familiar use of negative numbers is to describe temperatures. The freezing point of water is 0° Celsius (or centigrade) and a temperature of 5 °C below freezing point is written −5 °C.

Most people would call −5 °C 'minus 5 °C' but we will call it 'negative 5 °C' and there are good reasons for doing so because in mathematics 'minus' means 'take away'.

A temperature of 5 °C above freezing point is called 'positive 5 °C' and can be written as +5 °C. Most people would just call it 5 °C and write it without the positive symbol.

A number without any symbol in front of it is a positive number,

i.e.   2 means +2

and   +3 can be written as 3.

Exercise 1e

Draw a Celsius thermometer and mark a scale on it from −10° to +10°.
Use your drawing to write the following temperatures as positive or
negative numbers:

1 10° above freezing point   4 5° above zero

2 7° below freezing point   5 8° below zero

3 3° below zero   6 freezing point

Write down, in words, the meaning of the following
temperatures:

7 −2 °C   9 4 °C   11 +8 °C

8 +3 °C   10 −10 °C   12 0 °C

Which temperature is higher?

13 +8° or +10°   18 −2° or −5°

14 12° or 3°   19 1° or −1°

15 −2° or +4°   20 +3° or −5°

16 −3° or −5°   21 −7° or −10°

17 −8° or 2°   22 −2° or −9°

23 The contour lines on the map below show distances above sea level as
positive numbers and distances below sea level as negative numbers.

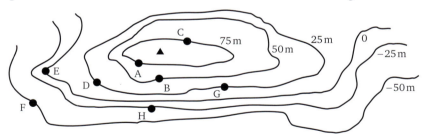

Write down in words the position relative to sea level of the points
A, B, C, D, E, F, G and H.

In questions **24** to **34** use positive or negative numbers to describe the
quantities.

A ball thrown up a distance of 5 m.

Up is above the start point so a positive number describes this.

+5 m

24 5 seconds before blast off of a rocket.

25 5 seconds after blast off of a rocket.

26 $50 in your purse.

27 $50 owed.

28 1 minute before the train leaves the station.

29 A win of $500 on a lottery.

30 A debt of $50.

31 Walking forwards five paces.

32 Walking backwards five paces.

33 The top of a hill that is 200 m above sea level.

34 A ball thrown down a distance of 5 m.

35 At midnight the temperature was −2 °C. One hour later it was 1° colder. What was the temperature then?

36 At midday the temperature was 18 °C. Two hours later it was 3° warmer. What was the temperature then?

37 A rockclimber started at +200 m and came a distance of 50 m down the rock face. How far above sea level was he then?

38 At midnight the temperature was −5 °C. One hour later it was 2° warmer. What was the temperature then?

39 At the end of the week my financial state could be described as $25. I was later given $50. How could I then describe my financial state?

40 Positive numbers are used to describe a number of paces forwards and negative numbers are used to describe a number of paces backwards. Describe where you are in relation to your starting point if you walk +10 paces followed by −4 paces.

Extending the number line

If a number line is extended beyond zero, negative numbers can be used to describe points to the left of zero and positive numbers are used to describe points to the right of zero.

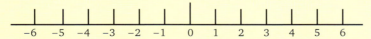

On this number line,   5 is to the *right* of 3

and we say that   5 is *greater* than 3.

    or   5 > 3   (> means 'is greater than')

Also   −2 is to the *right* of −4

and we say that   −2 is *greater* than −4

         or   −2 > −4

So 'greater' means 'higher up the scale'.

(A temperature of −2 °C is higher than a temperature of −4 °C.)

Now   2 is to the *left* of 6

and we say that   2 is *less* than 6

         or   2 < 6   (< means 'is smaller than')

Also   −3 is to the *left* of −1

and we say that   −3 is *less* than −1

         or   −3 < −1

So 'less than' means 'lower down the scale'.

Exercise 1f

Draw a number line.

In questions **1** to **12** write either > or < between the two numbers:

1	3 2	**4**	−3 −1	**7**	3 −2	**10**	−7 3
2	5 1	**5**	1 −2	**8**	5 −10	**11**	−1 0
3	−1 −4	**6**	−4 1	**9**	−3 −9	**12**	1 −1

In questions **13** to **24** write down the next two numbers in the pattern:

13	4, 6, 8	**16**	−4, −2, 0	**19**	5, 1, −3	**22**	−10, −8, −6
14	−4, −6, −8	**17**	9, 6, 3	**20**	2, 4, 8	**23**	−1, −2, −4
15	4, 2, 0	**18**	−4, −1, 2	**21**	36, 6, 1	**24**	1, 0, −1

Addition and subtraction of positive numbers

If you were asked to work out 5 − 7 you would probably say that it cannot be done. But if you were asked to work out where you would be if you walked 5 steps forwards and then 7 steps backwards, you would say that you were two steps behind your starting point.

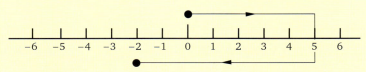

On the number line, $5 - 7$ means

start at 0 and go 5 places to the right

and then go 7 places to the left

So $5 - 7 = -2$

i.e. 'minus' a positive number means   move to the left

and 'plus' a positive number means   move to the right.

In this way $3 + 2 - 8 + 1$ can be shown on the number line as follows:

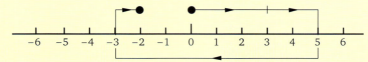

Therefore $3 + 2 - 8 + 1 = -2$

Exercise 1g

Find, using a number line if it helps:

1	$3 - 6$	**3**	$4 - 6$	**5**	$4 - 2$	**7**	$-2 + 3$	**9**	$-5 - 7$
2	$5 - 2$	**4**	$5 - 7$	**6**	$5 + 2$	**8**	$-3 + 5$	**10**	$-3 + 2$

$(+4) - (+3)$

$(+4) - (+3) = 4 - 3$ $(+4 = 4$ and $+3 = 3)$

$= 1$

11	$(+3) + (+2)$		**21**	$-4 + 2 + 5$
12	$(+2) - (+4)$		**22**	$-3 + 1 - 1$
13	$(+5) - (+7)$		**23**	$5 - 6 - 9$
14	$(-3) + (+2)$		**24**	$-3 - 4 + 2$
15	$(-1) + (+5)$		**25**	$-2 - 3 + 9$
16	$5 - 2 + 3$		**26**	$(+3) + (+4) - (+1)$
17	$7 - 9 + 4$		**27**	$(+2) - (+5) + (+6)$
18	$5 - 11 + 3$		**28**	$(+9) - (+7) - (+2)$
19	$10 - 4 - 9$		**29**	$(-3) + (+5) - (+5)$
20	$3 + 6 - 10$		**30**	$(-8) - (+4) + (+7)$

Remember that +3 is the same as 3.

What number does x represent if $x + 2 = 1$?

You can think of this as starting at 2 on the number line. What do you have to do to get to 1?

$$-1 + 2 \quad \text{is} \quad 1$$

so $$x = -1$$

Find the value of x:

31	$x + 4 = 5$	**36**	$3 + x = 1$
32	$x - 2 = 0$	**37**	$5 - x = 4$
33	$x + 2 = 4$	**38**	$5 + x = 7$
34	$x + 2 = 0$	**39**	$9 - x = 4$
35	$2 + x = 1$	**40**	$x - 6 = 10$

Addition and subtraction of negative numbers

Most of you will have some money of your own, from pocket money and other sources. Many of you will have borrowed money at some time.

At any one time you have a *balance* of money, i.e. the total sum that you own or owe!

If you own $2 and you borrow $4, your balance is a debt of $2. We can write this as

$$(+2) + (-4) = (-2)$$

or as $$2 + (-4) = -2$$

But $$2 - 4 = -2$$

$\therefore$ $$+(-4) \quad \text{means} \quad -4$$

If you owe $2 and then take away that debt, your balance is zero. We can write this as

$$(-2) - (-2) = 0$$

You can pay off a debt on your balance only if someone gives you $2. So subtracting a negative number is equivalent to adding a positive number, i.e. $-(-2)$ is equivalent to $+2$.

$$-(-2) \quad \text{means} \quad +2$$

Exercise 1h

Find:

1   $3 + (-1)$

2   $5 + (-8)$

3   $4 - (-3)$

4   $-1 - (-4)$

5   $-2 + (-7)$

6   $-2 - (-5)$

7   $4 + (-7)$

8   $-3 - (-9)$

9   $-4 + (-10)$

10   $2 - (-8)$

11   $-7 + (-7)$

12   $-3 - (-3)$

13   $+4 + (-4)$

14   $+2 - (-4)$

15   $-3 + (-3)$

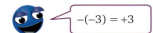

$+(-1) = -1$

$-(-3) = +3$

$$2 + (-1) - (-4)$$

$$+(-1) = -1 \text{ and } -(-4) = +4$$

$$2 + (-1) - (-4) = 2 - 1 + 4$$

$$= 5$$

16   $5 + (-1) - (-3)$

17   $(-1) + (-1) + (-1)$

18   $4 - (-2) + (-4)$

19   $-2 - (-2) + (-4)$

20   $6 - (-7) + (-8)$

21   $9 + (-5) - (-9)$

22   $8 - (-7) + (-2)$

23   $10 + (-9) + (-7)$

24   $12 + (-8) - (-4)$

25   $9 + (-12) - (-4)$

Addition and subtraction of directed numbers

We can now use the following rules:

$$+(+a) = +a \quad \text{and} \quad -(+a) = -a$$
$$+(-a) = -a \quad \text{and} \quad -(-a) = +a$$

Exercise 1i

Find:

1   $3 + (-2)$

2   $-3 - (+2)$

3   $6 - (-3)$

4   $4 + (+4)$

5   $-5 + (-7)$

6   $9 - (+2)$

7 $7+(-3)$

8 $8+(+2)$

9 $10-(-5)$

10 $-2-(-4)$

11 $12+(-7)$

12 $-4-(+8)$

13 $3-(-2)$

14 $-5+(-4)$

15 $8+(-7)$

16 $4-(-5)$

17 $7+(-3)-(+5)$

18 $2-(-4)+(-6)$

19 $5+(-2)-(+1)$

20 $8-(-3)+(+5)$

21 $7+(-4)-(-2)$

22 $3-(+2)+(-5)$

23 $-9+(-2)-(-3)$

24 $8+(+9)-(-2)$

25 $7+(-9)-(+2)$

26 $4+(-1)-(+7)$

27 $-3+(+5)-(-2)$

28 $-4+(+8)+(-7)$

29 $-9-(+4)-(-10)$

30 $-2-(+8)+(-9)$

$-8-(4-7)$

$$-8-(4-7) = -8-(-3) \qquad \text{(brackets first)}$$
$$= -8+3$$
$$= -5$$

31 $3-(4-3)$

32 $5+(7-9)$

33 $4+(8-12)$

34 $-3-(7-10)$

35 $6+(8-15)$

36 $(3-5)+2$

37 $5-(6-10)$

38 $(4-9)-2$

39 $(7+4)-15$

40 $8+(3-8)$

41 $(3-8)-(9-4)$

42 $(3-1)+(5-10)$

43 $(7-12)-(6-9)$

44 $(4-8)-(10-15)$

45   Add $(+7)$ to (-5).

46   Subtract 7 from -5.

47   Subtract (-2) from 1.

48   Find the value of 8 take away -10.

49   Add -5 to $+3$.

50   Find the sum of -3 and $+4$.

51   Find the sum of -8 and $+10$.

52   Subtract positive 8 from negative 7.

53   Find the sum of -3 and -3 and -3.

54   Find the value of twice negative 3.

55   Find the value of four times -2.

Puzzle

An early Greek mathematician set up a secret society in the 6th century BCE. Members swore never to give away any mathematical secrets. One member was killed because he told one of the secrets to a friend who was not a member. The first letters to complete each of the statements below taken in order will spell his name.

The distance around a plane figure is called its _____.

There are 365 days in a _____.

A plane figure with three sides is a _____.

−1 is _____ of −2.

The _____ of a triangle add up to 180 degrees.

> means 'is _____ than'.

The point where the x and y axes intersect is the _____.

$\frac{1}{4}$ of 360° is called a _____ angle.

To find the _____ of a rectangle we multiply the length by the width.

−8 is the _____ of −6 and −2.

Division of negative numbers by positive numbers

Because $2 \times 3 = 6$, $6 \div 3 = 2$.

In the same way, $(-3) \times 4 = -12$, so $(-12) \div 4 = -3$.

Notice that the order *does* matter in division, e.g.

$$(-12) \div 4 = -3$$

but we shall see that $4 \div (-12) = -\frac{1}{3}$

Exercise 1j

Find a $(-9) \div 3$ b $(-14) \div (+2)$ c $\frac{-10}{2}$

a $(-9) \div 3 = -3$

b $(-14) \div (+2) = -7$ $((-14) \div (+2) = -14 \div 2)$

c $\frac{-10}{2} = -5$ $\left(\frac{-10}{2} \text{ means } -10 \div 2\right)$

Find:

1	$(-6) \div 2$	**7**	$(-30) \div (+3)$	**13**	$\frac{-8}{4}$
2	$(-10) \div 5$	**8**	$(-36) \div 12$	**14**	$\frac{-12}{6}$
3	$(-15) \div 3$	**9**	$(-20) \div 4$	**15**	$\frac{-16}{4}$
4	$(-24) \div 6$	**10**	$(-28) \div (+7)$	**16**	$\frac{-27}{3}$
5	$(-12) \div (+3)$	**11**	$(-3) \div 3$	**17**	$\frac{-36}{9}$
6	$(-18) \div (+9)$	**12**	$(-10) \div (+5)$	**18**	$\frac{-30}{15}$

Multiplication of directed numbers

Consider the expression $6x - (x - 3)$.

From $6x$ we have to subtract the number in the bracket, which is 3 less than x.

If we start by writing $6x - x$ we have subtracted 3 too many.

To put it right we must add on 3.

Therefore $\qquad 6x - (x - 3) = 6x - x + 3$

Similarly $8x - 3(x - 2)$ means

'from $8x$ subtract three times the number that is 2 less than x', which is $8x - (3x - 6)$

If we write $8x - 3x$ we have subtracted 6 too many. So we must add 6 on again giving $8x - 3(x - 2) = 8x - 3x + 6$

From this, and from the previous exercise, we have

(a) $(+3) \times (+2) = +6$

This is just the multiplication of positive numbers,

i.e. $(+3) \times (+2) = 3 \times 2 = 6$

(b) $(-3) \times (+2) = -6$

Here we could write $(-3) \times (+2) = -3(2)$. This is equivalent to subtracting 3 twos, i.e. subtracting 6.

(c) $(+4) \times (-3) = -12$

This means four lots of -3,

i.e. $(-3) + (-3) + (-3) + (-3) = -12$

(d) $(-2) \times (-3) = +6$

This can be thought of as taking away two lots of -3,

i.e. $-2(-3) = -(-6)$

We have already seen that taking away a negative number is equivalent to adding a positive number, so $(-2) \times (-3) = +6$.

Calculate   **a** $(+2) \times (+4)$ **b** 2×4.
a $(+2) \times (+4) = 8$
b $2 \times 4 = 8$
This shows that $(+2) \times (+4)$ means the same as 2×4.

Calculate   **a** $(-3) \times (+4)$ **b** -3×4.
a $(-3) \times (+4) = -12$
b $-3 \times 4 = -12$
This shows that $(-3) \times (+4)$ means the same as -3×4.

Because order does not matter when two quantities are multiplied together, $(+4) \times (-3)$ gives the same answer of -12.

So -3 and $+4$ can be multiplied together in four different ways, but they all mean the same thing.

Calculate   **a** $(-5) \times (-2)$ **b** $-5(-2)$.
a $(-5) \times (-2) = 10$
b $-5(-2) = 10$ $-5(-2)$ means $-5 \times (-2)$

Calculate:

1	$(-3) \times (+5)$	**11**	$(-3) \times (-9)$	**21**	$3(-2)$
2	$(+4) \times (-2)$	**12**	$(-2) \times (+8)$	**22**	5×3
3	$(-7) \times (-2)$	**13**	$7 \times (-5)$	**23**	$6 \times (-3)$
4	$(+4) \times (+1)$	**14**	$-6(-4)$	**24**	$-5(-4)$
5	$(+6) \times (-7)$	**15**	-3×5	**25**	$6 \times (-4)$
6	$(-4) \times (-3)$	**16**	$5 \times (-9)$	**26**	$-3(+8)$
7	$(-6) \times (+3)$	**17**	$-6(4)$	**27**	$(+5) \times (+9)$
8	$(-8) \times (-2)$	**18**	$-2(-4)$	**28**	-4×5
9	$(+5) \times (-1)$	**19**	$-(-3)$	**29**	$7(-4)$
10	$(-6) \times (-3)$	**20**	$4 \times (-2)$	**30**	$(-4) \times (-9)$

Exercise 1l

Multiply out $-(x+2)$

$$-(x+2) \quad \text{means} \quad -1(x+2)$$
$$-(x+2) = -x-2$$

Multiply out the following brackets:

 1 $-6(x-5)$ 13 $3(2x-6)$

2   $-5(3c+3)$ 14 $-7(2+x)$

3   $-2(5e-3)$ 15 $-2(3x-1)$

4   $-(3x-4)$ <u>16</u> $-(3x+2)$ 25 $-(5a+5b)$

5   $-8(2-5x)$ <u>17</u> $8(2-3x)$ 26 $2(3x+2y+1)$

<u>6</u> $-7(x+4)$ <u>18</u> $-3(2y-4x)$ 27 $-5(5+2x)$

<u>7</u> $-3(2d-2)$ <u>19</u> $5(4x-1)$ 28 $4(x-y)$

<u>8</u> $-2(4+2x)$ <u>20</u> $-5(1-4x)$ 29 $-(4c-5)$

<u>9</u> $-7(2-3x)$ 21 $6(4+5x)$ 30 $9(2x-1)$

<u>10</u> $-(4-5x)$ 22 $-6(4+5x)$

 11   $4(3x+9)$ 23 $6(4-5x)$

12   $5(2+3x)$ 24 $-6(4-5x)$

> Multiply each term inside the bracket by –6.

> Multiply each term inside the bracket by 4.

Exercise 1m

Simplify the following expressions:

1   $5x+4(5x+3)$ 11 $7(3x+1)-2(2x+4)$ 21 $4(x-1)+5(2x+3)$

2   $42-3(2c+5)$ 12 $5(2x-3)-(x+3)$ 22 $4(x-1)-5(2x+3)$

3   $2m+4(3m-5)$ 13 $2(4x+3)+(x-5)$ 23 $4(x-1)+5(2x-3)$

4   $7-2(3x+2)$ 14 $7(3-x)-(6-2x)$ 24 $4(x-1)-5(2x-3)$

5   $x+(5x-4)$ 15 $5+3(4x+1)$ 25 $8(2x-1)-(x+1)$

<u>6</u> $9-2(4g-2)$ <u>16</u> $6x+2(3x-7)$ <u>26</u> $3x+2(4x+2)+3$

<u>7</u> $4-(6-x)$ <u>17</u> $20x-4(3+4x)$ <u>27</u> $5-4(2x+3)-7x$

8   $10f+3(4-2f)$ <u>18</u> $4(x+1)+5(x+3)$ <u>28</u> $3(x+6)-(x-3)$

<u>9</u> $7-2(5-2s)$ <u>19</u> $3(2x+3)-5(x+6)$ <u>29</u> $3(x+6)-(x+3)$

<u>10</u> $7x+3(4x-1)$ 20 $5(6x-3)+(x+4)$ <u>30</u> $7x+8x-2(5x+1)$

Exercise 1n

Solve the following equations.

Some equations may have negative answers.

$x+8=6$

$$x+8=6$$

Take 8 from both sides $\quad x=-2$

1	$x+4=2$	**3**	$3+a=1$	**5**	$4+w=2$
2	$x+6=1$	**4**	$s+3=2$	**6**	$c+6=2$

$x-6=2$

$$x-6=2$$

Add 6 to both sides $\quad x=8$

When there are several letter and number terms, deal with the letters first, then the numbers.

$5x+2=2x+9$

$$5x+2=2x+9$$

Take $2x$ from both sides $\quad 3x+2=9$

Take 2 from both sides $\quad\quad 3x=7$

Divide both sides by 3 $\quad\quad x=\dfrac{7}{3}=2\dfrac{1}{3}$

7	$3x+4=2x+8$	**11**	$7x+3=3x+31$
8	$x+7=4x+4$	**12**	$6z+4=2z+1$
9	$2x+5=5x-4$	**13**	$7x-25=3x-1$
10	$3x-1=5x-11$	**14**	$11x-6=8x+9$

Choose to take away the lower number of xs.

$9+x=4-4x$

$$9+x=4-4x$$

Add $4x$ to both sides $\quad 9+5x=4$

Take 9 from both sides $\quad\quad 5x=-5$

Divide both sides by 5 $\qquad x = -1$

Check: If $\quad x = -1$, left-hand side $= 9 + (-1)$

$$= 8$$

right-hand side $= 4 - (-4)$

$$= 8$$

So $x = -1$ is the solution.

Remember that negative numbers are lower down the number line than positive numbers.

15	$4x - 3 = 39 - 2x$	**18**	$24 - 2x = 5x + 3$	**21**	$32 - 6x = 8 + 2x$
16	$5 + x = 17 - 5x$	**19**	$5x - 6 = 3 - 4x$	**22**	$9 - 3x = -5 + 4x$
17	$7 - 2x = 4 + x$	**20**	$12 + 2x = 24 - 4x$		

$9 - 3x = 15 - 4x$

$$9 - 3x = 15 - 4x$$

$-4x < -3x$ so add $4x$ to both sides $\qquad 9 + x = 15$

Take 9 from both sides $\qquad\qquad\qquad x = 6$

Think of the number line to decide which is the lower number of xs.

23	$5 - 3x = 1 - x$	**29**	$4 - 2x = 8 - 5x$	**35**	$13 - 4x = 4x - 3$
24	$16 - 2x = 19 - 5x$	**30**	$3 - x = 5 - 3x$	**36**	$7x + 6 = x - 6$
25	$6 - x = 12 - 2x$	**31**	$6 - 3x = 4x - 1$	**37**	$6 - 2x = 9 - 5x$
26	$-2 - 4x = 6 - 2x$	**32**	$4z + 1 = 6z - 3$	**38**	$3 - 2x = 3 + x$
27	$16 - 6x = 1 - x$	**33**	$3 - 6x = 6x - 3$		
28	$4 - 3x = 1 - 4x$	**34**	$8 - 4x = 14 - 7x$		

$3 - 2x = 5$

$$3 - 2x = 5$$

There are zero x's on the right-hand side, and $-2x < 0x$

Add $2x$ to both sides $\qquad\qquad 3 = 5 + 2x$

Take 5 from both sides $\qquad\quad -2 = 2x$

Divide both sides by 2 $\qquad\quad x = -1$

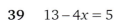

Remember that a negative number is less than 0.

39	$13 - 4x = 5$	**43**	$9x + 4 = 3x + 1$	**47**	$-4x - 5 = -2x - 10$
40	$6 = 2 - 2x$	**44**	$2x + 3 = 12x$	**48**	$5 - 3x = 2$
41	$6 = 8 - 3x$	**45**	$7 - 2x = 3 - 6x$	**49**	$6 + 3x = 7 - x$
42	$0 = 6 - 2x$	**46**	$3x - 6 = 6 - x$	**50**	$5 - 2x = 4x - 7$

Did you know?

If the sum of two numbers is 24 and the difference is 6, then one of the numbers is $\dfrac{(24+6)}{2} = 15$

or

If the sum of two numbers is 28 and the difference is 12, then one of them is $\dfrac{(28+12)}{2} = 20$

Use your algebra to prove that this is always so.

! Investigation

Try this on a group of pupils or friends.

Think of a number between 1 and 10.
Add 4.
Multiply the result by 5.
Double your answer.
Divide the result by 10.
Take away the number you first thought of.
Write down your answer.

However many times you try this the answer is always 4. Investigate what happens when you use numbers other than numbers between 1 and 10. Try, for example, larger whole numbers, decimals, negative whole numbers, fractions.

Is the answer always 4?

Mixed exercises

Exercise 1p

1 Which is the higher temperature, $-5°$ or $-8°$.

2 Write $<$ or $>$ between   **a** -3 2 **b** -2 -4.

Find:

3 $-4-(-6)$	**7** $-2+(-3)-(-5)$	**11** $(-5)\times(-2)$
4 $3+2-10$	**8** $4-(2-3)$	**12** $-2(7)$
5 $2+(-4)$	**9** $6\times(-4)$	**13** $(+4)\times(-6)$
6 $3-(-1)$	**10** $-36\div3$	**14** $-2(-3)+1$

15 Simplify   **a** $-3(2-5x)$ **b** $3(5-2x)-(x-7)$.

Exercise 1q

1 A is the point (2, −1). What are the coordinates of the point immediately

   a 2 above A

   b 3 to the left of A

   c 4 to the right of A and 2 below it?

Draw your own grid and point A on it.

2 Plot the points A(−2, −1), B(2, 2), C(4, 0), D(0, −3) on axes scaled from −4 to 4 for both x and y.

   Join the points to make the figure ABCD.

   a What name is given to this shape?

   b Write down the coordinates of the midpoint of   **i** BC   **ii** AC.

3 Simplify: **a** $-3(3b-4)-3$ **b** $3(2x-1)-5(2-x)$

4 Solve these equations.

   a $s+4=-1$

   b $27-5x=6+2x$

   c $-3-4x=3-2x$

In this chapter you have seen that...

✔ directed numbers is the collective name for positive and negative numbers

✔ directed numbers can be used to describe quantities that can be measured above or below a natural zero

✔ the rules for addition and subtraction are

$+(+a)$ and $-(-a)$ both give $+a$ and $+(-a)$ and $-(+a)$ both give $-a$

You can remember these as

SAME SIGNS GIVE POSITIVE, DIFFERENT SIGNS GIVE NEGATIVE

✔ $5\times(-3)$, $(+5)\times(-3)$, $(-3)\times5$, $(-3)\times(+5)$ ALL MEAN THE SAME

✔ when a negative number is divided by a positive number, the answer is negative

✔ negative coordinates allow us to plot points anywhere in the xy plane

✔ a negative number multiplied by a negative number gives a positive number

✔ you can solve equations involving directed numbers

2 Working with numbers

Ratio

A ratio is a comparison between two related quantities.

For example, Peter makes a model of his father's boat.
The model is 2 m long and the boat is 20 m long.

We say that the ratio of the length of the model to the length of the boat is 2 m to 20 m.

We write this as 2 m : 20 m. As the units are the same we can simply write the ratio as 2 : 20.

We can also write the ratio as the fraction $\frac{2}{20}$.

Simplifying ratios

As the ratio $2:20$ can be written as $\frac{2}{20}$, we can simplify $\frac{2}{20}$ to $\frac{1}{10}$ by dividing numerator and denominator by 2. So $2:20 = 1:10$

Ratios can be simplified in the same way by thinking of them as fractions. There is no need to write a ratio as a fraction, you can simplify a ratio by dividing each part by the same number.

Exercise 2a

Express the ratios **a** 24 to 72 **b** 2 cm to 1 m in their simplest form.

a $\frac{24}{72} = \frac{3}{9} = \frac{1}{3}$ (dividing both numbers by 8 and then by 3)

or $24:72 = 3:9 = 1:3$

so $24:72 = 1:3$

b (Before we can compare 2 cm and 1 m they must be expressed in the same unit.)

$\frac{2\,cm}{1\,m} = \frac{2\,cm}{100\,cm}$ or $2\,cm:1\,m = 2\,cm:100\,cm$

$= 2:100$

$= \frac{1}{50}$ $= 1:50$

so $2\,cm:1\,m = 1:50$

Express the following ratios in their simplest form:

1	$8:10$	**6**	$45\,g:1\,kg$
2	$20:16$	**7**	$\$4:75\,c$
3	$12:18$	**8**	$48\,c:\$2.88$
4	$2\,cm:8\,cm$	**9**	$288:306$
5	$32\,c:96\,c$	**10**	$10\,cm^2:1\,m^2$

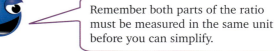

Remember both parts of the ratio must be measured in the same unit before you can simplify.

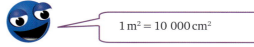

$1\,m^2 = 10\,000\,cm^2$

Simplify the ratio $24:18:12$

(As there are three numbers involved, this ratio cannot be expressed as a single fraction.)

$24:18:12 = 4:3:2$ (dividing each number by 6)

11	$4:6:10$	**15**	$20:24:32$	**19**	$144:12:24$
12	$18:24:36$	**16**	$7:56:49$	**20**	$98:63:14$
13	$2:10:20$	**17**	$15:20:35$		
14	$9:12:15$	**18**	$16:128:64$		

We know that we can produce equivalent fractions by multiplying or dividing both numerator and denominator by the same number,

so that $\frac{2}{3} = \frac{4}{6}$ or $\frac{12}{18}$ or $\frac{20}{30}$.

We can do the same with a ratio in the form $3:6$.

$\qquad 3:6 = 6:12 \qquad$ (multiplying both numbers by 2)

and $\qquad\qquad 2:\frac{1}{3} = 6:1 \qquad$ (multiplying both numbers by 3)

We can use this to simplify ratios containing fractions.

Exercise 2b

Express in their simplest form the ratios

a $\quad 3:\frac{1}{4}$ $\qquad$ **b** $\quad \frac{2}{3}:\frac{4}{5}$

a $\quad 3:\frac{1}{4} = 12:1 \qquad\qquad$ (multiplying both numbers by 4)

b $\quad \frac{2}{3}:\frac{4}{5} = \overset{5}{15} \times \frac{2}{\underset{1}{3}} : \overset{3}{15} \times \frac{4}{\underset{1}{5}} \qquad$ (multiplying both numbers by 15, which is the lowest common multiple of 3 and 5)

$\qquad\quad = 10:12$

$\qquad\quad = 5:6$

Express the following ratios in their simplest forms:

1	$5:\frac{1}{3}$	**5**	$\frac{1}{3}:\frac{3}{4}$	**9**	$2\frac{2}{3}:1\frac{1}{6}$	**13**	$4:\frac{9}{10}$	**17**	$1\frac{1}{2}:3:4\frac{1}{2}$
2	$2:\frac{1}{4}$	**6**	$\frac{7}{12}:\frac{5}{6}$	**10**	$\frac{2}{3}:\frac{7}{15}$	**14**	$\frac{4}{5}:6$	**18**	$6:4\frac{1}{2}$
3	$\frac{1}{2}:\frac{1}{3}$	**7**	$\frac{5}{4}:\frac{6}{7}$	**11**	$24:15:9$	**15**	$7\frac{1}{2}:9\frac{1}{2}$	**19**	$\frac{1}{6}:\frac{1}{8}:\frac{1}{12}$
4	$\frac{3}{4}:\frac{1}{4}$	**8**	$3:\frac{4}{3}$	**12**	$\frac{4}{9}:\frac{2}{3}$	**16**	$\frac{1}{4}:\frac{1}{5}$	**20**	$6:8:12$

Relative sizes

Exercise 2c

Which ratio is the larger, 6:5 or 7:6?

(We need to compare the sizes of $\frac{6}{5}$ and $\frac{7}{6}$ so we express both with the same denominator.)

$$\frac{6}{5} = \frac{36}{30} \text{ and } \frac{7}{6} = \frac{35}{30}$$

so   6:5   is larger than   7:6

1   Which ratio is the larger, 5:7 or 2:3?

2   Which ratio is the smaller, 7:4 or 13:8?

3   Which ratio is the larger, $\frac{5}{8}$ or $\frac{7}{12}$?

4   Which ratio is the smaller, $\frac{3}{4}$ or $\frac{7}{10}$?

In the following sets of ratios some are equal to one another. In each question identify the equal ratios.

5   6:8,   24:32, $\frac{3}{4}$:1

6   10:24, $\frac{5}{9}$:$\frac{4}{5}$, $\frac{5}{9}$:$\frac{4}{3}$

7   8:64,   2:14, $\frac{1}{16}$:$\frac{1}{2}$

8   $\frac{2}{3}$:3,   4:18,   2:6

Simplify each ratio, then you can see which are equal.

Problems

Exercise 2d

A family has 12 pets of which 6 are cats or kittens, 2 are dogs and the rest are birds. Find the ratio of the numbers of

a   birds to dogs   **b**   birds to pets.

a   There are 4 birds and 2 dogs

So number of birds : number of dogs = 4:2

= 2:1

b   There are 4 birds and 12 pets

So number of birds : number of pets = 4:12

= 1:3

In each question give your answer in its simplest form.

1   A couple have 6 grandsons and 4 granddaughters. Find
   a   the ratio of the number of grandsons to that of granddaughters.
   b   the ratio of the number of granddaughters to that of grandchildren.

2   Square A has side 6 cm and square B has side 8 cm. Find the ratio of
   a   the length of the side of square A to the length of the side of square B
   b   the area of square A to the area of square B.

3   Tom walks 2 km to school in 40 minutes and John cycles 5 km to school in 15 minutes. Find the ratio of
   a   Tom's distance to John's distance
   b   Tom's time to John's time.

4   Mary has 18 sweets and Jane has 12. As Mary has 6 sweets more than Jane she tries to even things out by giving Jane 6 sweets. What is the ratio of the number of sweets Mary has to the number Jane has
   a   to start with   **b**   to end with?

5   If $p : q = 2 : 3$,   find the ratio $6p : 2q$

6   Rectangle A has length 12 cm and width 6 cm while rectangle B has length 8 cm and width 5 cm. Find the ratio of
   a   the length of A to the length of B
   b   the area of A to the area of B
   c   the perimeter of A to the perimeter of B
   d   the size of an angle of A to the size of an angle of B.

7   A triangle has sides of lengths 3.2 cm, 4.8 cm and 3.6 cm. Find the ratio of the lengths of the sides to one another.

8   Two angles of a triangle are 54° and 72°. Find the ratio of the size of the third angle to the sum of the first two.

9   For a school fête, Mrs Jones and Mrs Brown make marmalade in 1 lb jars. Mrs Jones makes 5 jars of lemon marmalade and 3 jars of orange. Mrs Brown makes 7 jars of lemon marmalade and 5 of grapefruit. Find the ratio of the numbers of jars of
   a   lemon to orange to grapefruit
   b   Mrs Jones' to Mrs Brown's marmalade
   c   Mrs Jones' lemon to orange.

Finding missing quantities

Some missing numbers are fairly obvious.

Exercise 2e

Find the missing numbers in the following ratios:

a $6:5 = \quad :10$ 　　　　　　　　**b** $\frac{4}{3}:\frac{}{9}:\frac{24}{}$

a 10 is 5×2, so the missing number is 6×2.

$6:5 = 12:10$

b Think of these as equivalent fractions. $\frac{4}{3} = \frac{12}{9} = \frac{24}{18}$

Find the missing numbers in the following ratios:

1 $2:5 = 4:$ 　　　　　**5** $3:\quad = 12:32$ 　　　　　**9** $\frac{6}{8}:\frac{}{12}$

2 $\quad :6 = 12:18$ 　　　**6** $\quad :15 = 8:10$ 　　　　**10** $6:9 = 8:$

3 $24:14 = 12:$ 　　　　**7** $9:6 = \quad :4$

4 $\frac{6}{}:\frac{9}{3}$ 　　　　　　　**8** $\frac{}{4}:\frac{15}{10}$

Indices

The accepted shorthand way of writing $2\times2\times2\times2$ is 2^4.
We read this as '2 to the power of 4' or '2 to the four'.
The 4 is called the *index*.
Hence $16 = 2\times2\times2\times2 = 2^4$ and similarly $3^3 = 3\times3\times3 = 27$.

Expressing a number using indices gives a convenient way for writing large numbers. For example, it has been discovered that

$$2^{216091} - 1$$

is a prime number. This is a very large number that would fill two newspaper pages if it were written in full, as it contains 65 050 digits. Using indices we are able to write it in the short form shown above.

Exercise 2f

Write the following products in index form:

1 $2\times2\times2$ 　　　　　　　**3** $5\times5\times5\times5$

2 $3\times3\times3\times3$ 　　　　　**4** $7\times7\times7\times7\times7$

5	$2\times2\times2\times2\times2$	**8**	19×19
6	$3\times3\times3\times3\times3\times3$	**9**	$2\times2\times2\times2\times2\times2\times2$
7	$13\times13\times13$	**10**	$6\times6\times6\times6$

Find the value of:

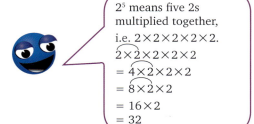

11	2^5	**15**	3^2	**19**	10^2
12	3^3	**16**	7^2	**20**	10^3
13	5^2	**17**	3^4	**21**	10^4
14	2^3	**18**	4^2	**22**	10^1

2^5 means five 2s multiplied together, i.e. $2\times2\times2\times2\times2$.
$2\times2\times2\times2\times2$
$= 4\times2\times2\times2$
$= 8\times2\times2$
$= 16\times2$
$= 32$

Express the following numbers in index form:

23 4   **24** 9   **25** 8   **26** 27   **27** 49   **28** 25   **29** 32   **30** 64

Find $3^2\times5^3$

$3^2\times5^3 = 3\times3\times5\times5\times5$
$\qquad\ = 9\times5\times5\times5$
$\qquad\ = 45\times5\times5$
$\qquad\ = 225\times5$
$\qquad\ = 1125$

Find the value of:

31 $2^3\times3$ **32** $2^4\times5^2$ **33** $4^2\times5$ **34** $2^5\times3^2$ **35** $2^6\times10^2$

Recurring decimals

Consider the calculation

$$3\div4 = 0.75 \qquad\qquad 4\overline{)3.00}$$
$$0.75$$

By adding two zeros after the point we are able to finish the division and give an exact answer. Now consider

$$2\div3 = 0.666\ldots \qquad\qquad 3\overline{)2.0000\ldots}$$
$$0.6666\ldots$$

We can see that we will continue to obtain 6s for ever and we say that the 6 *recurs*.

Consider

$$31 \div 11 = 2.8181\ldots \qquad \qquad 11\overline{)31.0000\ldots}$$
$$2.8181\ldots$$

Here 81 recurs.

Sometimes it is one figure which is repeated and sometimes it is a group of figures. If one figure or a group continues to *recur* we have a *recurring decimal*.

Exercise 2g

Calculate $0.2 \div 7$

$$7\overline{)0.200\,000\,000\,000\ldots}$$
$$0.2 \div 7 = 0.028\,571\,428\,571\ldots \qquad \qquad 0.028\,571\,428\,571\ldots$$

Calculate:

1 $1.4 \div 6$ **2** $0.03 \div 11$ **3** $4 \div 7$ **4** $0.43 \div 3$ **5** $0.03 \div 7$ **6** $1.1 \div 9$

Express the following fractions as decimals:

$\frac{4}{3}$

$$\frac{4}{3} = 4 \div 3 = 1.333\ldots \qquad \qquad 3\overline{)4.00\ldots}$$
$$1.33\ldots$$

7 $\frac{4}{9}$ **8** $\frac{2}{3}$ **9** $\frac{2}{11}$ **10** $\frac{5}{7}$ **11** $\frac{7}{9}$ **12** $\frac{8}{7}$

To save writing so many figures we use a dot notation for recurring decimals.

For example
$$\frac{1}{6} = 1 \div 6 = 0.1666\ldots \qquad \qquad 6\overline{)1.000\ldots}$$
$$0.166\ldots$$
$$= 0.1\dot{6}$$

and $\qquad \qquad 0.2 \div 7 = 0.0\dot{2}8\,571\,\dot{4} \qquad \qquad 7\overline{)0.2}$
$$0.028\,571\,428\,571\,428$$

The dots are placed over the single recurring number or over the first and last figures of the recurring group.

13 Write the answers **1** to **12** above using the dot notation.

Correcting to a given number of decimal places

Often we need to know only the first few figures of a decimal. For instance, if we measure a length with an ordinary ruler we usually need an answer to the nearest $\frac{1}{10}$ cm and are not interested, or cannot see, how many $\frac{1}{100}$ cm are involved.

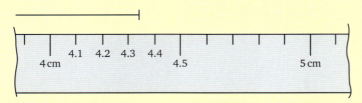

Look at this enlarged view of the end of a line that is being measured. We can see that with a more accurate measure we might be able to give the length as 4.34 cm. However on the given ruler we would probably measure it as 4.3 cm because we can see that the end of the line is nearer 4.3 than 4.4. We cannot give the exact length of the line but we can say that it is 4.3 cm long to the nearest $\frac{1}{10}$ cm. We write this as 4.3 cm correct to 1 decimal place.

Consider the numbers 0.62, 0.622, 0.625, 0.627 and 0.63. To compare them we write 0.62 as 0.620 and 0.63 as 0.630 so that each number has 3 figures after the point. When we write them in order in a column:

0.620
0.622
0.625
0.627
0.630

we can see that 0.622 is nearer to 0.620 than to 0.630 while 0.627 is nearer to 0.630 so we write

0.62|2 = 0.62   (correct to 2 decimal places)

0.62|7 = 0.63   (correct to 2 decimal places)

It is not so obvious what to do with 0.625 as it is halfway between 0.62 and 0.63. To save arguments, if the figure after the cut-off line is 5 or more we add 1 to the figure before the cut-off line, i.e. we round the number *up*, so we write

0.62|5 = 0.63   (correct to 2 decimal places)

Exercise 2h

Give 10.9315 correct to:

a the nearest whole number   **b** 1 decimal place   **c** 3 decimal places.

a   10.9315 = 11   (correct to the nearest whole number)
b   10.9315 = 10.9   (correct to 1 decimal place)
c   10.9315 = 10.932   (correct to 3 decimal places)

Give 4.699 and 0.007 correct to 2 decimal places.
4.699 = 4.70   (correct to 2 decimal place.)
0.007 = 0.01   (correct to 2 decimal place.)

Give the following numbers correct to 2 decimal places:

1	0.328		**6**	0.6947
2	0.322		**7**	0.8351
3	1.2671		**8**	3.927
4	2.345		**9**	0.0084
5	0.0416		**10**	3.9999

Mark in the vertical cut-off line similar to that shown in the worked examples.

Give the following numbers correct to the nearest whole number:

11	13.9	**13**	26.5	**15**	4.45	**17**	109.7	**19**	74.09
12	6.34	**14**	2.78	**16**	6.783	**18**	6.145	**20**	3.9999

Give the following numbers correct to 3 decimal places:

21	0.3627		**26**	0.0843
22	0.026 234		**27**	0.084 47
23	0.007 14		**28**	0.3251
24	0.0695		**29**	0.032 51
25	0.000 98		**30**	3.9999

Don't forget the cut-off line.

Give the following numbers correct to the number of the decimals places indicated in the brackets:

31	1.784	(1)	**36**	1.639	(2)
32	42.64	(1)	**37**	1.639	(1)
33	1.0092	(2)	**38**	1.689	(nearest whole number)
34	0.009 42	(4)	**39**	3.4984	(2)
35	0.7345	(3)	**40**	3.4984	(1)

Don't forget the cut-off line.

If we are asked to give an answer correct to a certain number of decimal places, we work out one more decimal place than is asked for. Then we can find the size of the last figure required.

Exercise 2i

Find $4.28 \div 6$ giving your answer correct to 2 decimal places.

$4.28 \div 6 = 0.71|3\ldots$

$\qquad = 0.71 \qquad$ (correct to 2 decimal places)

Calculate $302 \div 14$ correct to 1 decimal place.

$302 \div 14 = 21.5|7\ldots$

$\qquad = 21.6 \qquad$ (correct 1 decimal place)

Remember to work to one more decimal place than asked for.

$$
\begin{array}{r}
\overset{2}{}\\
6\overline{)4.280}\\
0.713\ldots
\end{array}
$$

$$
\begin{array}{r}
21.57\ldots\\
14\overline{)302.00}\\
28\\
\overline{22}\\
14\\
\overline{80}\\
70\\
\overline{100}\\
98
\end{array}
$$

Calculate, giving your answers correct to 2 decimal places:

1	$0.496 \div 3$	**5**	$25.68 \div 9$	**9**	$5.68 \div 24$
2	$6.49 \div 7$	**6**	$2.35 \div 15$	**10**	$3.85 \div 101$
3	$3.12 \div 9$	**7**	$0.68 \div 16$	**11**	$1.73 \div 8$
4	$12.2 \div 6$	**8**	$0.99 \div 21$	**12**	$48.4 \div 51$

Calculate, giving your answers correct to 1 decimal place:

13	$32.9 \div 8$	**17**	$124 \div 17$	**21**	$213 \div 22$
14	$402 \div 7$	**18**	$16.2 \div 14$	**22**	$8.4 \div 13$
15	$15.3 \div 6$	**19**	$45 \div 21$	**23**	$26 \div 15$
16	$9.76 \div 11$	**20**	$15.1 \div 16$	**24**	$519 \div 19$

Find, giving your answers correct to 3 decimal places:

25	$0.023 \div 4$	**29**	$0.45 \div 12$	**33**	$0.2584 \div 16$
26	$0.123 \div 7$	**30**	$0.012 \div 13$	**34**	$0.321 \div 17$
27	$1.25 \div 3$	**31**	$0.654 \div 23$	**35**	$1.26 \div 32$
28	$0.23 \div 11$	**32**	$0.98 \div 32$	**36**	$0.88 \div 24$

Changing fractions to decimals

Exercise 2j

Give $\frac{4}{25}$ as a decimal.

$\frac{4}{25} = 4 \div 25 = 0.16$

(This is an exact answer.)

$$\begin{array}{r} 0.16 \\ 25\overline{)4.00} \\ 25 \\ \overline{150} \end{array}$$

Give $\frac{4}{7}$ as a decimal correct to 3 decimal places.

$\frac{4}{7} = 4 \div 7 = 0.5714...$

$= 0.571$ (correct to 3 decimal places)

$$\begin{array}{r} 7\overline{)4.0000} \\ 0.5714... \end{array}$$

(This is an approximate answer.)

Give the following fractions as exact decimals:

1	$\frac{5}{8}$	6	$\frac{7}{50}$
2	$\frac{3}{40}$	7	$\frac{1}{16}$
3	$\frac{3}{16}$	8	$\frac{11}{8}$
4	$\frac{3}{5}$	9	$\frac{13}{25}$
5	$\frac{9}{25}$	10	$\frac{3}{80}$

Remember, you change a fraction to a decimal by dividing the bottom into the top.

Give the following fractions as decimals correct to 3 decimal places:

11	$\frac{3}{7}$	21	$\frac{3}{14}$
12	$\frac{4}{9}$	22	$\frac{4}{17}$
13	$\frac{1}{6}$	23	$\frac{6}{13}$
14	$\frac{2}{3}$	24	$\frac{4}{21}$
15	$\frac{9}{11}$	25	$\frac{3}{19}$
16	$\frac{6}{7}$	26	$\frac{3}{17}$
17	$\frac{8}{7}$	27	$\frac{4}{15}$
18	$\frac{1}{9}$	28	$\frac{7}{18}$
19	$\frac{1}{3}$	29	$\frac{3}{22}$
20	$\frac{4}{11}$	30	$\frac{4}{33}$

Puzzle

Number Game

This is a game for two or more players.

a Get someone who is not playing to call out a fraction.

b Each player now estimates the decimal equivalent of the fraction to two decimal places and keeps a record in a table like the one shown below.

c The player or players with the nearest estimate scores 1 point.

d The game ends after an agreed number of fractions – say 6 or 10. The person with the highest score wins.

Use a score sheet like this.

Fraction	Estimated decimal value	Actual decimal value (to 3 d.p.)	Difference between estimated value and actual value	Score
$\frac{7}{12}$	0.6	0.583	0.017	

Division by decimals

$0.012 \div 0.06$ can be written as $\frac{0.012}{0.06}$. We know how to divide by a whole number so we need to find an equivalent fraction with denominator 6 instead of 0.06. Now $0.06 \times 100 = 6$. Therefore we multiply the numerator and denominator by 100.

$$\frac{0.012}{0.06} = \frac{0.012 \times 100}{0.06 \times 100} = \frac{1.2}{6}$$
$$= 0.2$$

To divide by a decimal, the denominator must be made into a whole number but the numerator need not be. We can write, for short,

$$0.012 \div 0.06 = \frac{0.01{\vdots}2}{0.06{\vdots}} = \frac{1.2}{6}$$

(keeping the points in line)

the dotted line indicating where we want the point to be so as to make the denominator a whole number.

Exercise 2k

Find $0.024 \div 0.6$

$0.024 \div 0.6 = \dfrac{0.0|24}{0.6} = \dfrac{0.24}{6}$

$ = 0.04$

(See that the decimal points are one beneath the other.

Draw a line through the fraction where you want the decimal point to be.)

$$6\overline{)0.24}$$
$$0.04$$

$64 \div 0.08$

$64 \div 0.08 = \dfrac{64.00|}{0.08|} = \dfrac{6400}{8}$

$ = 800$

(Multiplying the top and bottom by 100 makes the bottom a whole number.)

$$8\overline{)6400}$$
$$800$$

Find the exact answers to the following questions:

1	$0.04 \div 0.2$	**14**	$1.08 \div 0.003$	**27**	$0.496 \div 1.6$
2	$0.0006 \div 0.03$	**15**	$0.0012 \div 0.1$	**28**	$0.0288 \div 0.18$
3	$4 \div 0.5$	**16**	$0.009 \div 0.9$	**29**	$34.3 \div 1.4$
4	$0.8 \div 0.04$	**17**	$0.9 \div 0.009$	**30**	$10.24 \div 3.2$
5	$90 \div 0.02$	**18**	$0.92 \div 0.4$	**31**	$0.0204 \div 0.017$
6	$0.48 \div 0.04$	**19**	$16.8 \div 0.8$	**32**	$102.5 \div 2.5$
7	$0.032 \div 0.2$	**20**	$0.00132 \div 0.11$	**33**	$9.8 \div 1.4$
8	$3.6 \div 0.6$	**21**	$0.000\,068\,4 \div 0.04$	**34**	$0.168 \div 0.14$
9	$3.6 \div 0.06$	**22**	$20.8 \div 0.0004$	**35**	$1.35 \div 0.15$
10	$3 \div 0.6$	**23**	$0.0012 \div 0.3$	**36**	$0.192 \div 2.4$
11	$6.5 \div 0.5$	**24**	$4.8 \div 0.08$		
12	$8.4 \div 0.07$	**25**	$1.76 \div 2.2$		
13	$72 \div 0.09$	**26**	$144 \div 0.16$		

Use long division.

Exercise 2l

Find the value of $16.9 \div 0.3$ giving your answer correct to 1 decimal place.

$16.9 \div 0.3 = \dfrac{16.9|}{0.3|} = \dfrac{169}{3}$

$ = 56.3|3...$

$ = 56.3$ (correct to 1 decimal place)

$$3\overline{)169.00}$$
$$56.33...$$

Calculate, giving your answers correct to 2 decimal places:

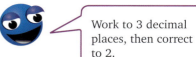

Work to 3 decimal places, then correct to 2.

1   3.8 ÷ 0.6   **6**   1.25 ÷ 0.03

2   0.59 ÷ 0.07   **7**   0.0024 ÷ 0.09

3   15 ÷ 0.9   **8**   0.65 ÷ 0.7

4   5.633 ÷ 0.2   **9**   0.0072 ÷ 0.007

5   0.796 ÷ 1.1   **10**   5 ÷ 7

Calculate, giving your answers correct to the number of decimal places indicated in the brackets:

11   0.123 ÷ 6   (2)

12   2.3 ÷ 0.8   (1)

13   90 ÷ 11   (1)

14   0.0078 ÷ 0.09   (3)

15   12 ÷ 9   (4)

16   0.23 ÷ 0.007   (1)

17   16.2 ÷ 0.8   (1)

18   0.21 ÷ 6.5   (3)

19   85 ÷ 0.3   (3)

20   1.37 ÷ 0.8   (1)

21   56.9 ÷ 1.6   (nearest whole number)

22   0.89 ÷ 0.23   (1)

23   0.75 ÷ 4.5   (3)

24   0.023 ÷ 0.021   (1)

25   3.2 ÷ 1.4   (1)

26   0.045 ÷ 0.012   (nearest whole number)

27   12.3 ÷ 17   (2)

28   0.0054 ÷ 0.021   (4)

29   0.012 ÷ 0.021   (2)

30   0.52 ÷ 0.21   (1)

Remember to work to one more decimal place than you need in your answer. For an answer correct to 1 decimal place you must work to 2 decimal places.

Mixed multiplication and division

Exercise 2m

Calculate, giving your answers exactly:

1   0.48 × 0.3   **7**   0.0042 × 0.03

2   0.48 ÷ 0.3   **8**   0.0042 ÷ 0.03

3   2.56 × 0.02   **9**   16.8 × 0.4

4   2.56 ÷ 0.02   **10**   1.68 ÷ 0.4

5   3.6 × 0.8   **11**   20.4 × 0.6

6   9.6 × 0.6   **12**   5.04 ÷ 0.06

Ignore the decimal points and multiply the numbers. The number of decimal places in the answer in equal to the total number of decimal places in the numbers multiplied.

Find $\frac{0.12 \times 3}{0.006}$

$\frac{0.12 \times 3}{0.006} = \frac{0.36}{0.006}$ (Multiply 0.12 by 3 first.) $12 \times 3 = 36$

$\qquad\qquad = \frac{360}{6}$ (Multiply top and bottom by 1000.)

$\qquad\qquad = 60$

Find the value of:

13 $\frac{0.2 \times 0.6}{0.4}$ **16** $\frac{3.2}{4 \times 0.2}$ **19** $\frac{2.5 \times 0.7}{3.5 \times 4}$

14 $\frac{1.2 \times 0.04}{0.3}$ **17** $\frac{3}{0.6 \times 0.5}$ **20** $\frac{5.6 \times 0.8}{6.4}$

15 $\frac{4.8 \times 0.2}{0.6 \times 0.4}$ **18** $\frac{4.4 \times 0.3}{11}$ **21** $\frac{0.9 \times 4}{0.5 \times 0.6}$

Relative sizes

To compare the sizes of numbers they need to be in the same form, either as fractions with the same denominators, or as decimals.

Exercise 2n

Express 0.82, $\frac{4}{5}$, $\frac{9}{11}$ as decimals where necessary and write them in order of size with the smallest first.

$\frac{4}{5} = 0.8$

$\frac{9}{11} = 0.8181...$ $11\overline{)9.000}$
                                              0.8181...

In order of size: $\frac{4}{5}$, $\frac{9}{11}$, 0.82

Express the following sets of numbers as decimals or as fractions and write them in order of size with the smallest first:

1 $\frac{1}{4}$, 0.2   **4** $\frac{1}{3}$, 0.3, $\frac{3}{11}$ **7** $\frac{3}{8}$, $\frac{9}{25}$, 0.35   **10** $0.\dot{7}$, $\frac{8}{11}$

2 $\frac{2}{5}$, $\frac{4}{9}$ **5** $\frac{8}{9}$, 0.9, $\frac{7}{8}$ **8** $\frac{3}{5}$, $\frac{4}{7}$, 0.59   **11** $0.\dot{3}$, $\frac{5}{12}$

3 $\frac{1}{2}$, $\frac{4}{9}$ **6** $\frac{3}{4}$, $\frac{17}{20}$ **9** $\frac{3}{7}$, $\frac{5}{11}$, $\frac{6}{13}$ **12** $\frac{1}{2}$, 0.45, $\frac{9}{19}$

Changing between fractions, decimals and percentages

To change a fraction to a decimal, divide the numerator by the denominator.

The fraction $\frac{3}{8}$ means $3 \div 8$

You can calculate $3 \div 8$:

$$8\overline{)3.000}^{\,0.375}$$

So $\frac{3}{8} = 0.375$

Any fraction can be treated like this.

To change a decimal to a fraction express it as a number of tenths or hundredths, etc. and, if possible, simplify.

The decimal 0.6 can be written $\frac{6}{10}$, which simplifies to $\frac{3}{5}$ and the decimal 1.85 can be written $1\frac{85}{100}$, which simplifies to $1\frac{17}{20}$

To change a percentage to a decimal divide the percentage by 100.

To express a percentage as a decimal, start by expressing it as a fraction, but *do not simplify,* because dividing by 100, or by a multiple of 100, is easy.

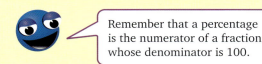

Remember that a percentage is the numerator of a fraction whose denominator is 100.

For example $44\% = \frac{44}{100} = 44 \div 100 = 0.44$ and $12.5\% = \frac{12.5}{100} = 12.5 \div 100 = 0.125$

To change a decimal to a percentage simply multiply by 100

For example $0.34 = 34\%$ and $1.55 = 155\%$

To change a percentage to a fraction divide by 100 and simplify.

We know that 20% of the cars in a car park means $\frac{20}{100}$ of the cars there.

Now $\frac{20}{100}$ can be simplified to the equivalent fraction $\frac{1}{5}$, i.e. $20\% = \frac{1}{5}$.

Similarly 45% of the sweets in a bag means the same as $\frac{45}{100}$ of them and $\frac{45}{100} = \frac{9}{20}$ i.e. $45\% = \frac{9}{20}$

To change a fraction to a percentage change it to a decimal, then multiply by 100.

You can write a fraction as a percentage in two steps.

First write the fraction as a decimal; you do this by dividing the top by the bottom.

For example, $\frac{4}{5} = 4 \div 5 = 0.8$

then change the decimal to a percentage: $0.8 = 80\%$.

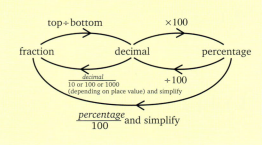

Exercise 2p

1 Work out each fraction as a decimal.

a $\frac{3}{4}$ b $\frac{3}{5}$ c $\frac{3}{10}$ d $\frac{3}{20}$ e $\frac{7}{8}$ f $\frac{6}{25}$

2 Work out $\frac{9}{20}$ as a decimal.

Now decide which is larger, $\frac{9}{20}$ or 0.47?

3 Write each decimal as a fraction in its lowest terms, using mixed numbers where necessary.

a 0.06   b 0.004   c 15.5   d 2.01   e 3.25

In questions **4** and **5** write each decimal as a fraction in its lowest terms.

4 It is estimated that 0.86 of the families in Northgate Street own a car.

5 There were 360 seats on the aircraft and only 0.05% of them were vacant.

6 Write these decimals as percentages.

a 0.3   b 0.2   c 0.7   d 0.035   e 0.925

7 Write these decimals as percentages.

a 1.32   b 1.5   c 2.4   d 1.05   e 2.555

(Remember that 1 is 100%, so $1.66 = 100\% + 66\% = 166\%$)

8 Write these percentages as decimals.

a 45%   b 60%   c 95%   d 5.5%   e 12.5%

9 Express each percentage as a fraction in its lowest terms.

a 40%   b 65%   c 54%   d 25%

10 Express each fraction as a percentage.

a $\frac{2}{5}$ b $\frac{3}{20}$ c $\frac{21}{50}$ d $\frac{15}{50}$

In questions **11** to **14** express the given percentage as a fraction in its lowest terms.

11 Last summer 60% of the pupils in my class went on holiday.

12 At my youth club only 35% of the members are boys.

13 The Post Office claims that 95% of the letters posted arrive the following day.

14 A survey showed that 32% of the pupils in a year group needed to wear glasses.

In each question from **15** to **18** express the fraction as a percentage.

15 At a youth club $\frac{17}{20}$ of those present took part in at least one sporting activity.

16 About $\frac{17}{50}$ of first-year pupils watch more than 20 hours of television a week.

17 Approximately $\frac{3}{5}$ of sixteen-year-olds have a Saturday job.

18 Recently, at the local garage, $\frac{1}{8}$ of the cars tested failed to get a test certificate.

19 Copy and complete the following table:

Fraction	Percentage	Decimal
$\frac{3}{4}$	60%	0.6
$\frac{4}{5}$		
	75%	
		0.7
$\frac{11}{20}$		
	44%	

20 The registers showed that only 0.05 of the pupils in the first year had 100% attendance last term.

   a What fraction is this?

   b What percentage of the first-year pupils had 100% attendance last term?

21 Marion spends $\frac{21}{50}$ of her income on food and lodgings.

   a What percentage is this?

   b As a decimal, what part of her total income does she spend on food and lodging?

22 Marmalade consists of 28% fruit, $\frac{3}{5}$ sugar and the remainder water.

   a What fraction of the marmalade is fruit?

   b What percentage of the marmalade is sugar?

   c What percentage is water?

23 An alloy is 60% copper, $\frac{7}{20}$ nickel and the remainder is tin.

   a What fraction is copper?

   b What percentage is   **(i)** nickel   **(ii)** either nickel or copper?

   c Express the part that is tin as a decimal.

   d What is the ratio of the amount of copper to the amount of tin?

? Puzzle

Everton Giles stands on the middle rung of a ladder. He climbs 3 rungs higher but has forgotten something so descends 7 rungs to get it. He now goes up 16 rungs and reaches the top of the ladder. How many rungs are there to the ladder?

The first significant figure

Rea wants to paint this door but does not know if she has enough paint. A full tin covers 5 square metres but her tin is half full. She measures this door: it is 2.18 m high and 0.83 m wide.

She only needs a rough value of the area of the door. Rea approximates the measurements as 2 m by 0.8 m. This gives an area of approximately 1.6 square metres. This is well under half the area that a full tin of paint will cover so she probably has enough paint.

By approximating 2.18 to 2 and 0.83 to 0.8, Rea has corrected each number to its highest non-zero value digit.
The highest non-zero value digit in a number is called the first significant figure.

The first significant figure in 38.65 is 3, and the first significant figure in 0.007 02 is 7.

Rough estimates

If you were asked to find 1.397×62.54 you could do it by long multiplication or you could use a calculator. Whichever method you choose, it is essential first to make a rough estimate of the answer. You will then know whether the actual answer you get is reasonable or not.

One way of estimating the answer to a calculation is to write each number correct to the highest non-zero value digit. This is called correcting the number to one significant figure (1 s.f.).

So
$$1.397 \times 62.57 \approx 1 \times 60 = 60$$

Correct each number to 1 s.f. and hence give a rough answer to

a 9.524×0.0837 **b** $54.72 \div 0.761$

a $9.524 \times 0.0837 \approx 10 \times 0.08 = 0.8$

b $\dfrac{54.72}{0.761} \approx \dfrac{50}{0.8} = \dfrac{500}{8}$

           $= 60$ (giving $500 \div 8$ to 1 s.f.)

Correct each number to 1 s.f. and hence give a rough answer to each of the following calculations:

1 4.78×23.7

2 56.3×0.573

3 $0.0674 \div 5.24$

4 354.6×0.0475

5 576×256

6 $82.8 \div 146$

7 0.632×0.845

8 0.0062×574

9 $7.835 \div 6.493$

10 4736×729

11 34.7×21

12 8.63×0.523

13 $34.9 \div 15.8$

14 $0.47 \div 0.714$

15 $985 \div 57.2$

16 $0.0326 \div 12.4$

17 0.00724×0.783

18 $3581 \div 45$

19 1097×94

20 45.07×0.0327

Correct each number to 1 s.f. and hence estimate $\dfrac{0.048 \times 3.275}{0.367}$ to 1 s.f.

$$\frac{0.04\vert8 \times 3.\vert275}{0.3\vert67} \approx \frac{0.05 \times 3}{0.4} = \frac{0.15}{0.4} = \frac{1.5}{4}$$

$$= 0.4 \text{ (to 1 s.f.)}$$

21 $\dfrac{3.87 \times 5.24}{2.13}$

22 $\dfrac{0.636 \times 2.63}{5.47}$

23 $\dfrac{21.78 \times 4.278}{7.96}$

24 $\dfrac{6.38 \times 0.185}{0.0628}$

25 $\dfrac{43.8 \times 3.62}{4.72}$

26 $\dfrac{89.03 \times 0.07937}{5.92}$

27 $\dfrac{975 \times 0.636}{40.78}$

28 $\dfrac{8.735}{5.72 \times 5.94}$

29 $\dfrac{0.527}{6.41 \times 0.738}$

30 $\dfrac{57.8}{0.057 \times 6.93}$

Calculations: multiplication and division

When you key in a number on your calculator it appears on the display. Check that the number on display is the number that you intended to enter.

Also check that you press the correct operator, i.e. press $\times$ to multiply and $\div$ to divide.

Always make an estimate and use it to check that the answer given by the calculator is reasonable.

Exercise 2r

First make a rough estimate of the answer. Then use your calculator to give the answer correct to the number of decimal places shown in brackets.

1 2.16×3.28 (2)
2 2.63×2.87 (2)
3 1.48×4.74 (2)
4 4.035×2.116 (2)
5 3.142×2.925 (2)
6 6.053×1.274 (2)
7 2.304×3.251 (2)
8 8.426×1.086 (2)
9 $5.839 \div 3.618$ (2)
10 $6.834 \div 4.382$ (2)
11 $9.571 \div 2.518$ (2)
12 $5.393 \div 3.593$ (2)
13 $7.384 \div 2.51$ (2)
14 $4.931 \div 3.204$ (2)
15 $8.362 \div 5.823$ (2)
16 23.4×56.7 (1)
17 384×21.8 (1)
18 45.8×143.7 (1)
19 $537.8 \div 34.6$ (1)
20 $45.35 \div 6.82$ (2)
21 63.8×2.701 (1)
22 $40.3 \div 2.74$ (1)
23 $400 \div 35.7$ (1)
24 $(34.2)^2$ (1)
25 5007×2.51 (1)

26 $5703 \div 154.8$ (1)
27 39.03×49.94 (1)
28 $2000 \div 52.66$ (1)
29 $(36.8)^2$ (1)
30 $29\,006 \div 2.015$ (1)
31 0.366×7.37 (2)
32 0.0526×0.372 (4)
33 $6.924 \times 0.007\,93$ (4)
34 0.638×825 (1)
35 52×0.0895 (2)
36 0.0826×0.582 (4)
37 24.78×0.0724 (2)
38 $0.008\,35 \times 0.617$ (5)
39 0.5824×6.813 (2)
40 $(0.74)^2$ (3)
41 $0.583 \div 4.82$ (3)
42 $0.628 \div 7.61$ (4)
43 $0.493 \div 1.253$ (3)
44 $0.518 \div 5.047$ (3)
45 $82.7 \div 593$ (3)
46 $89.5 \div 0.724$ (1)
47 $38.07 \div 0.682$ (1)
48 $5.71 \div 0.0623$ (1)
49 $7.045 \div 0.0378$ (1)
50 $6.888 \div 0.0072$ (1)

51 $45.37 \div 0.925$ (1)
52 $8.41 \div 0.000\,748$ (1)
53 $6.934 \div 0.0829$ (1)
54 $0.824 \div 0.362$ (2)
55 $0.572 \div 0.851$ (3)
56 $0.528 \div 0.0537$ (2)
57 $0.571 \div 0.824$ (3)
58 $0.0455 \div 0.0613$ (3)
59 $0.006 \div 0.047\,03$ (3)
60 $0.824 \div 0.000\,008$ (exact)
61 $5000 \div 0.789$ (1)
62 $(0.078)^2$ (5)
63 0.0608×573 (1)
64 $(78.5)^3$ (1)
65 $\dfrac{3.782 \times 0.467}{4.89}$ (3)
66 $4.88 \times 0.004\,17$ (4)
67 $0.9467 \div 7683$ (6)
68 $0.0467 \div 0.000\,074$ (1)
69 $(0.000\,31)^2$ (10)
70 $\dfrac{54.9 \times 36.6}{0.406}$ (1)
71 $68.41 \div 392.9$ (3)
72 $0.0482 \div 0.002\,89$ (1)
73 $(0.0527)^3$ (6)
74 $\dfrac{0.857 \times 8.109}{0.5188}$ (1)

Mixed exercises

Exercise 2s

1   Simplify the ratio $96:216$.

2   Simplify the ratio $\frac{1}{4}:\frac{2}{5}$.

3   Divide 0.0432 by 0.9

4   Write $\frac{2}{3}$ as a decimal correct to 3 d.p.

5   Express 0.095 as a fraction in its lowest terms.

6   Give 57 934 correct to 1 s.f.

7   Give 0.061 374 correct to 4 d.p.

8   Find 0.582×6.382, giving your answer correct to 3 d.p.

9   Find $45.823 \div 15.89$, giving your answer correct to 2 d.p.

10   Find the value of $2^4 \times 3^2$.

Exercise 2t

1   Simplify the ratio $10\,\text{mm}:2\,\text{cm}$.

2   Which is smaller, $32:24$ or $30:22$?

3   Which is larger, 6.6 or $6\frac{2}{3}$? Give a reason for your answer.

4   Find $6.43 \div 0.7$ correct to 3 d.p.

5   Express $\frac{7}{8}$ as a decimal.

6   Give 45 823 correct to 1 s.f.

7   The organisers of a calypso show hope that, to the nearest thousand, 8000 people will buy tickets. What is the minimum number of tickets that they hope to sell?

8   Estimate the value of $12.07 \div 0.008\,97$ giving your answer correct to 1 s.f.

9   Give 7.7815 correct to
   a   the nearest whole number   **b**   1 d.p.   **c**   3 d.p.

10   Change 35% into
   a   a fraction in its lowest terms   **b**   a decimal.

Exercise 2u

1   Two squares have sides 8 m and 12 cm. Find the ratio of the lengths of their sides.

2   Which is larger, 16 : 13 or 9 : 7?

3   Give $\frac{5}{7}$ as a recurring decimal.

4   Write 75% as   **a**   a decimal   **b**   a fraction.

5   Give 9764 correct to 1 s.f.

6   How many pieces of ribbon of length 0.3 m can be cut from a piece 7.5 m long?

7   Find $\frac{0.6 \times 0.3}{0.09}$

8   Estimate, then find $0.0468 \div 0.004\,73$ giving your answer correct to 1 d.p.

9   Find $\frac{56.82 \times 0.714}{8.625}$ giving your answer correct to 2 d.p.

10 Change $\frac{5}{8}$ into   **a**   a percentage   **b**   a decimal.

11 Find the value of $5^2 \times 2^3$.

In this chapter you have seen that...

✔ the units in all parts of a ratio must be the same before they can be omitted

✔ a ratio can be written as a fraction, e.g. $5 : 8 = \frac{5}{8}$

✔ a ratio can be simplified by dividing (or multiplying) all parts in the ratio by the same number

✔ 5^3 means $5 \times 5 \times 5$ and $4 \times 4 \times 4 \times 4 \times 4$ can be shortened to 4^5

✔ you can interchange fractions, percentages and decimals using these rules

  • to divide by a decimal, multiply top and bottom by the same number so that the bottom becomes a whole number

  • the first significant figure is the highest non-zero value digit in a number

  • to correct a number to a given degree of accuracy, place a cut-off line after the place value required and look at the next figure – if it is 5 or more, round up, otherwise round down

✔ you can make a rough estimate of a calculation by correcting each number to one significant figure

✔ you need to be careful when you use a calculator to work out accurate answers.

3 Number patterns

Finding further terms in a number pattern

In Grade 7, Chapter 2, we studied simple number patterns. This chapter revises and extends that work.

Reminders

A square number can be represented by a number of dots arranged in a square.

The first five square numbers are 1, 4, 9, 16 and 25.

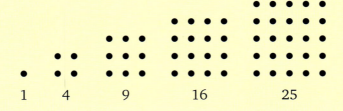

A rectangular number can be shown as a rectangular pattern of dots.

For example 12 can be arranged as 2×6 dots

or as 3×4 dots.

11 is not a rectangular number as it cannot be arranged to give a rectangle.

11 is a prime number because it has no factors other than itself and 1.

Triangular numbers can be shown as dots arranged in rows so that each row is one dot longer than the row above it.

The first five triangular numbers are

1, 3, 6, 10 and 15.

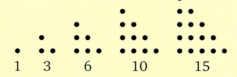

Other patterns may involve numbers increasing or decreasing by the same amount each time. For example in 3, 7, 11, 14, ... the numbers increase by 4. Some patterns involve multiples of a number. For example in 2, 6, 18, 54, ... each number is 3 times the number before it.

Exercise 3a

1 Which of these numbers are rectangular numbers?

11, 20, 31, 40

2 Which of these numbers are square numbers?

30, 36, 64, 80, 100, 120, 169

3 Square numbers, rectangular numbers and triangular numbers are different types of numbers. Which of the numbers 9, 12, 24, 36 and 64 belongs to all three categories?

Find the next two numbers in the number pattern 7, 2, −3, −8,

Each term is 5 less than the one before it.

The next two terms are $-8 - 5 = -13$ and $-13 - 5 = -18$

In questions **4** to **20** write down the next two numbers in the number pattern.

4 3, 6, 12, 24, ...

5 5, 9, 13, 17, ...

6 9, 5, 1, −3, ...

7 2, 5, 10, 17, ...

8 $\frac{1}{2}, \frac{1}{4}, \frac{1}{8}, \frac{1}{16}, \ldots$

9 $\frac{1}{2}, \frac{2}{3}, \frac{3}{4}, \frac{4}{5}, \ldots$

10 1, 0.1, 0.01, 0.001, …

11 8, −1, −10, −19, …

12 32, 16, 8, 4, …

13 4, 9, 19, 34, …

14 −3, −5, −8, −12, …

15 52, 42, 33, 25, …

16 9, −2, −13, −24, …

17 1, −4, 9, −16, …

18 2, −4, 8, −16, …

19 0.5, 0.1, 0.02, …

20 −2, 6, −12, 20, −30, …

So far we have been asked to find the next two terms in a number pattern.

Sometimes terms are missing in a pattern and we need to find them.

Work out the missing numbers in the pattern 4, …, −2, −5, −8, …, −14, …

The terms −2, −5, and −8 show that each term is 3 less than the term before it.

The first missing term is therefore $4 - 3 = 1$

and the last missing term is $-8 - 3 = -11$

In questions **21** to **25** copy the number patterns and write the missing numbers in the spaces.

21 5, 10, …, 40, 80, …, …

22 20, 17, 14, …, …, 5, …

23 …, …, 4, 8, 16.

24 …, 8, …, 2, −1, −4, …

25 $\frac{1}{16}, \frac{1}{8}, …, \frac{1}{2}, …, 2$

26 Copy the diagram and table onto squared paper.

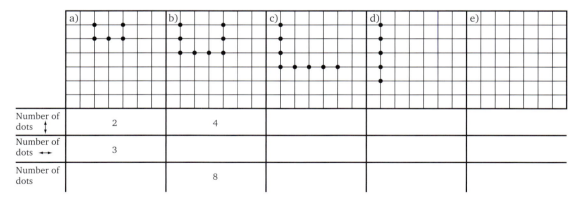

	a)		b)		c)		d)		e)
Number of dots ↕	2		4						
Number of dots ↔	3								
Number of dots			8						

Complete the 'U' shapes in (c), (d) and (e) and then fill in the table.

Fill in the fifth column in the table without drawing another diagram.

27 Copy the diagram onto squared paper.

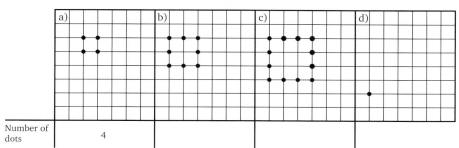

Complete the third and fourth squares and fill in the number of dots under each diagram.

How many dots will there be in the fifth diagram?

28 Continue the pattern of triangular numbers by writing the next four.

1, 3, 6, 10, …

29 The first six terms of a number pattern are 2, 6, 12, 20, 30, 42, …

   a Write down the next two terms.

   b Explain why all the numbers in this number pattern are even.

30 This diagram shows that the triangular numbers 6 and 10 add together to give a square number.

Draw similar diagrams to show that the following triangular numbers give square numbers   **a** $3 + 6$ **b** $10 + 15$

$$\times \times \times \bullet$$
$$\times \times \bullet \bullet$$
$$\times \bullet \bullet \bullet$$
$$\bullet \bullet \bullet \bullet$$

31 Continue each number pattern for another three terms:

   a $1 \times 2, 2 \times 3, 3 \times 4, \ldots$

   b $1 + (2 \times 3), 2 - (3 \times 4), 3 + (4 \times 5), 4 - (5 \times 6), \ldots$

32 Use three different diagrams to show that 24 is a rectangular number.

33 Copy and complete the first six rows of this pattern.

$(1 + 2) \times 3 = 9$

$(2 + 3) \times 4 = 20$

$(3 + 4) \times 5 = ?$

$(4 + 5) \times ? = ?$

$(5 + ?) \times ? = ?$

$(? + ?) \times ? = ?$

34 Copy and complete the first six rows of this pattern.

$(1 \times 2) + 3 = 5$

$(2 \times 3) + 4 = 10$

$(3 \times 4) + 5 = ?$

$(4 \times 5) + ? = ?$

$(5 \times ?) + ? = ?$

$(? \times ?) + ? = ?$

35 Write down the next two rows of this pattern.

$3 - (1 \times 2)$

$4 - (2 \times 3)$

$5 - (3 \times 4)$

$6 - (4 \times 5)$

36 Write down the next two rows of this pattern.

$(1 \times 2) - 4$

$(2 \times 3) - 5$

$(3 \times 4) - 6$

$(4 \times 5) - 7$

37 The number of dots in each of these diagrams are arranged as a diamond.

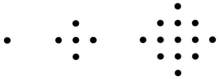

a Draw the next two diagrams in the pattern.

b Write down the number of dots in the first five patterns.

38

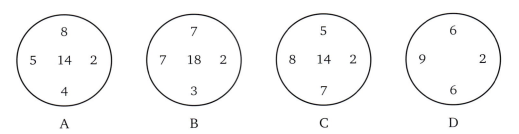

For each circle the number in the centre is the sum found by multiplying two opposite numbers together then using the difference between the remaining two. Find the number at the centre of D.

? Puzzle

a Copy and complete this magic square, using the digits 1 to 9 once each, so that the total in every row and column, and in both diagonals is 15.

6	1	8
7		
2		

b Copy and complete this magic square, using the numbers 1 to 16 once each, so that the total in every row, column and in both diagonals is 34.

1			4
	6		
		11	
13			16

Number patterns from problems

Sometimes number patterns arise in games and practical situations.

Exercise 3b

1 In a restaurant chairs can be arranged around one or more square tables as shown below.

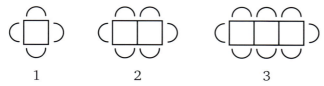

1   2   3

How many chairs would be required when the number of tables arranged in this way is **a** 4   **b** 5   **c** 8?

2 Sally uses matchsticks to make patterns. One of her patterns is shown below.

1   2   3

a How many matchsticks does she use to make
   i pattern 1   **ii** pattern 2   **iii** pattern 3?
b How many matchsticks does she need to make pattern 10?

3 The first three terms of a number pattern are 2, 3, 5, …
a If these numbers are the first three terms of the number pattern of prime numbers, write down the next six terms.
b If other numbers in the pattern do not have to be prime numbers write down another way of continuing the pattern. State the rule you are using.

4

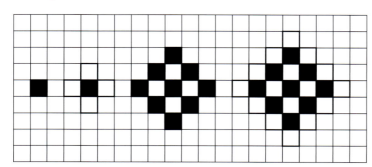

a Look at this tiling pattern, then copy and complete the following table.

Pattern number	1	2	3	4
Number of black tiles	1	1		
Number of white tiles	0	4		
Total number of tiles	1	5		

b Without drawing pattern number 5 add column 5 to the table and go on to add column 6.

c Is it true to say that the total number of tiles used is equal to the sum of the squares of two consecutive whole numbers?

d Which two consecutive whole numbers need to be squared to give the total number of tiles used

    i in pattern number 7   **ii** in pattern number 10?

5

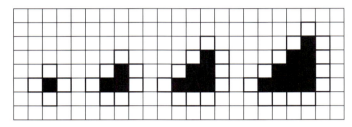

The diagrams show a pattern of black tiles surrounded by white tiles.

a Draw the next two diagrams in the pattern.

b Copy and complete the following table which shows the number of each type of tile needed for the first six diagrams in the pattern.

Diagram number	1	2	3	4	5	6
Number of black tiles	1	3				
Number of white tiles	8	12				

c Without drawing a diagram, determine the number of white tiles in diagram number 8.

6

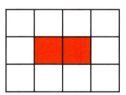

The diagram shows a pattern of two red tiles surrounded by ten white tiles.

How many white tiles are required to surround a row of:

a 3 red tiles   **b** 4 red tiles   **c** 10 red tiles?

7 **a** Look at this number pattern and write down the next five lines

$$1^2 + 2^2 = 3^2 - 2^2$$

$$2^2 + 3^2 = 7^2 - 6^2$$

$$3^2 + 4^2 = 13^2 - 12^2$$

Verify that each line is correct.

b Use the pattern to complete the following lines

  i $9^2 + 10^2 = \quad - \quad$

  ii $\quad + \quad = 57^2 - 56^2$

8 A number pattern is formed using the following rules: The first two terms are 2 and 3.

To find the third term multiply the first two terms together and add 1.

To find the fourth term multiply the first three terms together and add 1.

Subsequent terms are found by continuing this process.

  a Write down the first six terms of this sequence.

  b Are all the numbers in the pattern prime? Give a reason for your answer

9 Nine snooker balls are put into a square frame as shown in the diagram.

Another 4 can be placed on these at the points marked with stars, and a further ball can be placed on these 4.

This gives a square pyramid with three layers

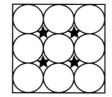

The total number of snooker balls in a two-layer square pyramid is 5 and the total number in a three-layer square pyramid is 14.

How many are needed to build a pyramid with

  a four layers   **b** five layers?

10 Gary agrees his pocket money with his father. He is to get $100 a week rising by $10 a week every six months. The agreement his brother Frank negotiates with their father is different. Frank's pocket money is to be $100 a week to start, increasing by $20 every year.

  a For each son write down the pattern of the different weekly amounts they will receive.

  b Which son has the better deal, Gary or Frank? Justify your answer.

The nth term of a pattern

So far we have been asked to find the next or missing terms in a number pattern.

Sometimes we are given a formula for the nth term.

For example, if we are told that the nth term is T_n and $T_n = 4n + 1$ then we can make a table.

1st term $n = 1$	2nd term $n = 2$	3rd term $n = 3$		nth term	....
$4 \times 1 + 1 = 5$	$4 \times 2 + 1 = 9$	$4 \times 3 + 1 = 13$	....	$4n + 1$	

By substituting the position number into the formula, we can find any term.

In the example, the 10th term is the value of $4n + 1$ when $n = 10$,

so 10th term $= 4 \times 10 + 1 = 41$

Exercise 3c

The nth term, T_n, of a number pattern is given by the formula $T_n = 3n - 2$.

    a Write down the first four terms.

    b Work out the 20th term.

    c Which term is 43?

    a $T_1 = 3 \times 1 - 2 = 1$, $T_2 = 3 \times 2 - 2 = 4$, $T_3 = 3 \times 3 - 2 = 7$ and
        $T_4 = 3 \times 4 - 2 = 10$
        so the number pattern is 1, 4, 7, 10, ...

    b $T_{20} = 3 \times 20 - 2 = 58$

    c $T_n = 3n - 2$ so for some value of n, $3n - 2 = 43$
           $\therefore 3n = 43 + 2 = 45$
        i.e. $n = 15$ so 43 is the 15th term.

1 The nth term, T_n, of a number pattern is given by the formula $T_n = 5n$.
Write down the first 6 terms.

2 The nth term, T_n, of a number pattern is given by the formula $T_n = 4n + 3$.
Write down the first 5 terms.

3 In a number pattern $T_n = 3n - 4$. Write down the first 6 terms.

4 In a number pattern $T_n = 7n - 3$. Write down the first 5 terms.

5 The nth term, T_n, of a number pattern is given by the formula $T_n = \frac{1}{2}n + 2$.
Write down the first 4 terms.

6 The nth term, T_n, of a number pattern is given by the formula $T_n = 2n - 7$.
Write down the first 5 terms.

7 In a number pattern $T_n = 12 - 3n$. Write down the first 5 terms.

8 In a number pattern $T_n = 3(n + 1)$. Write down the first 6 terms.

9 The nth term, T_n, of a number pattern is given by the formula $T_n = 2(4n - 1)$.
Write down the first 5 terms.

10 The nth term, T_n, of a number pattern is given by the formula $T_n = 2n(n + 1)$
Write down the first 6 terms.

11 The nth term, T_n, of a number pattern is given by the formula $T_n = 4n - 3$.
    a Write down the first 6 terms.
    b Work out the 15th term.
    c Which term has a value of 81?

12 The nth term, T_n, of a number pattern is given by the formula $T_n = 3n - 2$.
    a Write down the first 6 terms.
    b Work out the 12th term.
    c Find the value of n if T_n is the first term in the pattern bigger
      than 200.

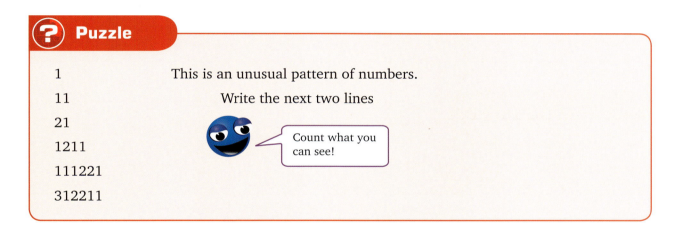

? Puzzle

1
11
21
1211
111221
312211

This is an unusual pattern of numbers.

Write the next two lines

Count what you can see!

In this chapter you have seen that...

✔ you can find other numbers in a number pattern by recognising the pattern.

✔ you can find the numbers in a number pattern by substituting
the values 1, 2, 3, 4, 5, … into an expression for the nth term, T_n.

4 Triangles and angles

At the end of this chapter you should be able to...

1. Identify supplementary angles and complementary angles.
2. Know the relationship between vertically opposite angles.
3. Name the sides and angles of a given triangle.
4. Draw rough copies of given triangles.
5. State the sum of the angles of a triangle.
6. Calculate the third angle of a triangle given the other two angles.
7. Classify triangles as isosceles, equilateral, right-angled or scalene.
8. Identify equal sides (angles) of an isosceles triangle, given the equal angles (sides).
9. Calculate angles of isosceles triangles from necessary data.

You need to know...

✔ how to add and subtract whole numbers
✔ that angles on a straight line add up to 180°

In Grade 7 we saw that one complete revolution is divided into 360°.

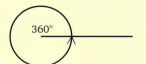

We also saw that one quarter of a revolution is called a right angle,

and that angles on a straight line add up to 180°.

Exercise 4a

In questions **1** and **2** first measure the angle marked *r*. Then estimate the size of the angle marked *s*. Check your estimate by measuring angle *s*.

1

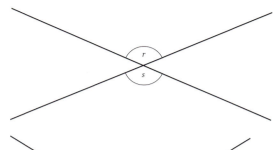

2

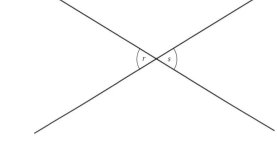

3 Draw some more similar diagrams and repeat questions **1** and **2**.

In each of the following questions, write down the size of the angle marked *t*, without measuring it:

4

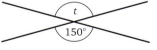

7

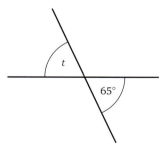

5

8

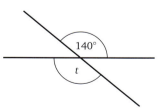

6

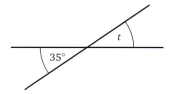

9

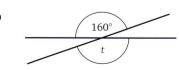

Vertically opposite angles

When two straight lines cross, four angles are formed.

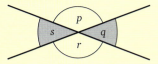

The two angles that are opposite each other are called *vertically opposite angles*. After working through the last exercise you should now be convinced that

vertically opposite angles are equal

i.e. $p = r$ and $s = q$.

Complementary angles

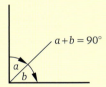

Angles in a right angle add up to 90°.

Two angles that add up to 90° are called complementary angles.

Supplementary angles

Angles on a straight line add up to 180°.

Two angles that add up to 180° are called supplementary angles.

Exercise 4b

In questions **1** to **6** write down the pairs of angles that are supplementary:

1

2

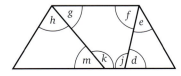

3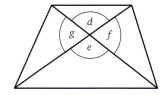

5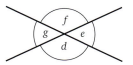

4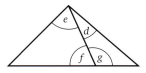

6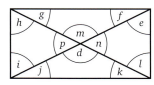

In questions **7** to **14** calculate the size of the angles marked with a letter:

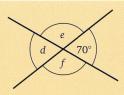

d and 70° are equal (they are vertically opposite)

∴   d = 70°

e and 70° add up 180° (they are angles on a straight line)

∴   e = 110°

f and e are equal (they are vertically opposite)

∴   f = 110°

7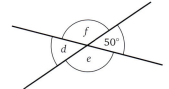

10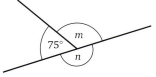

8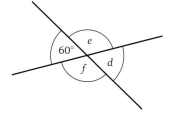

11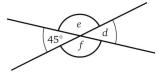

9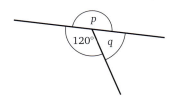

12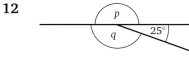

13

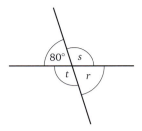

14

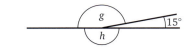

Angles at a point

When several angles make a complete revolution they are called *angles at a point*.

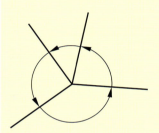

Angles at a point add up to 360°.

Exercise 4c

In questions **1** to **10** find the size of the angle marked with a letter:

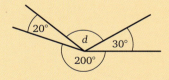

The four angles at the points add up to 360°

The three given angles add up to 250°.

∴ $d = 360° - 250°$

$d = 110°$

$$\begin{array}{r} 30 \\ 200 \\ +20 \\ \hline 250 \\ \hline \end{array}$$

1

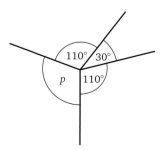

2

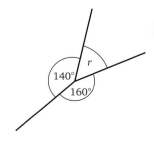

Remember that, however many angles there are at a point, their sum is 360°.

3

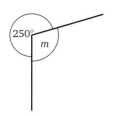

7

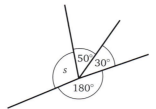

4

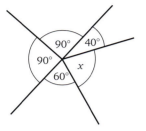

8

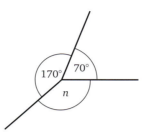

5

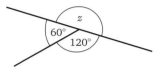

9

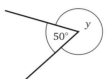

6

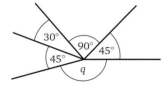

10

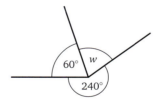

Problems

Exercise 4d

1 Find each of the equal angles marked *s*.

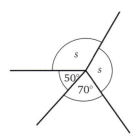

Do not always expect to see immediately how to find the angles asked for. Copy the diagram. On your copy, mark the sizes of any angles that you know or can work out. This may prompt you where to go next.

2   The angle marked f is twice the angle marked g. Find angles f and g.

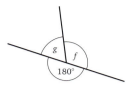

3   Find each of the equal angles marked d.

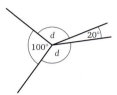

4   Each of the equal angles marked p is 25°. Find the reflex angle q.

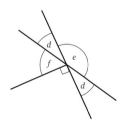

5   Each of the equal angles marked d is 30°. Angle d and angle e are supplementary. Find angles e and f. (An angle marked with a square is a right angle.)

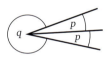

6   Angle s is twice angle t. Find angle r.

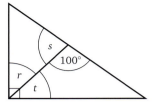

7   The angle marked d is 70°. Find angle e.

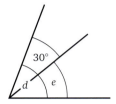

8   Find the angles marked p, q, r and s.

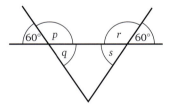

Triangles

A triangle has three sides and three angles.

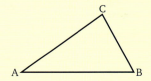

The corners of the triangle are called vertices. (One corner is called a vertex.) So that we can refer to one particular side, or to one particular angle, we label the vertices using capital letters. In the diagram above we use the letters A, B and C so we can now talk about 'the triangle ABC' or '△ABC'.

The side between A and B is called 'the side AB' or AB.
The side between A and C is called 'the side AC' or AC.
The side between B and C is called 'the side BC' or BC.

The angle at the corner A is called 'angle A' or $\widehat{A}$ for short.

Exercise 4e

1   Write down the name of the side that is 4 cm long.

Write down the name of the side that is 2 cm long.

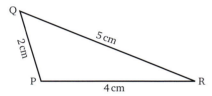

2   Write down the name of
   a   the side that is 2.5 cm long
   b   the side that is 2 cm long
   c   the angle that is 70°

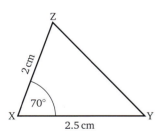

In the following questions, draw a rough copy of the triangle and mark the given measurements on your drawing:

3   In △ABC, AB = 4 cm, $\widehat{B}$ = 60°, $\widehat{C}$ = 50°.

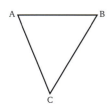

4   In △DEF, $\widehat{E}$ = 90°, $\widehat{F}$ = 70°, EF = 3 cm.

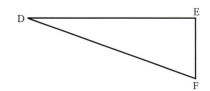

5   In △LMN, $\hat{L}$ = 100°, $\hat{N}$ = 30°, NL = 2.5 cm.

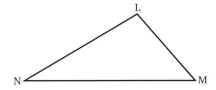

6   In △FGH, FG = 3.5 cm, GH = 3 cm, $\hat{H}$ = 35°.

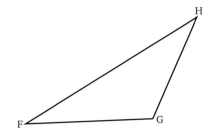

Make a rough drawing of the following triangles.

Label each one and mark the measurements given:

7   △ABC in which AB = 10 cm, BC = 8 cm and $\hat{B}$ = 60°.

8   △PQR in which $\hat{P}$ = 90°, $\hat{Q}$ = 30° and PQ = 6 cm.

9   △DEF in which DE = 8 cm, $\hat{D}$ = 50° and DF = 6 cm.

10   △XYZ in which XY = 10 cm, $\hat{X}$ = 30° and $\hat{Y}$ = 80°.

(!) Investigation

Five different shaped triangles can be drawn on a 2×2 grid.

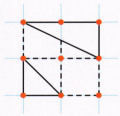   Remember that is the same as

Two of them are shown. Sketch the other three.

Investigate the number of different triangles that can be drawn on a 3×3 grid.

Extend your investigation to a 4×4 grid and then a 5×5 grid.

Can you find a connection between the number of different triangles that can be drawn and the length of the side of the grid?

Angles of a triangle

? Practical work

Draw a large triangle of any shape. Use a straight edge to draw the sides. Measure each angle in this triangle, turning your page to a convenient position when necessary. Add up the sizes of the three angles.

Draw another triangle of a different shape. Again measure each angle and then add up their sizes.

Now try this: on a piece of paper draw a triangle of any shape and cut it out. Next tear off each corner and place the three corners together.

They should look like this:

The three angles of a triangle add up to 180°.

Can you tell why this is so? Hint: the sum of the angles on a straight line is 180°.

Exercise 4f

Find the size of angle A (an angle marked with a square is a right angle):

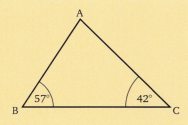

$A + 57° + 42° = 180°$ (angles of $\triangle$ add up to 180°)

$A = 180° - 99°$

$\quad = 81°$

$$\begin{array}{r} 57 \\ +42 \\ \hline 99 \end{array}$$

1

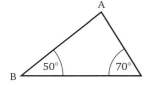

2

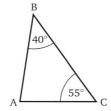

3

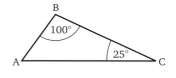

4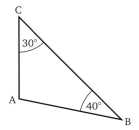

8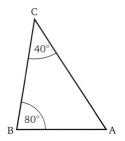

12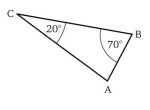

5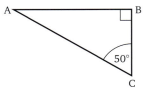

9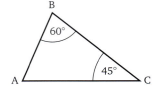

13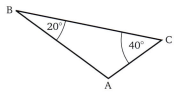

6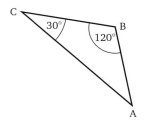

10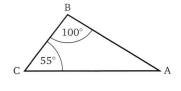

14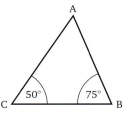

7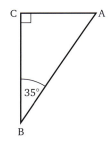

11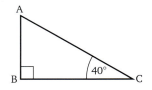

15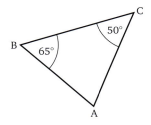

Problems

Reminder

- Vertically opposite angles are equal.
- Angles on a straight line add up to 180°.

You will need these facts in the next exercise.

Exercise 4g

1 Find angles *d* and *f*.

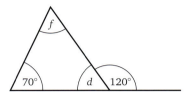

In each question make a rough copy of the diagram and mark the sizes of the angles that you are asked to find. You do not need to find them in alphabetical order. You can also mark in any other angles that you know. This may help you find the angles asked for.

2 Find angles *s* and *t*.

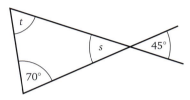

6 Find each of the equal angles *g*.

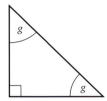

3 Find each of the equal angles *x*.

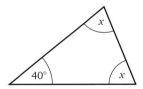

7 Find each of the equal angles *x*.

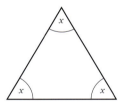

4 Find angles *p* and *q*.

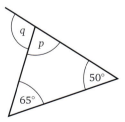

8 Angle *h* is twice angle *j*. Find angles *h* and *j*.

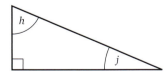

5 Find angles *s* and *t*.

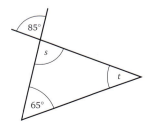

9 Find angles *p* and each of the equal angles *q*.

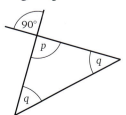

Quadrilaterals

A quadrilateral is bounded by four straight sides. These shapes are examples of quadrilaterals:

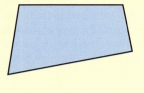 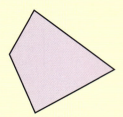 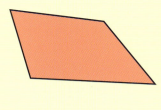

The following diagrams are also quadrilaterals, but each one is a 'special' quadrilateral with its own name:

 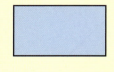   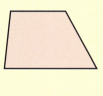

    square   rectangle   parallelogram   rhombus   trapezium

Draw yourself a large quadrilateral, but do not make it one of the special cases. Measure each angle and then add up the sizes of the four angles.

Do this again with another three quadrilaterals.

Now try this: on a piece of paper draw a quadrilateral. Tear off each corner and place the vertices together. It should look like this:

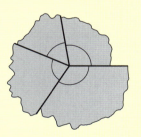

The sum of the four angles of a quadrilateral is 360°.

This is true of any quadrilateral whatever its shape or size.

Can you tell why? Hint: draw one of its diagonals and use what you know about the angles of a triangle.

Exercise 4h

In questions **1** to **10** find the size of the angle marked d:

1

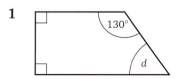

 Make a rough copy of the following diagrams and mark on your diagram the sizes of the required angles. You can also write in the sizes of any other angles that you can: this may help you find the angles you need.

2

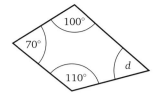

5

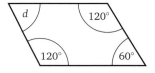

8

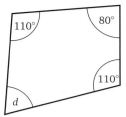

3

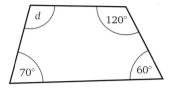

6

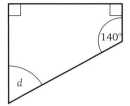

9

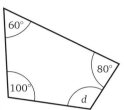

4

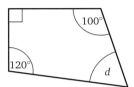

7

10

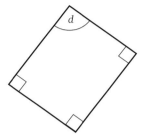

11   Find each of the equal angles *d*.

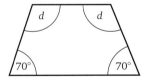

12   Find each of the equal angles *d*.

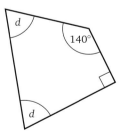

13   Angle *e* is twice angle *d*. Find angles *d* and *e*.

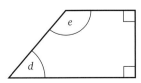

14   Find angles *d* and *e*.

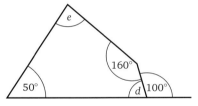

15   Find *d* and each of the equal angles *e*.

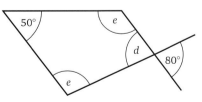

16   Angles *d* and *e* are supplementary. Find *d* and each of the equal angles *e*.

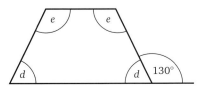

Puzzle

Eight square serviettes are placed flat but overlapping on a table and give the outlines shown in the diagram. In which order must they be removed if the top one is always next.

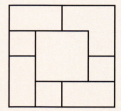

Some special triangles: equilateral, isosceles and right-angled

A triangle in which all three sides are the same length is called an *equilateral triangle*.

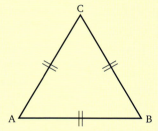

In an equilateral triangle all three sides are the same length and each of the three angles is 60°.

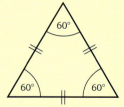

A triangle in which two sides are equal is called an *isosceles triangle*.

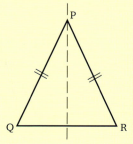

An isosceles triangle has one line of symmetry.

In an isosceles triangles two sides are equal and the two angles opposite the equal sides are equal.

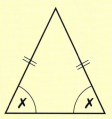

A right-angled triangle has one angle equal to 90º.

A scalene triangle has no equal sides or angles and does not contain a right angle.

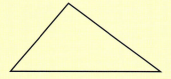

Exercise 4i

In questions **1** to **10** make a rough sketch of the triangle and mark angles that are equal:

1

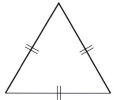

2

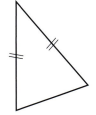

3

4

5

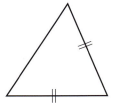

6

7

8

9

10

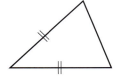

11 Classify these triangles as scalene, right-angled, isosceles or equilateral. If you are not sure, measure the lengths of sides and the sizes of angles.

One or more triangles may fit two classifications.

a

b

c
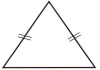

d

e

12 Use GeoGebra to draw
   a a right-angled triangle   **b** an isosceles triangle
   c an equilateral triangle   **d** a right-angled isosceles triangle.

In questions **13** to **24** find angle d:

13

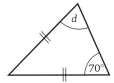

14

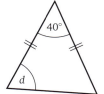

15

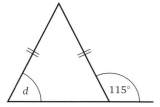

16

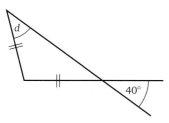

17

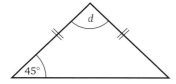

18

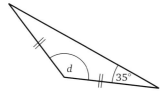

19

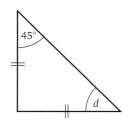

20

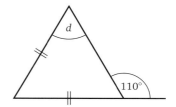

<u>**21**</u>

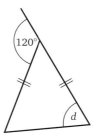

<u>**22**</u>

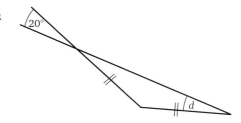

<u>**23**</u>

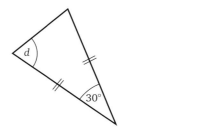

<u>**24**</u>

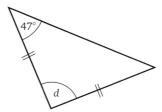

In questions **25** to **28** mark the equal sides:

25

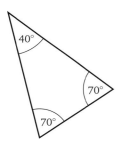

26

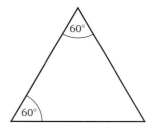

27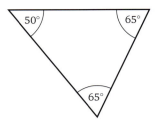

28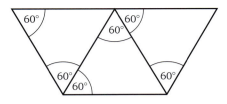

In questions **29** to **34** find angles *d* and *e*:

29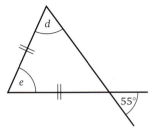

30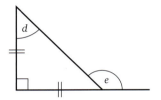

31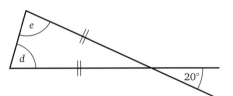

32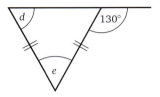

33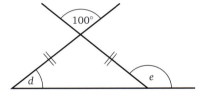

34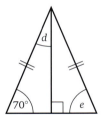

? **Practical work**

On a piece of paper draw an equilateral triangle of side 4 cm.

Draw an equilateral triangle, again of side 4 cm, on each of the three sides of the first triangle.

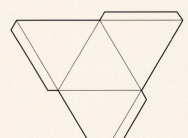

Cut out the complete diagram. Fold the outer triangles up so that the corners meet. Stick the edges together using the tabs. You have made a tetrahedron. (These make good Christmas tree decorations if made out of foil-covered paper.)

Mixed exercises

Exercise 4j

1   Find the size of the angle marked *x*.

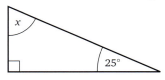

2   Find the size of the angle marked *y*.

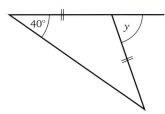

3   Find the size of the angle marked *t*.

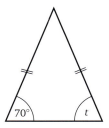

Exercise 4k

1   Find the size of the angles marked *p* and *q*.

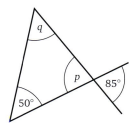

2   Find the size of the angles marked *x* and *y*.

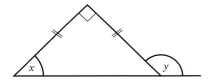

3   Find the size of the angles marked *u* and *v*.

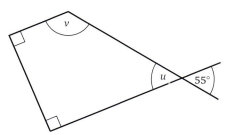

Exercise 4I

1 All three sides of the large triangle are equal. Find angles r and s.

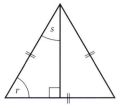

2 Find angles x, y and z.

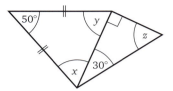

3 Find angles f and g.

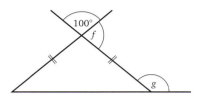

? Puzzle

Find a way of cutting up an equilateral triangle into four pieces so that the pieces fit together to form a square.

In this chapter you have seen that...

✔ complementary angles add up to 90°

✔ supplementary angles add up to 180°

✔ vertically opposite angle are equal

✔ angles at a point add up to 360°

✔ the three angles of a triangle add up to 180°

✔ the four angles of a quadrilateral add up to 360°

✔ an equilateral triangle has three equal sides and each of its angles is 60°

✔ an isosceles triangle has two equal sides and the two angles at the base of these sides are equal

✔ a right-angled triangle has one angle equal to 90°

✔ a scalene triangle has no equal sides, no equal angles and no angle equal to 90°.

5 Parallel lines and angles

You need to know...

✔ that the three angles of a triangle add up to 180°

✔ that angles on a straight line add up to 180°

✔ that vertically opposite angles are equal

✔ that angles at a point add up to 360°

✔ that the sum of the four angles in a quadrilateral add up to 360°

✔ that isosceles triangles have two equal sides and the base angles are equal

✔ that equilateral triangles have three equal sides and three equal angles

✔ how to read a protractor

Key words

alternate angles, base angle, construct, corresponding angles, horizontal, interior angle, isosceles, midpoint, parallel lines, parallelogram, protractor, rotational symmetry, transversal, vertically opposite angles

Parallel lines

Two straight lines that are always the same distance apart, however far they are drawn, are called parallel lines.

The lines in your exercise books are parallel. You can probably find many other examples of parallel lines.

Exercise 5a

1   Using the lines in your exercise book, draw three lines that are parallel. Do not make them all the same distance apart. For example

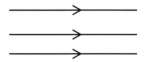

(We use arrows to mark lines that are parallel.)

2   Using the lines in your exercise book, draw two parallel lines. Make them fairly far apart. Now draw a slanting line across them. For example

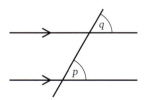

Mark the angles in your drawing that are in the same position as those in the diagram. Are they acute or obtuse angles? Measure your angles marked p and q.

3   Draw a grid of parallel lines like the diagram below. Use the lines in your book for one set of parallels and use the two sides of your ruler to draw the slanting parallels.

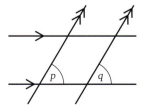

Mark your drawing like the diagram. Are your angles p and q acute or obtuse? Measure your angles p and q.

4   Repeat question 3 but change the direction of your slanting lines.

5   Draw three slanting parallel lines like the diagram below, with a horizontal line cutting them. Use the two sides of your ruler and move it along to draw the third parallel line.

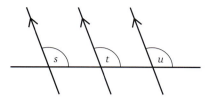

       Mark your drawing like the diagram. Decide whether angles *s*, *t* and *u* are acute or obtuse and then measure them.

6   Repeat question **5** but change the slope of your slanting lines.

Corresponding angles

In the exercise above, lines were drawn that crossed a set of parallel lines.

A line that crosses a set of parallel lines is called a *transversal*.

When you have drawn several parallel lines you should notice that two parallel lines on the same flat surface will never meet however far apart they are drawn.

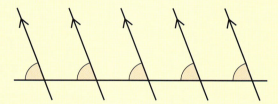

If you draw the diagram above by moving your ruler along you can see that all the shaded angles are equal. These angles are all in corresponding positions: they are all above the transversal and to the left of the parallel lines. Angles like these are called *corresponding angles*.

When two or more parallel lines are cut by a transversal, the corresponding angles are equal.

Exercise 5b

In the diagrams below write down the letter that corresponds to the shaded angle:

1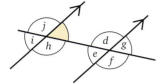

5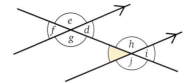

2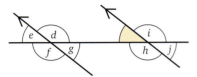

6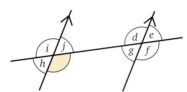

3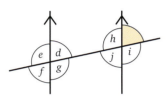

7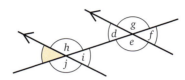

4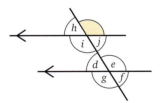

8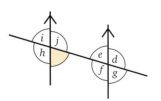

Drawing parallel lines (using a protractor)

The fact that the corresponding angles are equal gives us a method for drawing parallel lines.

If you need to draw a line through the point C that is parallel to the line AB, first draw a line through C to cut AB.

Use your protractor to measure the shaded angle.
Place your protractor at C as shown in the diagram.
Make an angle at C the same size as the shaded angle and in the corresponding position.

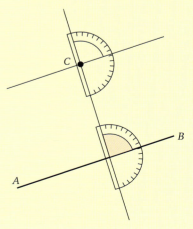

You can now extend the arm of your angle both ways, to give the parallel line.

Exercise 5c

1 Using your protractor draw a grid of parallel lines like the one in the diagram. (It does not have to be an exact copy.)

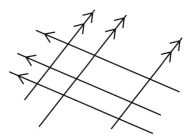

2 Trace the diagram below.

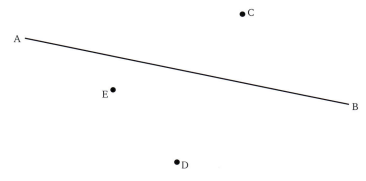

Now draw lines through the points C, D and E so that each line is parallel to AB.

3 Draw a sloping line on your exercise book. Mark a point C above the line. Use your protractor to draw a line through C parallel to your first line.

4 Trace the diagram below.

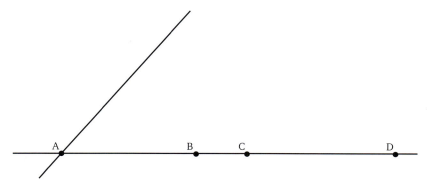

Measure the acute angle at A. Draw the corresponding angles at B, C and D. Extend the arms of your angles so that you have a set of four parallel lines.

In questions **5** to **8** remember to draw a rough sketch before doing the accurate drawing.

5 Draw an equilateral triangle with the equal sides each 8 cm long. Label the corners A, B and C. Draw a line through C that is parallel to the side AB.

6 Draw an isosceles triangle ABC with base AB which is 10 cm long and base angles at A and B which are each 30°. Draw a line through C that is parallel to AB.

7 Draw the triangle as given in question **5** again and this time draw a line through A that is parallel to the side BC.

8 Make an accurate drawing of the figure below where the side AB is 7 cm, the side AD is 4 cm and $\hat{A} = 60°$.

(A figure like this is called a *parallelogram*.)

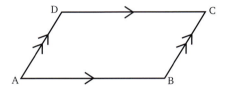

Copy this grid.

How many different-shaped parallelograms can you draw on this grid?

Each vertex must be on a dot. One has been drawn for you.

Do not include squares and rectangles.

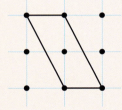

Problems involving corresponding angles

The simplest diagram for a pair of corresponding angles is an F shape.

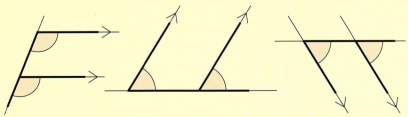

Looking for an F shape may help you to recognise the corresponding angles.

Exercise 5d

Write down the size of the angle marked *d* in each of the following diagrams:

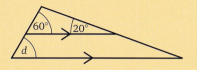

$d = 60°$ (*d* and the angle of 60° are corresponding angles.)

1
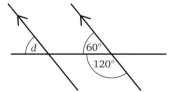

Look for an F shape round the angle you need to find.

2

6
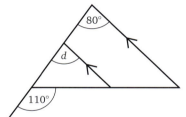

3
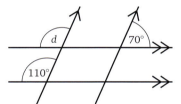

7
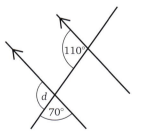

4
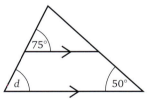

8
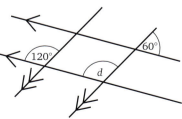

5
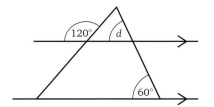

9
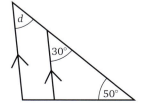

10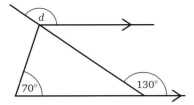

11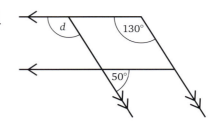

Reminder

Vertically opposite angles are equal.   Angles on a straight line add up to 180°.

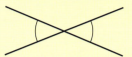   

Angles at a point add up to 360°.   The angles of a triangle add up to 180°.

   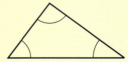

You will need these facts in the next exercise. If you cannot see immediately the angle you want, copy the diagram. On your diagram, write down the size of any angles you can, including those that are not marked. This should help you to find the size of other angles in the diagram, including those that you need. Remember you can use any facts you know about angles.

Exercise 5e

Find the size of each marked angle:

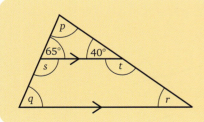

$p = 75°$ (angles of △ add up to 180°)
$q = 65°$ (corresponding angles)
$s = 115°$ (s and 65° add up to 180°)
$r = 40°$ (corresponding angles)
$t = 140°$ (t and 40° add up to 180°)

1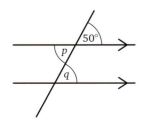

2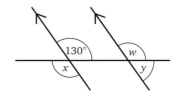

3

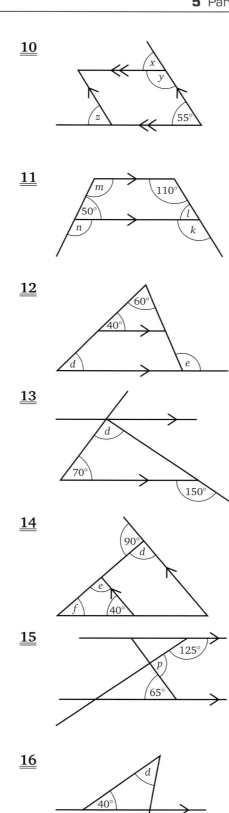

4

5

6

7

8

9

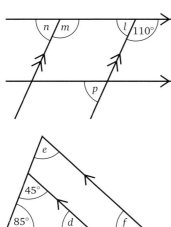

Find the size of angle *d* in questions **17** to **24**:

17

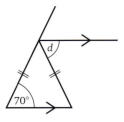

21

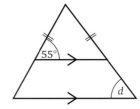

18

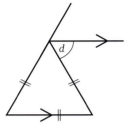

22

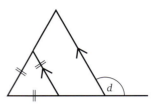

19

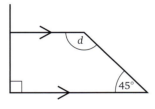

23

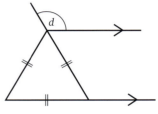

20

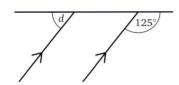

24

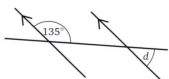

Alternate angles

Draw a large letter Z. Use the lines of your exercise book to make sure that the outer arms of the Z are parallel.

This letter has rotational symmetry about the point marked with a cross. This means that the two shaded angles are equal. Measure them to make sure.

Draw a large N, making sure that the outer arms are parallel.

This letter also has rotational symmetry about the point marked with a cross, so once again the shaded angles are equal. Measure them to make sure.

The pairs of shaded angles like those in the Z and N are between the parallel lines and on alternate sides of the transversal.

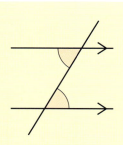

Angles like these are called *alternate angles*.

When two parallel lines are cut by a transversal, the alternate angles are equal.

The simplest diagram for a pair of alternate angles is a Z shape.

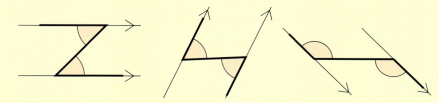

Looking for a Z shape may help you to recognise the alternate angles.

Exercise 5f

Write down the angle that is alternate to the shaded angle in the following diagrams:

Look for a Z shape around the angle you want to find.

1
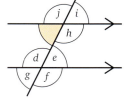

4
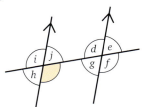

7
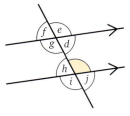

2
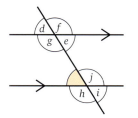

5
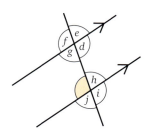

8
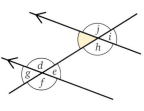

3
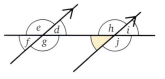

6
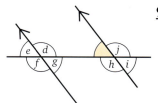

9
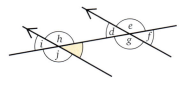

Problems involving alternate angles

Without doing any measuring we can show that alternate angles are equal by using the facts that we already know:

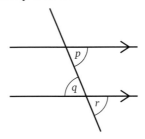

$p = r$ because they are corresponding angles

$q = r$ because they are vertically opposite angles

∴ $p = q$ and these are alternate angles

Proof that the sum of the angles in a triangle is **180°**

On page 67 we showed how to tear off the three angles of a triangle to demonstrate that they fit together to make 180°. It shows that the angles of just those triangles add up to 180°. We can now use the facts we know about angles and parallel line to prove that it is true for *all* triangles.

In the diagram, △ABC is any triangle and LM is parallel to BC. $\angle d + \angle e + \angle f = 180°$ because they are angles on a straight line. But $\angle d = \angle g$ and $\angle f = \angle h$ because they are alternate angles. Therefore $\angle g + \angle e + \angle h = 180°$.

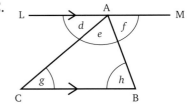

We can use this fact to prove another fact about angles and triangles. In the diagram below, the side CB is extended to D.
The angle k between AB and BD is called an exterior angle of triangle ABC.

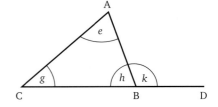

We know that $\angle g + \angle e + \angle h = 180°$.
Also, $\angle k + \angle h = 180°$ because they are angles on a straight line.
Therefore $\angle k = \angle g + \angle e$
This means that an exterior angle of any triangle is equal to the sum of the two interior opposite angles.

Exercise 5g

Find the size of each marked angle:

1

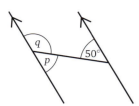

Remember that you can use any angle you know.

2

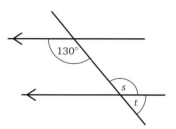

3

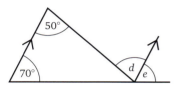

4

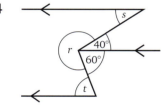

5

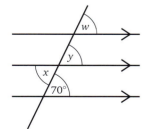

6

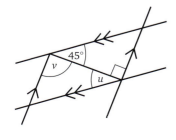

7

8

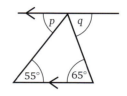

9

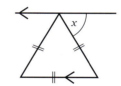

10

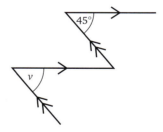

11

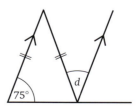

Investigation

This diagram represents a child's billiards table.

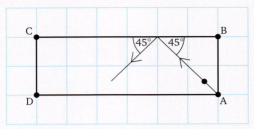

There is a pocket at each corner.

The ball is projected from the corner A at 45° to the sides of the table. It carries on bouncing off the sides at 45° until it goes down a pocket. (This is a very superior toy – the ball does not lose speed however many times it bounces!)

a How many bounces are there before the ball goes down a pocket?

b Which pocket does it go down?

c What happens if the table is 2 squares by 8 squares?

d Can you predict what happens for a 2 by 20 table?

e Now try a 2 by 3 table.

f Investigate for other sizes of tables. Start by keeping the width at 2 squares, then try other widths. Copy this table and fill in the results.

Size of table	Number of bounces	Pocket
2×6		
2×8		
2×3		
2×5		

g Can you predict what happens with a 3×12 table?

Co-interior angles

In the diagram on the right, f and g are on the same side of the transversal and 'inside' the parallel lines.

Pairs of angles like f and g are called *co-interior angles*.

Co-interior angles are often called just interior angles.

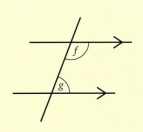

Exercise 5h

In the following diagrams, two of the marked angles are a pair of interior angles.

You may find it helpful to look for a U shape.

Name them:

1

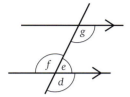

3

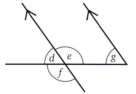

5

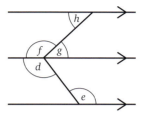

2

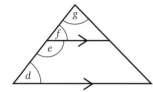

4

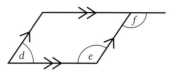

6

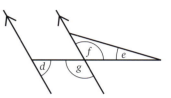

In the following diagrams, use the information given to find the size of p and of q. Then find the sum of p and q:

7

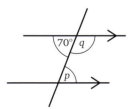

9

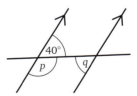

8

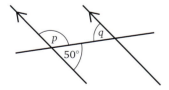

10

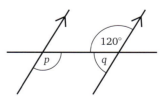

11 Make a large copy of the diagram below. Use the lines of your book to make sure that the outer arms of the 'U' are parallel.

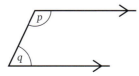

Measure each of the interior angles p and q. Add them together.

The sum of a pair of interior angles is 180°.

You will probably have realised this fact by now.
We can show that it is true from the following diagram.

$d + f = 180°$ because they are angles on a straight line

$d = e$ because they are alternate angles

So $e + f = 180°$

The simplest diagram for a pair of interior angles is a U shape.

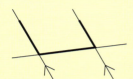 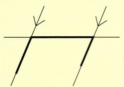 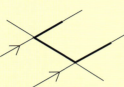

Looking for a U shape may help you to recognise a pair of interior angles.

Exercise 5i

Find the size of each marked angle:

1

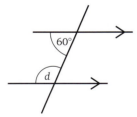

4

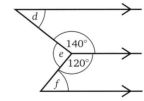

2

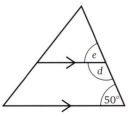

5

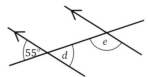

3

6

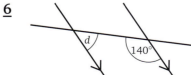

7

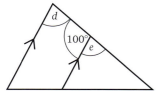

9

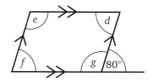

8

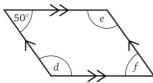

10

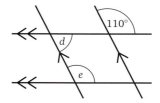

Mixed exercises

You now know that when a transversal cuts a pair of parallel lines

the corresponding (F) angles are equal

the alternate (Z) angles are equal

the interior (U) angles add up to 180°.

You can use any of these facts, together with the other angle facts you know, to answer the questions in the following exercises.

Exercise 5j

Find the size of each marked angle:

1

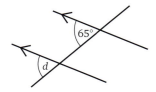

3

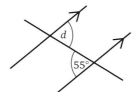

5

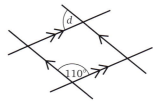

2

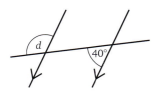

4

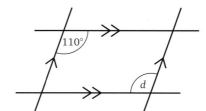

6

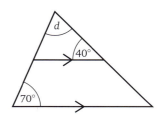

7

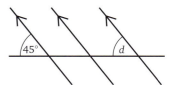

Find the size of each marked angle:

1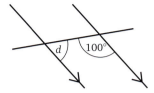

4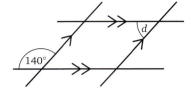

7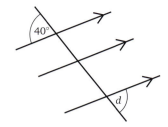

2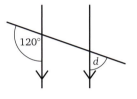

5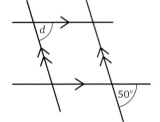

8   Draw the parallelogram below, making it full size.

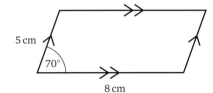

3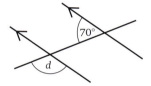

6

In this chapter you have seen that...

✔ you can draw parallel lines

✔ parallel lines cut by a transversal give different types of angles – some are called corresponding angles, some alternate angles and others co-interior angles

✔ corresponding angles are equal; they can be recognised by an F shape

✔ alternate angles are equal; they can be recognised by a Z shape

✔ co-interior angles add up to 180°; they can be recognised by a U shape

✔ geometry problems can often be solved by starting with a copy of the diagram and filling in the sizes of the angles you know.

6 Algebra

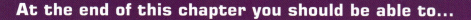

Equations

You worked with equations in Grade 7. This section revises that work.

Imagine a set of scales.

Imagine a balance.

On the left-hand side there are two bags each containing the same (but unknown) number of cricket balls and three loose cricket balls.

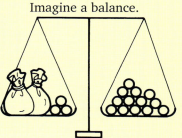

On the right-hand side there are thirteen cricket balls.

The scales balance.

Using the letter x to stand for the unknown number of cricket balls in each bag we can write this as an equation:

$$2x + 3 = 13$$

We can solve this equation (i.e. find the number that x represents) as follows:

take three cricket balls off each pan $2x = 10$

halve the contents of each pan $\quad\quad x = 5$

Remember that we want to isolate x and that we can do anything as long as we do it to both sides of the equation. (The imaginary set of scales must always balance.)

Exercise 6a

Solve the equation $5x - 4 = 6$

$$5x - 4 = 6$$

Add 4 to each side $\quad\quad\quad 5x = 10$

Divide each side by 5 $\quad\quad\quad x = 2$

Check: LHS $= 5 \times 2 - 4 = 6$ RHS $= 6$

Solve the following equations:

1	$2x = 8$	**6**	$3x - 2 = 10$
2	$x - 3 = 1$	**7**	$5 + 2x = 7$
3	$x + 4 = 16$	**8**	$5x - 4 = 11$
4	$2x + 3 = 7$	**9**	$3 + 6x = 15$
5	$3x + 5 = 14$	**10**	$7x - 6 = 15$

Solve the equation $3x + 4 = 12 - x$

(We need to start by getting the terms containing x on one side of the equation and the terms without x on the other side. The left-hand side has the greater number of xs, so we will collect them on this side.)

$$3x + 4 = 12 - x$$

Add x to each side $\quad\quad 4x + 4 = 12$

Take 4 from each side $\quad\quad\quad 4x = 8$

Divide each side by 4 $\quad\quad\quad x = \dfrac{8}{4} = 2$

Solve the equation $4-x=6-3x$

(There are fewer xs missing from the LHS so we will collect them on this side.)

$$4-x=6-3x$$

Add $3x$ to each side $\qquad 4-x+3x=6$

$$4+2x=6$$

Take 4 from each side $\qquad\qquad 2x=2$

Divide each side by 2 $\qquad\qquad\; x=1$

Solve the following equations:

11	$2x+5=x+9$	**16**	$x+4=4x+1$
12	$3x+2=2x+7$	**17**	$3x-2=2x+1$
13	$x-4=2-x$	**18**	$1-3x=9-4x$
14	$3-2x=7-3x$	**19**	$2-5x=6-3x$
15	$2x+1=4-x$	**20**	$5-3x=1+x$

Solve the equation $4x+2-x=7+x-3$

$$4x+2-x=7+x-3$$

Collect like terms $\qquad\quad\; 3x+2=4+x$

Take x from each side $\qquad\; 2x+2=4$

Take 2 from each side $\qquad\quad 2x=2$

Divide each side by 2 $\qquad\quad\; x=1$

Solve the following equations:

21	$x+2+2x=8$	**31**	$5x-8=2$
22	$x-4=3-x+1$	**32**	$4-x=3x$
23	$3x+1-x=5$	**33**	$5-x=7+2x-4$
24	$4+3x-1=6$	**34**	$4-2x=8-4x$
25	$7+4x=2-x+10$	**35**	$15=21-2x$
26	$3+x-1=3x$	**36**	$x+4-3x=2-x$
27	$x-4+2x=5+x-1$	**37**	$3x-7=9-x+6$
28	$x+5-2x=3+x$	**38**	$x+4=6x$
29	$x+17-4x=2-x+6$	**39**	$8-3x=5x$
30	$8-3x-3=x-4+2x$	**40**	$5-4x+7=2x$

Brackets

Reminder: If we want to multiply both x and 3 by 4, we group x and 3 together in a bracket and write $4(x+3)$.

So $4(x+3)$ means that *both x and* 3 are to be multiplied by 4. (Note that the multiplication sign is invisible.)

i.e. $4(x+3) = 4x+12$

($4x$ and 12 are unlike terms so $4x+12$ cannot be simplified)

Exercise 6b

Multiply out the following brackets:

1	$6(x+4)$	**3**	$4(x-3)$	**5**	$4(3-2x)$	**7**	$3(2-3x)$	**9**	$2(5x-7)$
2	$3(2x+1)$	**4**	$2(3x-5)$	**6**	$5(4x+2)$	**8**	$7(5-4x)$	**10**	$6(7+2x)$

Simplify:

11 $2(3+x)+3(2x+4)$

12 $7(2x+3)+4(3x-2)$

First multiply the brackets, then collect like terms.

13 $4(6x+3)+5(2x-5)$

14 $2(2x-4)+4(x+3)$

15 $5(3x-2)+3(2x+5)$

16 $3(3x+1)+4(x+4)$

17 $5(2x+3)+6(3x+2)$

18 $6(2x-5)+2(3x-7)$

19 $8(2-x)+3(3+4x)$

20 $5(7-2x)+4(3-5x)$

21 $3(2x-1)+4(x+2)$

22 $5(2-x)+2(2x+1)$

23 $3(x-4)+7(2x-3)$

24 $2(2x+1)+4(3-2x)$

25 $6(2-x)+2(1-2x)$

26 $5(4+3x)+3(2+7x)$

27 $4(3+2x)+5(4-3x)$

28 $8(x+1)+7(2-x)$

29 $3(2x+7)+5(3x-8)$

30 $9(x-2)+5(4-3x)$

Solve the following equations:

31 $2(x+2) = 8$

32 $4(2-x) = 2$

Start by multiplying out the brackets.

33 $5(3x+1) = 20$

34 $2(2x-1) = 6$

35 $3(2x+5) = 18$

36 $3(3-2x) = 3$

37 $2(x+4) = 3(2x+1)$

38 $4(2x-3) = 2(3x-5)$

39 $6(3x+5) = 12$

40 $6(x+3) = 2(2x+5)$

41 $8(x-1) = 4$

42 $3(1-4x) = 11$

43 $5(3-2x) = 3(4-3x)$

44 $7(1+2x) = 21$

45 $7(2x-1) = 5(3x-2)$

46 $4(3x+2) = 14$

(?) Puzzle

James bought seventeen pens, some black and some red, for $720. He paid $10 more for each red pen than for each black pen.

How many of each colour did he buy?

Multiplication and division of fractions

Remember that, to multiply fractions, the numerators are multiplied together and the denominators are multiplied together:

i.e. $\frac{3}{4} \times \frac{5}{7} = \frac{3 \times 5}{4 \times 7} = \frac{15}{28}$

Also $\frac{1}{6}$ of x means $\frac{1}{6} \times x = \frac{1}{6} \times \frac{x}{1} = \frac{x}{6}$ (1)

Remember that, to divide by a fraction, that fraction is turned upside down and multiplied:

i.e. $\frac{2}{3} \div \frac{5}{7} = \frac{2}{3} \times \frac{7}{5} = \frac{14}{15}$

and $x \div 6 = \frac{x}{1} \div \frac{6}{1} = \frac{x}{1} \times \frac{1}{6} = \frac{x}{6}$ (2)

Comparing (1) and (2) we see that

$\frac{1}{6}$ of x, $\frac{1}{6}x$, $x \div 6$ and $\frac{x}{6}$ are all equivalent

Equations containing fractions

Exercise 6c

Solve the equation $\frac{x}{3} = 2$

(As $\frac{x}{3}$ means $\frac{1}{3}$ of x, to find x we need to make $\frac{x}{3}$ three times larger.)

$$\frac{x}{3} = 2$$

Multiply each side by 3 $\frac{x}{3} \times \frac{3}{1} = 2 \times 3$

$$x = 6$$

Solve the following equations:

| **1** $\frac{x}{5} = 3$ | **3** $\frac{x}{6} = 8$ | **5** $16 = \frac{9x}{2}$ | **7** $\frac{4x}{7} = 8$ |
| **2** $\frac{x}{2} = 4$ | **4** $\frac{2x}{3} = 8$ | **6** $\frac{2x}{5} = 9$ | **8** $\frac{6x}{5} = 10$ |

Solve the equation $\frac{2x}{5} = \frac{1}{3}$

$$\frac{2x}{5} = \frac{1}{3}$$

Multiply each side by 5
to get rid of the fraction $\frac{2x}{5} \times \frac{5^{1}}{1} = \frac{1}{3} \times \frac{5}{1}$
on the left-hand side.

$$2x = \frac{5}{3}$$

Divide each side by 2 $x = \frac{5}{3} \div 2$

$$x = \frac{5}{3} \times \frac{1}{2}$$

$$x = \frac{5}{6}$$

Solve the following equations:

| **9** $\frac{3x}{2} = \frac{1}{4}$ | **11** $\frac{2x}{9} = \frac{1}{3}$ | **13** $\frac{3x}{8} = \frac{1}{2}$ | **15** $\frac{3x}{5} = \frac{1}{4}$ |
| **10** $\frac{4x}{3} = \frac{1}{5}$ | **12** $\frac{6x}{5} = \frac{2}{3}$ | **14** $\frac{5x}{7} = \frac{3}{4}$ | **16** $\frac{4x}{7} = \frac{2}{5}$ |

Solve the equation $\frac{x}{5} + \frac{1}{2} = 1$
(Both 5 and 2 divide into 10, so by multiplying each side by 10 we can
eliminate all fractions from this equation before we start to solve for x.)

$$\frac{x}{5} + \frac{1}{2} = 1$$

Multiply both sides by 10 $10\left(\frac{x}{5} + \frac{1}{2}\right) = 10 \times 1$

$$\frac{\overset{2}{10}}{1} \times \frac{x}{5} + \frac{\overset{5}{10}}{1} \times \frac{1}{2} = 10$$

$$2x + 5 = 10$$

Take 5 from each side $2x = 5$

Divide each side by 2 $x = 2\frac{1}{2}$

Solve the following equations:

| **17** $\frac{x}{3} + \frac{1}{4} = 1$ | **19** $\frac{x}{5} + \frac{2x}{3} = 3$ | **21** $\frac{2x}{3} - \frac{1}{2} = 4$ | **23** $\frac{x}{3} - \frac{2}{9} = 4$ | **25** $\frac{3}{4} - \frac{x}{5} = 1$ |
| **18** $\frac{x}{5} - \frac{3}{4} = 2$ | **20** $\frac{5x}{7} + \frac{x}{2} = 2$ | **22** $\frac{x}{3} + \frac{5}{6} = 2$ | **24** $\frac{3x}{4} - \frac{x}{2} = 5$ | **26** $\frac{5}{7} + \frac{3x}{4} = 2$ |

Solve the following equations:

27 $\frac{x}{3} + \frac{1}{4} = \frac{1}{2}$

28 $\frac{x}{5} + \frac{2}{3} = \frac{14}{15}$

29 $\frac{x}{4} - \frac{1}{2} = \frac{9}{4}$

30 $\frac{2x}{3} + \frac{2}{7} = \frac{1}{3}$

31 $\frac{x}{2} - \frac{3}{7} = \frac{1}{2}$

32 $\frac{3x}{5} + \frac{2}{9} = \frac{11}{15}$

33 $\frac{5x}{6} + \frac{x}{8} = \frac{3}{4}$

34 $\frac{3x}{4} + \frac{1}{8} = \frac{1}{2}$

35 $\frac{5x}{12} - \frac{1}{3} = \frac{x}{8}$

36 $\frac{2x}{5} - \frac{x}{15} = \frac{5}{9}$

37 $\frac{3x}{4} + \frac{1}{3} = \frac{x}{2} + \frac{5}{8}$

38 $\frac{2x}{7} - \frac{3}{4} = \frac{x}{14} + \frac{1}{2}$

39 $\frac{5x}{7} - \frac{2}{3} = \frac{3}{7} - \frac{x}{3}$

40 $\frac{2x}{9} - \frac{3}{4} = \frac{7}{18} - \frac{5x}{12}$

41 $\frac{3}{11} - \frac{x}{2} = \frac{2x}{11} + \frac{1}{4}$

42 $\frac{3}{5} - \frac{x}{9} = \frac{2}{15} - \frac{2x}{45}$

43 $\frac{4}{7} + \frac{2x}{9} = \frac{15}{9} - \frac{4x}{21}$

44 $\frac{x}{3} + \frac{1}{4} - \frac{x}{6} = \frac{7}{12}$

45 $\frac{5}{8} - \frac{x}{6} + \frac{1}{12} = \frac{3}{4}$

46 $\frac{5}{9} - \frac{7x}{12} = \frac{1}{6} - \frac{x}{8}$

> Multiply both sides by the lowest common multiple of the denominators.

Problems

Exercise 6d

Form an equation for each of the following problems and then solve the equation.

> A bag of sweets was divided into three equal shares. David had one share and he got 8 sweets. How many sweets were there in the bag?
>
> Let x stand for the number of sweets in the bag.
>
> One share is $\frac{1}{3}$ of x $\therefore \frac{1}{3}$ of $x = 8$
>
> $$\frac{x}{3} = 8$$
>
> Multiply each side by 3 $x = 24$
>
> Therefore there were 24 sweets in the bag.

1 Tracy Brown came first in a golf tournament and won \$100. This was $\frac{2}{3}$ of the total prize money paid out. Find the total prize money.

> Start by letting x or some other letter stand for the number you need to find.

2 Peter lost 8 marbles in a game. This number was one-fifth of the number that he started with. Find how many he started with.

3 The width of a rectangle is 12 cm. This is two-fifths of its length. Find the length of the rectangle.

4   I think of a number, halve it and the result is 6. Find the number that I first thought of.

5   The length of a rectangle is 8 cm and this is $\frac{1}{3}$ of its perimeter. Find its perimeter.

6   In an equilateral triangle, the perimeter is 15 cm. Find the length of one side of the triangle.

7   I think of a number, take $\frac{1}{3}$ of it and then add 4. The result is 7. Find the number I first thought of.

8   I think of a number and divide it by 3. The result is 2 less than the number I first thought of. Find the number I first thought of.

9   I think of a number and add $\frac{1}{3}$ of it to $\frac{1}{2}$ of it. The result is 10. Find the number I first thought of.

10   John Smith won the singles competition of a local tennis tournament, for which he got $\frac{1}{5}$ of the total prize money. He also won the doubles competition, for which he got $\frac{1}{20}$ of the prize money. He got \$250 altogether. How much was the total prize money?

Ratios and fractions

When two ratios are equal but one number is missing, you can sometimes find it because it is obvious.

Some missing numbers are not so obvious. You can find them by letting x represent the missing number.

Exercise 6e

Find the missing numbers in

a $:4 = 3:5$ **b** $6: = 5:3$

(Fill the gap with an x to start with.)

a $x:4 = 3:5$ **b** $6:x = 5:3$

$$\frac{x}{4} = \frac{3}{5}$$

or $x:6 = 3:5$

Now solve the equation   (turning **both** ratios round)

$$4 \times \frac{x}{4} = 4 \times \frac{3}{5} \qquad\qquad \frac{x}{6} = \frac{3}{5}$$

$$x = \frac{12}{5} \qquad\qquad\qquad\quad 6 \times \frac{x}{6} = 6 \times \frac{3}{5}$$

$$= 2\frac{2}{5} \qquad\qquad\qquad\qquad x = \frac{18}{5}$$

$$2\frac{2}{5}:4 = 3:5 \qquad\qquad\qquad\qquad = 3\frac{3}{5}$$

$$6:3\frac{3}{5} = 5:3$$

Find x in questions **1** to **12**.

1	$\frac{x}{3} = \frac{4}{5}$	**4**	$x:5 = 4:3$	**7**	$3:5 = x:6$	**10**	$5:1 = 3:x$
2	$\frac{x}{4} = \frac{1}{3}$	**5**	$x:4 = 1:3$	**8**	$7:3 = 3:x$	**11**	$6:5 = 12:x$
3	$x:7 = 3:4$	**6**	$4:x = 3:5$	**9**	$3:x = 2:5$	**12**	$x:3 = 7:15$

Find the missing numbers in questions **13** to **22**.

13	$\quad:9 = 3:5$	**16**	$3:\quad = 5:1$	**19**	$10:3 = \quad:5$	**22**	$12:\quad = 10:3$
14	$\quad:3 = 5:2$	**17**	$4:\quad = 6:5$	**20**	$4:3 = 5:\quad$		
15	$\quad:3 = 3:4$	**18**	$9:5 = \quad:4$	**21**	$\quad:6 = 5:8$		

Two speeds are in the ratio $12:5$. If the first speed is $8\,$km/h, what is the second speed?

Let the second speed be $x\,$km/h. Then we can write the ratio of the speeds as $8:x$; but we know this ratio is equal to $12:5$

$$\therefore 8:x = 12:5$$
$$\frac{x}{8} = \frac{5}{12}$$
$$\overset{1}{8} \times \frac{x}{\underset{1}{8}} = \overset{2}{8} \times \frac{5}{\underset{3}{12}}$$
$$x = \frac{10}{3}$$
$$= 3\frac{1}{3}$$

The second speed is $3\frac{1}{3}\,$km/h.

23 The ratio of the amount of money in David's pocket to that in Indira's pocket is $9:10$. Indira has $25. How much has David got?

24 Two lengths are in the ratio $3:7$. The second length is $42\,$cm. Find the first length.

25 If the ratio in question **24** were $7:3$, what would the first length be?

26 In a rectangle, the ratio of length to width is $9:4$. The length is $24\,$cm. Find the width.

27 The ratio of the perimeter of a triangle to its shortest side is $10:3$. The perimeter is $35\,$cm. What is the length of the shortest side?

28 A length, originally $6\,$cm, is increased so that the ratio of the new length to the old length is $9:2$. What is the new length?

29 A class is making a model of the school building and the ratio of the lengths of the model to the lengths of the actual building is $1:20$. The gym is $6\,$m high. How high, in centimetres, should the model of the gym be?

30 The ratio of lengths of a model boat to those of the actual boat is $3:50$. Find the length of the actual boat if the model is $72\,$cm long.

Directed numbers

Reminder

- $(+2) \times (+3) = +6 \qquad (+2) \times (-3) = -6$
- $(-2) \times (+3) = -6 \qquad (-2) \times (-3) = +6$

Another way to remember these rules when *multiplying* directed numbers is like signs give positive, unlike signs give negative.

Division is the reverse of multiplication, e.g. $4 \div 4 = 1$, so $-4 \div (-4) = 1$

This means that dividing a negative number by a negative number gives a positive number,

e.g. $-8 \div (-2) = +4$

In the same way $(+8) \div (-2) = -4$

and $\qquad\qquad (-8) \div (+2) = -4$

Hence we can use the rules of multiplication of directed numbers for the division of directed numbers.

Exercise 6f

Evaluate:

1	$(+2) \times (-4)$	**4**	$(-\frac{1}{2}) \times (+6)$	**7**	$(+12) \div (-3)$
2	$(-3) \times (-5)$	**5**	$(-16) \div (-8)$	**8**	$(+\frac{1}{2}) \times (+\frac{2}{3})$
3	$(-6) \times (+4)$	**6**	$(-4) \times (-7)$	**9**	$(-6) \div (-2)$

Remember that the positive sign is often omitted, i.e. 6 means +6.

Simplify $4(x-3) - 3(2-3x)$

Multiply out the brackets. Remember that $-3(2-3x) = -3 \times 2 - 3 \times (-3x)$

$$4(x-3) - 3(2-3x) = 4x - 12 - 6 + 9x$$

Collect like terms $\qquad\qquad\qquad\qquad = 13x - 18$

Simplify:

10	$7 - 2(x-5)$	**13**	$4 - 7(2x-3)$	**16**	$2(3x+5) - 2(4+3x)$
11	$2x + 5(3x-4)$	**14**	$3x - 4(5-3x)$	**17**	$5(2x-8) - 3(2-5x)$
12	$3x - 6(3x+5)$	**15**	$3(x-4) + 6(3-2x)$	**18**	$7(x-2) - (2x+3)$

Solve the following equations:

19 $4x - 2(x - 3) = 8$

20 $7 - 3(5 - 2x) = 10$

21 $4x + 2(2x - 5) = 6$ **25** $2x - \frac{1}{2}(6 + 2x) = 7$

22 $3(x - 4) - 7 = 2(x - 3)$ **26** $10 - \frac{1}{4}(4x - 8) = 5$

23 $4 - 3x = 3 + 4(2x - 3)$ **27** $3 - \frac{2}{3}(6x + 9) = 5 - 2x$

24 $3x - 2(4 - 5x) = 5 - 3x$ **28** $\frac{3}{4}(4 - 8x) = 2x - \frac{2}{3}(6 - 12x)$

Start by multiplying out brackets, then collect like terms.

❓ Puzzle

My dog weighs nine-tenths of its weight plus nine-tenths of a kilogram. What does it weigh?

Formulae

A formula is a rule for finding one quantity in terms of other quantities. For example, the rule for finding the area of any triangle is find half the length of the base multiplied by the perpendicular height.

We can write this using letters to represent the area, the length of the base and the perpendicular height as $A = \frac{1}{2}bh$.

We can use this formula to find the area of the triangle shown by substituting the numerical values for b and h.

$A = \frac{1}{2} \times 7 \times 6 \, \text{cm}^2 = 21 \, \text{cm}^2$

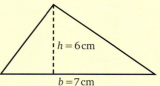

$h = 6\,\text{cm}$

$b = 7\,\text{cm}$

We can substitute numbers for letters in any formula. We do not need to know what the letters stand for.

Exercise 6g

If $v = u - at$, find v when $u = 5$, $a = -2$, $t = -3$

$$v = u - at$$

When $u = 5$, $a = -2$, $t = -3$, $v = 5 - (-2) \times (-3)$

$$= 5 - (+6)$$
$$= 5 - 6$$
$$= -1$$

(Notice that where negative numbers are substituted for letters they have been put in brackets. This makes sure that only one operation at a time is carried out.)

1   If $N = p + q$, find N when $p = 4$ and $q = -5$.

2   If $C = RT$, find C when $R = 4$ and $T = -3$.

3   If $z = w + x - y$, find z when $w = 4$, $x = -3$ and $y = -4$.

4   If $r = u(v - w)$, find r when $u = -3$, $v = -6$ and $w = 5$.

5   Given that $X = 5(T - R)$, find X when $T = 4$ and $R = -6$.

6   Given that $P = d - rt$, find P when $d = 3$, $r = -8$ and $t = 2$.

7   Given that $v = l(a + n)$, find v when $l = -8$, $a = 4$ and $n = -6$.

8   If $D = \frac{a - b}{c}$, find D when $a = -4$, $b = -8$ and $c = 2$.

9   If $Q = abc$, find Q when $a = 3$, $b = -7$ and $c = -5$.

10   If $l = \frac{2}{3}(x + y - z)$, find l when $x = 4$, $y = -5$ and $z = -6$.

11   If $v^2 = u^2 + 2as$, find a when $v = 3$, $u = 2$ and $s = 12$.

12   If $d = \frac{1}{2}(a + b + c)$, find a when $d = 16$, $b = 4$ and $c = -3$.

13   If $H = P(Q - R)$, find Q when $H = 12$, $P = 4$ and $R = -6$.

> Put negative numbers in brackets.

> Subsititute the numbers for the letters, then solve the equation for the letter you need to find.

14   Given that $v = at$, find the value of

   a v when $a = 4$ and $t = 12$ **b** v when $a = -3$ and $t = 6$

   c t when $v = 18$ and $a = 3$ **d** a when $v = 25$ and $t = 5$.

15   Given that $N = 2(n - m)$, find the value of

   a N when $n = 6$ and $m = 4$ **b** N when $n = 7$ and $m = -3$

   c n when $N = 12$ and $m = 2$ **d** m when $N = 16$ and $n = -4$.

16   If $A = P + QT$, find the value of

   a A when $P = 50$, $Q = \frac{1}{2}$ and $T = 4$ **b** A when $P = 70$, $Q = 5$ and $T = -10$

   c P when $A = 100$, $Q = \frac{1}{4}$ and $T = 16$ **d** T when $A = 25$, $P = -15$ and $Q = -10$.

17 Given that $s = \frac{1}{2}(a - b)$, find the value of

   a s when $a = 16$ and $b = 6$ **b** s when $a = -4$ and $b = -10$

   c a when $s = 15$ and $b = 8$ **d** b when $s = 10$ and $a = -4$.

18 Given that $z = x - 3y$, find the value of

   a z when $x = 3\frac{1}{2}$ and $y = \frac{3}{4}$ **b** z when $x = \frac{3}{8}$ and $y = -1\frac{1}{2}$

   c x when $z = 5\frac{1}{3}$ and $y = 2\frac{1}{2}$ **d** y when $z = \frac{1}{4}$ and $x = \frac{7}{8}$

19 If $P = 100r - t$, find the value of

   a P when $r = 0.25$ and $t = 10$ **b** P when $r = 0.145$ and $t = 15.6$

   c t when $P = 18.5$ and $r = 0.026$ **d** t when $P = 50$ and $t = -12$.

Expressions, equations and formulae

In this chapter you have worked with expressions, equations and formulae.

Remember that:

An **expression** is a collection of one or more algebraic terms, for example $2x$, $5x + 2y$, $a^2 - 4b$ and $6(2x - 3)$ are expressions.

An **equation** is an equality between two expressions, for example $2x = 4$ and $y + 2 = 3x + 1$ are equations.

A **formula** is a general rule for finding one quantity in terms of other quantities, for example the formula for finding the area, $A\,\text{cm}^2$, of a rectangle measuring $l\,\text{cm}$ by $b\,\text{cm}$ is $A = l \times b$. A is called the subject of the formula.

($A = l \times b$ is also an equation.)

Exercise 6h

For each question write down whether what is given is an expression, an equation or a formula.

1 $P = 2(l + b)$	**5** $y = \frac{1}{2}(3x + 3z)$	**9** $3x + 2y = 6$
2 $5x - 1 = 4x + 2$	**6** $3(x - 2) + 8$	**10** $A = \pi r^2$
3 $5x - 9 - \frac{1}{2}x = 0$	**7** $5a + 2b - 3(a - b)$	**11** $r = C/2\pi$
4 $4(a - 3) + 2(b + 4)$	**8** $\frac{1}{3}(x + 3) = 4$	**12** $4(7a + 4)$

Mixed exercises

Exercise 6i

1   Solve the equation $8 = 3 + 2x$.

2   Solve the equation $x - 4 = 5 - 2x + 1$.

3   Multiply out $3(2x - 8)$.

4   Solve the equation $\frac{2x}{3} = 8$.

5   Find the value of x if $\frac{x}{2} + \frac{1}{6} = \frac{1}{3}$.

6   Simplify the expression $3x - 2(4 - x)$.

7   Write down a formula for P if P cm is the perimeter
    of the figure in the diagram. (Each letter stands for a
    number of centimetres.)

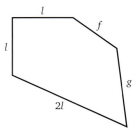

8   If $x = a - b$, find the value of x when $a = 2$ and $b = -5$.

Exercise 6j

1   Solve the equation $5(3 - 4x) = x - 2(3x - 5)$.

2   Simplify $5 - (x - 2y) - 3(x - 6)$.

3   Solve the equation $\frac{3}{8} - \frac{5x}{6} = \frac{2}{3}$.

4   Given that $r = s - vt$, find the value of r when $s = 4$, $v = 3$ and $t = -2$.

5   A rectangle is twice as long as it is wide. If it is a cm wide, write down
    a formula for P where P cm is the perimeter of the rectangle.

6   $L = 3pg$. Find the value of p when $L = 18$ and $g = 2$.

7   Solve the equation $5 - \frac{x}{2} = \frac{1}{4}$.

In this chapter you have seen that...

✔ to solve a linear equation where the unknown quantity is on both sides
  of the equals sign, collect the terms containing the unknown on the side
  where there are more of them

✔ when an equation contains brackets, multiply these out first

✔ fractions can be eliminated from an equation by multiplying both sides by
  the lowest common multiple of the denominators

✔ when you substitute numerical values into a formula, place negative
  numbers in brackets.

7 Probability

Single events

If we roll an unbiased die, each of the six numbers is equally likely to appear.
We say that there are six equally likely *outcomes* to this experiment. The
probability of scoring 2 is $\frac{1}{6}$, because out of six equally likely outcomes,
only one is 'successful', i.e. is a 2

We write $\qquad P(2) = \frac{1}{6}$

Similarly $\qquad P(\text{even number}) = \frac{3}{6} = \frac{1}{2}$ $\qquad$ (There are three equally likely
possibilities: 2, 4, 6)

$$P(10) = 0 \qquad \text{(The number of outcomes that}$$
give a score of 10 is zero.)

$$P(1 \text{ or } 2 \text{ or } 3 \text{ or } 4 \text{ or } 5 \text{ or } 6) = \frac{6}{6} = 1$$

$$P(\text{not } 2) = 1 - \frac{1}{6} \qquad \text{('not 2' is all the possible}$$
outcomes except one.)

111

To summarise, a successful event can be denoted by A and the probability of it happening is then $P(A)$. If A is 'rolling a 2' then $P(A) = \frac{1}{6}$.

$$P(\text{successful event}) = \frac{\text{number of successful outcomes}}{\text{total number of equally likely outcomes}}$$

$P(\text{certainty}) = 1$

$P(\text{impossibility}) = 0$

$0 \leqslant P(A) \leqslant 1$

$P(\text{not } A) = 1 - P(A)$

Exercise 7a

1 A letter is chosen at random from the letters of the word POPULAR. Find the probability that it is

  a L   **b** P   **c** a vowel   **d** not R.

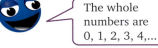

Chosen at random means that any letter is as likely to be chosen as any other letter.

2 A number is chosen at random from the set {2, 3, 5, 6 , 9, 10}. Find the probability that it

  a has two digits   **b** is prime   **c** is not even   **d** a multiple of 7.

3 A card is drawn at random from an ordinary pack of 52 playing cards. Find the probability that it is

  a an ace   **b** a club   **c** not an ace.

4 A number is chosen at random from the first 10 whole numbers. What is the probability that it is

  a an even number   **b** a prime number

The whole numbers are 0, 1, 2, 3, 4,...

  c a factor of 12?

5 A bag contains 4 blue discs and 5 white discs. One disc is removed at random. What is the probability that it is

  a blue   **b** white   **c** red   **d** blue or white?

Possibility space for two events

Suppose a 5c coin and a 10c coin are tossed together. One possibility is that the 5c coin will land head up and that the 10c coin will also land head up.

If we use H for a head on the 5c coin and H for a head on the 10c coin, we can write this possibility more briefly as the ordered pair (H, H).

To list all the possibilities, an organised approach is necessary, otherwise we may miss some. We use a table called a *possibility space*. The possibilities for the 10c coin are written across the top and the possibilities for the 5c coin are written down the side:

	10c coin	
	H	*T*
5c coin H		
T		

When both coins are tossed we can see all the combinations of heads and tails that are possible and then fill in the table.

	10c coin	
	H	*T*
5c coin H	(H, *H*)	(H, *T*)
T	(T, *H*)	(T, *T*)

Exercise 7b

1   Two bags each contain 3 white counters and 2 black counters. One counter is removed at random from each bag. Copy and complete the following possibility space for the possible combination of two counters.

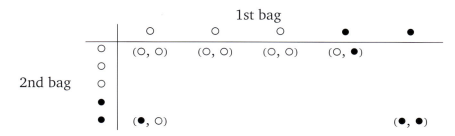

2   An ordinary six-sided die is rolled and a 10c coin is tossed. Copy and complete the following possibility space.

	Die					
	1	2	3	4	5	6
10c coin H		(H, 2)				
T				(T, 4)		

3 One bag contains 2 red counters, 1 yellow counter and 1 blue counter. Another bag contains 2 yellow counters, 1 red counter and 1 blue counter. One counter is taken at random from each bag. Copy and complete the following possibility space.

		1st bag			
		R	R	Y	B
2nd bag	R		(R, R)		
	Y				(Y, B)
	Y				
	B	(B, R)			

4 A top like the one in the diagram is spun twice. Copy and complete the possibility space.

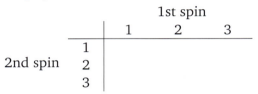

		1st spin		
		1	2	3
2nd spin	1			
	2			
	3			

5 A boy goes into a shop to buy a pencil and a rubber. He has a choice of a red, a green or a yellow pencil and a round, a square or a triangular shaped rubber. Make your own possibility space for the possible combinations of one pencil and one rubber that he could buy.

Using a possibility space

When there are several entries in a possibility space it can take a long time to fill in the ordered pairs. To save time we use a cross in place of each ordered pair. We can see which ordered pair a particular cross represents by looking at the edges of the table.

Exercise 7c

Two ordinary six-sided dice are rolled. Draw up a possibility space showing all the possible combinations in which the dice may land.

Use the possibility space to find the probability that a total score of at least 10 is obtained.

1st die

2nd die		1	2	3	4	5	6
	1	×	×	×	×	×	×
	2	×	×	×	×	×	×
	3	×	×	×	×	×	×
	4	×	×	×	×	×	⊗
	5	×	×	×	×	⊗	⊗
	6	×	×	×	⊗	⊗	⊗

(There are 36 entries in the table and 6 of these give a score of 10 or more.)

$P(\text{score of at least } 10) = \dfrac{6}{36} = \dfrac{1}{6}$

1 Use the possibility space in the example above to find the probability of getting a score of

   a 4 or less   **b** 9   **c** a double.

2 Use the possibility space for question **1** of Exercise **7b** to find the probability that the two counters removed

   a are both black   **b** contain at least one black.

3 Use the possibility space for question **2** of Exercise **7b** to find the probability that the coin lands head up and the die gives a score that is less than 3.

4 Use the possibility space for question **3** of Exercise **7b** to find the probability that the two counters removed are

   a both blue   **b** both red

   c one blue and one red   **d** such that at least one is red.

5 A 5 c coin and a 1 c coin are tossed together. Make your own possibility space for the combinations in which they can land. Find the probability of getting two heads.

6 A six-sided die has two of its faces blank and the other faces are numbered 1, 3, 4 and 6. This die is rolled with an ordinary six-sided die (faces numbered 1, 2, 3, 4, 5, 6). Make a possibility space for the ways in which the two dice can land and use it to find the probability of getting a total score of

   a 6   **b** 10   **c** 1   **d** at least 6.

7 One bag of coins contains three 10 c coins and two 25 c coins. Another bag contains one 10 c coin and one 25 c coin. One coin is removed at random from each bag. Make a possibility space and use it to find the probability that a 25 c coin is taken from each bag.

8   One bookshelf contains two storybooks and three textbooks. The next shelf holds three storybooks and one textbook. Draw a possibility space showing the various ways in which you could pick up a pair of books, one from each shelf. Use this to find the probability that
   a   both books are storybooks   **b**   both are textbooks.

9   The four aces and the four kings are removed from an ordinary pack of playing cards. One card is taken from the set of four aces and one card is taken from the set of four kings. Make a possibility space for the possible combinations of two cards and use it to find the probability that the two cards
   a   are both black   **b**   are both spades
   c   include at least one black card   **d**   are both of the same suit.

(?) Puzzle

A man asked a farmer how many animals he had. The farmer replied, "They're all cows but 2, all sheep but 2 and all pigs but 2."

How many animals did the farmer own?

Finding probability by experiment

We have assumed that if you toss a coin it is equally likely to land head up or tail up so that $P(\text{a head}) = \frac{1}{2}$. Coins like this are called 'fair' or 'unbiased'.

Most coins are likely to be unbiased but it is not necessarily true of all coins. A particular coin may be slightly bent or even deliberately biased so that there is not an equal chance of getting a head or a tail.

The only way to find out if a particular coin is unbiased is to toss it several times and count the number of times that it lands head up.

Then for that coin

$$P(\text{a head}) \approx \frac{\text{number of heads}}{\text{total number of tosses}}$$

The approximation gets nearer to the truth as the number of tosses gets larger.

Exercise 7e

Work with a partner or collect information from the whole class.

1 Toss a $10 coin 100 times and count the number of times it lands head up and the number of times it lands tail up.

Use tally marks, in groups of five, to count as you toss.

Find, approximately, the probability of getting a head with this coin.

2 Repeat question 1 with a $1 coin.

3 Repeat question 1 with the $10 coin that you used first but this time stick a small piece of plasticine on one side.

4 Choose two $5 coins and toss them both once. What do you think is the probability of getting two heads? Now toss the two coins 100 times and count the number of times that both coins land head up together. Use tally marks to count as you go: you will need to keep two tallies, one to count the total number of tosses and one to count the number of times you get two heads. Use your results to find approximately the probability of getting two heads.

5 Take an ordinary pack of playing cards and keep them well shuffled. If the pack is cut, what do you think is the probability of getting a red card? Cut the pack 100 times and keep count, using tally marks as before, of the number of times that you get a red card. Now find an approximate value for the probability of getting a red card.

6 Using the pack of cards again, what do you think is the probability of getting a spade? Now find this probability by experiment.

7 Use an ordinary six-sided die. Roll it 25 times and keep count of the number of times that you get a six. Use your results to find an approximate value for the probability of getting a six. Now roll the die another 25 times and add the results to the last set. Use these to find again the probability of getting a six. Now do another 25 rolls and add the results to the last two sets to find another value for the probability. Carry on doing this in groups of 25 rolls until you have done 200 rolls altogether.

You know that the probability of getting a six is $\frac{1}{6}$. Now look at the sequence of results obtained from your experiment. What do you notice? (It is easier to compare your results if you use your calculator to change the fractions into decimals correct to 2 d.p.)

8 Remove all the diamonds from an ordinary pack of playing cards. Shuffle the remaining cards well and then cut the pack. What do you think is the probability of getting a black card? Shuffle and cut the pack 100 times and use the results to find approximately the probability of cutting a black card.

9 Roll two dice together 100 times. Find the total for each roll and put a mark beside a number each time you roll that number. Use a copy of the chart below.

a What is the probability of getting a total of 7?

b What is the probability of getting

  i a total 2

  ii a total 12?

Total	Number of times
2	
3	
4	
5	
.	
.	
.	
12	

(!) **Investigation**

This is an experiment to find out if you can see into the future!

You need to work in pairs and you need one coin. One of you tosses the coin and records the results; the other is the guesser.

a The guesser predicts whether the coin will land head up or tail up. The tosser then tosses the coin.

  If the guesser has no psychic powers,
  i what is the probability that he/she guesses the actual outcome

  ii when this experiment is repeated 100 times, about how many times do you expect the guesser to predict the actual outcome?

b Now perform the experiment described at least 100 times and record each result 'right' or 'wrong' as appropriate.

c Compare what you expected to happen with what did happen, using appropriate diagrams as illustrations. Comment on the likelihood of the guesser being able to predict which way the coin will land.

d How could you make your results more reliable?

Did you know?

Did you know that much of the theory of probability was the work of the French philosopher and mathematician, Blaise Pascal (1623–1662). He also constructed, between 1642 and 1645, the first calculating machines, several of which still survive. His machine would add and subtract.

In this chapter you have seen that...

✔ a possibility space is a rectangular array showing all the possible ways in which the events being studied can occur. It can be used to find the probability that a particular event happens

✔ to estimate a probability by experiment find

$$\frac{\text{the number of times the event happens}}{\text{the number of times the experiment is repeated}}$$

✔ the probability that an event happens is

$$\frac{\text{the number of ways the event can happen}}{\text{the total number of equally likely events}}$$

✔ the probability that an event will happen is between zero and one.

impossible   certain

REVIEW TEST 1: CHAPTERS 1–7

In questions 1 to 13 choose the letter for the correct answer.

1 The value of $7 + (-5) - (-4)$ is

   A -6 **B** -2 **C** 2 **D** 6

2 Given that $3a + 2b = a + 5b$

   A $a = b$ **B** $2a = 3b$ **C** $3a = 5b$ **D** $4a = 7b$

3 $7(2x - 3) - (x - 1)$ simplifies to

   A $15x - 20$ **B** $15x - 22$ **C** $13x - 1$ **D** $13x - 20$

4 The value of x that satisfies the equation $8 - 5x = 4 + 7x$ is

   A $\frac{1}{3}$ **B** 1 **C** 2 **D** 3

5 The next two numbers in the number pattern 3, 8, 18, 33, ... are

   A 43 and 53   **B** 43 and 58   **C** 48 and 63   **D** 53 and 78

6 $(1 \times 3) - 2 = 1$ $(2 \times 4) - 3 = 5$ $(3 \times 5) - 4 = 11$ $(4 \times 6) - 5 = 19$

The next number on the right-hand side of this number pattern is

   A 27   **B** 28   **C** 29   **D** 41

7 The nth term T_n of a number pattern is given by the formula
$T_n = 3n(2n - 1)$.

The third and fourth terms are

   A 18 and 35   **B** 18 and 45   **C** 45 and 74   **D** 45 and 84

8 The value of $3^2 \times 4^2$ is

   A 12   **B** 48   **C** 144   **D** 20 736

9 The value of x that satisfies the equation $5(4x - 5) = 3(2x + 1)$ is

   A 1.1   **B** 1.5   **C** 2   **D** 3

10 The solution of the equation $\frac{3x}{4} = \frac{4}{3}$ is

   A $x = \frac{9}{16}$ **B** $x = \frac{7}{12}$ **C** $x = 1$ **D** $x = 1\frac{7}{9}$

11 A letter is chosen at random from the letters of the word
APPLICABLE. The probability that it is P is

   A $\frac{1}{10}$ **B** $\frac{1}{9}$ **C** $\frac{2}{9}$ **D** $\frac{1}{5}$

12 Which of these pairs do not have exactly the same value?

A $\frac{3}{5}$, 60% **B** 62.5%, $\frac{5}{8}$ **C** 66%, $\frac{8}{12}$ **D** $\frac{7}{10}$, 70%

13 Given that $p = q - 3r$, if $q = 5$ and $r = -2$ then $p =$

A 11 **B** 9 **C** 1 **D** −4

14 Find the size of each angle marked with a letter.

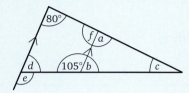

15 Given that $d = \frac{a}{2}(2b - c)$, find the value of d when $a = 3$, $b = 6$ and $c = -1$.

16 a In the figure below calculate the sizes of the angles marked x, y and z.

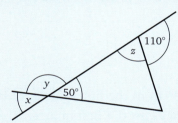

b A triangle ABC has three axes of symmetry. If AB = 6 cm, what are the lengths of BC and AC? What is the measure of the angle ABC? Explain your answers.

17 A bag contains 5 green discs, 7 yellow discs and 3 black discs.
Karen draws a disc at random from the bag.
What is the probability that

a it is green **b** it is not yellow **c** it is either green or black?

18 Two four-sided dice, one red and one blue, with faces labelled 1, 2, 3 and 4 are rolled together.

a Draw a possibility space diagram showing all the combinations in which the dice may land.

b How many outcomes are there?

c What is the probability that the number on the red die is one more than the number on the blue die?

8 Percentage increase and decrease

At the end of this chapter you should be able to...

1 Increase or decrease a quantity by a given percentage.

2 Solve problems involving percentage increases or decreases.

You need to know...

✔ how to find the percentage of a quantity

✔ how to multiply by a fraction

Key words

depreciates, multiplying factor, net pay, percentage decrease, percentage increase

Percentage increase

My telephone bill is to be increased by 8% from the first quarter of the year to the second quarter. It amounted to $6450 for the first quarter. From this information I can find the value of the bill for the second quarter.

If $6450 is increased by 8%, the increase is 8% of $6450,

i.e. $$\$\frac{8}{100} \times 6450 = \$516$$

The bill for the second quarter is therefore

$$\$6450 + \$516 = \$6966$$

The same result is obtained if we take the original sum to be 100%. The increased amount is $(100+8)\%$, or $\frac{108}{100}$, of the original sum,

i.e. the bill for the second quarter is $\$\frac{108}{100} \times 6450 = \$6966$

The quantity $\frac{108}{100}$ is called the multiplying factor. To increase a quantity by 12%, the multiplying factor would be $\frac{112}{100}$.

Percentage decrease

Similarly if we wish to decrease a quantity by 8%, the decreased amount is $(100-8)\%$, or $\frac{92}{100}$, of the original sum.

If we wish to decrease a quantity by 15%, the new quantity is 85% of the original quantity, and the multiplying factor is $\frac{85}{100}$.

Exercise 8a

If a number is increased by 40%, what percentage is the new number of the original number?

The new number is $100\% + 40\% = 140\%$ of the original.

If a number is increased by the given percentage, what percentage is the new number of the original number?

1	50%	**3**	20%	**5**	75%	**7**	48%	**9**	175%	**11**	57%
2	25%	**4**	60%	**6**	35%	**8**	300%	**10**	$12\frac{1}{2}\%$	**12**	15%

What multiplying factor increases a number by 44%?

The multiplying factor is $\frac{100+44}{100} = \frac{144}{100}$

Give the multiplying factor which increases a number by:

13	30%	**14**	80%	**15**	65%	**16**	130%.

If a number is decreased by 65%, what percentage is the new number of the original number?

The new number is $100\% - 65\% = 35\%$ of the original.

If a number is decreased by the given percentage what percentage is the new number of the original number?

17	50%	**19**	70%	**21**	35%	**23**	4%	**25**	$62\frac{1}{2}\%$	**27**	53%
18	25%	**20**	85%	**22**	42%	**24**	66%	**26**	$33\frac{1}{3}\%$	**28**	10%

What multiplying factor decreases a number by 30%?

The multiplying factor is $\frac{100-30}{100} = \frac{70}{100}$

What multiplying factor decreases a number by:

29 40% **30** 75% **31** 34% **32** 12%?

Increase 180 by 30%.

The new value is 130% of the old

i.e. the new value is $\frac{130}{100} \times 180 = 234$

Increase:

33	100 by 40%	**37**	1600 by 73%	**40**	111 by $66\frac{2}{3}$%
34	200 by 85%	**38**	745 by 14%	**41**	145 by 120%
35	340 by 45%	**39**	64 by $62\frac{1}{2}$%	**42**	644 by 275%.
36	550 by 36%				

Decrease 250 by 70%.

The new value is 100% − 70% = 30% of the original value

i.e. the new value is $\frac{30}{100} \times 250 = 75$

Decrease:

43	100 by 30%	**47**	3400 by 28%	**51**	208 by $87\frac{1}{2}$%
44	200 by 15%	**48**	3450 by 4%	**52**	248 by $37\frac{1}{2}$%.
45	350 by 46%	**49**	93 by $33\frac{1}{3}$%		
46	750 by 13%	**50**	273 by $66\frac{2}{3}$%		

Problems

Exercise 8b

The number of cases of mumps reported this year is 4% lower than the number of cases reported last year. There were 250 cases last year. How many cases were reported this year?

There were 4% fewer cases this year so to find the number of cases this year, we need to decrease 250 by 4%.

The multiplying factor is $\frac{100-4}{100} = \frac{96}{100}$.

The number of cases this year is $\frac{96}{100} \times 250 = 240$.

1   A boy's weight increased by 15% between his fifteenth and sixteenth birthdays. If he weighed 55 kg on his fifteenth birthday, what did he weigh on his sixteenth birthday?

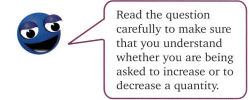

Read the question carefully to make sure that you understand whether you are being asked to increase or to decrease a quantity.

2   The water rates due on my house this year are 8% more than they were last year. Last year I paid $32 000. What must I pay this year?

3   There are 80 teachers in a school. It is anticipated that the number of staff next year will increase by 5%. How many staff should there be next year?

4   Pierre is 20% taller now than he was 2 years ago. If he was 150 cm tall then, how tall is he now?

5   A factory employs 220 workers. Next year this number will increase by 15%. How many extra workers will be taken on?

6   A living room suite is priced at $260 000 plus general consumption tax (GCT) at 15%. How much does the suite actually cost the customer?

7   A CD costs $3500 plus general consumption tax at 20%. How much does the CD actually cost?

8   The cost of a meal is $3200 plus a service charge of 15%. How much must I pay for the meal?

9   Miss Kendall earns $12 000 per week from which income tax is deducted at 30%. Find how much she actually gets. (This is called her *net* pay.)

10 In a certain week a factory worker earns $15 000 from which income tax is deducted at 30%. Find his net income after tax, i.e. how much he actually gets.

11 Mr Hall earns $1 000 000 per month. If income tax is deducted at 25%, find his net pay after tax.

12 As a result of using Alphamix fertilizer, my potato crop increased by 32% compared with last year. If I grew 150 kg of potatoes last year, what weight of potatoes did I grow this year?

13 The number of children attending White Sands village school is 8% fewer this year than last year. If 450 attended last year, how many are attending this year?

14 The marked price of a man's suit is $37 500. In a sale the price is reduced by 12%. Find the sale price.

15 In a sale all prices are reduced by 10%. What is the sale price of an article marked
    a $4000
    b $8500?

16 Last year in a school there were 75 reported cases of measles. This year the number of reported cases has dropped by 16%. How many cases have been reported this year?

17 Mr Connah was 115 kg when he decided to go on a diet. He lost 10% of his weight in the first month and a further 8% of his original weight in the second month. How much did he weigh after 2 months of dieting?

18 A car is valued at $24 000 000. It depreciates by 20% in the first year and thereafter each year by 15% of its value at the beginning of that year. Find its value

    a   after 2 years   **b**   after 3 years.

19 In any year the value of a motorcycle depreciates by 10% of its value at the beginning of that year. What is its value after two years if the purchase price was $7 200 000?

20 When John Short increases the speed at which he motors from an average of 40 km/h to 50 km/h, the number of km travelled per litre decreases by 25%. If he travels 9 km on each litre when his average speed is 40 km, how many km per litre can he expect at an average speed of 50 km/h?

21 When petrol was $100 per litre I used 700 litres in a year. The price rose by 12% so I reduced my yearly consumption by 12%. Find
    a   the new price of a litre of petrol
    b   my reduced annual petrol consumption
    c   how much more (or less) my petrol bill is for the year.

! Investigation

Ed bought a watch for $5000. He marked it up by 30% and put it in the window of his shop. He could not sell it, so in a sale marked it '30% off'. Molly bought the watch and claimed she had paid less than Ed bought it at. Was this true? Investigate.

Mixed exercises

Exercise 8c

1 Express 170 cm as a percentage of 4 m.

2 Find 62% of 3.5 m.

3 If a number is increased by 25%, what percentage is the new number of the original number?

4 What multiplying factor would increase a quantity by 45%?

5 **a** Increase 56 cm by 75%. **b** Decrease 1200 sheep by 20%.

6 If a number is decreased by 42%, what percentage is the new number of the original number?

7 What multiplying factor would decrease a quantity by 18%?

8 **a** Increase 70 m by 35%. **b** Decrease 55 miles by 84%.

9 In a sale a shopkeeper reduces the prices of his goods by 10%. Find the sale price of goods marked
   a $97 **b** $492.

⚠ Investigation

These references to percentages were in a newspaper. Investigate what each means.

a *'The annual rate of inflation has fallen from 3.5% last month to 3.46% this month.'*
Does this mean that prices are rising, falling or standing still?

b *'Everyone in the company is to get a raise of 4.5%. This means that the annual increase ranges from $235 000 to $50 000.'*
How can this be true if everybody gets the same raise? Can you explain?

In this chapter you have seen that...

✔ to increase a quantity by a%, multiply it by $\frac{100+a}{100}$; $\frac{100+a}{100}$ is called the multiplying factor

✔ to decrease a quantity by a%, multiply it by $\frac{100-a}{100}$; $\frac{100-a}{100}$ is called the multiplying factor.

9 Consumer arithmetic

At the end of this chapter you should be able to...

1 Total supermarket and other bills, and find the change from a given amount.

2 Calculate commission, bonuses and income tax deductions.

3 Calculate sales and other taxes.

4 Calculate the amount due on telephone, water and electricity bills, given the necessary information.

You need to know...

✔ how to work with decimals and fractions

✔ how to use a calculator

✔ the meaning of percentages

✔ how to find a percentage of a quantity

✔ how to find one quantity as a percentage of another

Key words

amount, bonus, commission, compound interest, general consumption tax (GCT), income tax, inflation, kilowatt hour, principal, salary, simple interest, standing charge

? Puzzle

Four married couples met at a party. Everyone shook hands with everyone else except each husband with his own wife.

How many handshakes were there?

Exercise 9a

Total the following supermarket bills. In each case find the change from two $5000 notes.

	$		$		$		$		$
1	288	**2**	269	**3**	255	**4**	239	**5**	126
	185		375		143		272		249
	344		377		143		142		255
	117		375		227		142		275
	138		429		164		293		245
	224		85		259		245		145
	329		239		319		245		145
	533		105		519		139		145
	134		770		255		139		145
	95		640		162		89		69
	129		225		173		420		345
	229		229		280		89		185
	659		225		334		165		375
	843		225		137		214		249
	129		489		252		225		561
	132		373		249		225		175
	132		89		126		372		175
	228		332		237		164		175
	316		276		409		85		99
	375		349		199		289		79
	243		123		175		289		495
	99		<u>149</u>		545		289		245
	242				<u>99</u>		199		75
	<u>118</u>						<u>75</u>		85
									<u>525</u>

Copy and complete the following bills: $

6   2 tins of paint at $1440 per tin
       3 brushes at $120 each
       24 ceramic tiles at $65 each   _____
                                                        _____

7   2 kg butter at $260 per kg
       3 litres milk at $180 per litre
       1 jar of nutmeg jam at $440   _____
                                                        _____

8 7 oranges at $50 each
8 grapefruit at $79 each
3 kg apples at $220 per kg
2 kg bananas at $265 per kg
_____
_____

9 2 jars of guava jelly at $475 per jar
3 jars of jam at $450 per jar
2 jars of marmalade at $520 per jar
3 jars of honey at $710 per jar
_____
_____

10 2 cans ackee at $725 each
3 cans of grape juice at $445 per can
2 kg sugar at $150 per kg
2 packets of peas at $135 each
1 packet of Jello at $245
1 bottle of bitters at $699
_____
_____

11 An importer of car batteries pays the following costs:
Price of battery $5599
Freight $1200
Duty $3239
Consumption tax $1044
Stamp duty $864
Calculate the total cost of a battery to the importer.

12 Residential lots of land are advertised as follows:
Down payment $420 000
Monthly payment $85 000

How much will it cost a person who pays for his land by making a down payment followed by 24 monthly payments?

Commission and bonus incentives

Some workers, such as salesmen and representatives, do not receive a fixed regular income. They are given a fairly low basic wage but they also get commission on every order they secure. The commission is usually a percentage of the value of the order.

Other workers get paid a fixed wage plus an amount that depends on the amount of work they do.

For example, Pete gets paid $12 000 a week plus $40 for every article he produces after the first 30.

Exercise 9b

1   In addition to a basic weekly wage of $14 000, Miss Black receives a commission of 1% for selling second-hand cars. Calculate her gross wage for a week when she sells cars to the value of $5 000 000.

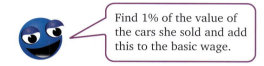

Find 1% of the value of the cars she sold and add this to the basic wage.

2   A salesman receives a basic wage of $5000 per week plus commission at 6% on the value of the goods he sells. Find his income in a week when sales amount to $530 000.

3   Tom Hannah receives a basic wage of $8500 per week and receives a commission of 2% on all sales over $100 000. Find his income for a week when he sells goods to the value of $1 880 000.

4   Sue Renner receives a basic wage of $12 000 per week plus a commission of 2% on her sales. Find her income for a week when she sells goods to the value of $2 120 000.

5   Penny George is paid a basic wage of $10 500 per week plus a commission of $1\frac{1}{2}$% on her sales over $150 000. Find her income for a week when she sells goods to the value of $2 130 000.

6   Alan McKay is paid a basic wage of $15 000 per week plus a commission of 3% on all sales over $240 000. Find his income for a week when he sells goods to the value of $1 740 000.

7   In addition to a weekly wage of $17 000, Olive MacCarthy receives commission of $1\frac{1}{2}$% on the sales of antique furniture. Calculate her gross wage in a week when she sells furniture to the value of $1 550 000.

8   Don Smith receives a guaranteed weekly wage of $26 000 plus a bonus of $40 for every circuit board he completes each day after the first 20. During a particular week the number of boards he produced are as follows:

First find the total number on which the bonus is paid: 33 the first day, 28 the second, and so on. Next calculate the total bonus and add it to the gross wage.

Monday 53, Tuesday 48, Wednesday 55, Thursday 51, Friday 47.
Calculate his gross wage for the week.

9   Audley Davis gets paid $40 for each article he completes up to 100
per day. For every article above this figure he receives $45. In a
particular week his production figures are

Mon	Tues	Wed	Thurs	Fri
216	192	234	264	219

a   How many articles does he produce in the week?
b   For how many of these is he paid $40 each?
c   For how many of these is he paid $45 each?
d   Find his earnings for the week.

10   The table shows the number of electric light fittings produced by five
factory workers each day for a week.

	Mon	Tues	Wed	Thurs	Fri
Ms Arnold	34	38	34	39	41
Mr Beynon	37	40	37	44	–
Miss Capstick	35	40	43	37	39
Mr Davis	42	45	40	52	46
Mrs Edmunds	39	38	37	35	42

The rate of payment is: $95 for each fitting up to 20 per day and
$145 for each fitting above 20 per day.

a   How many fittings does each person produce in the week?
b   For each person find
  i   how many fittings are paid at $95 each
  ii how many fittings are paid at $145 each.
c   Find each person's income for the week.
d   On which day of the week does this group of workers produce the
greatest number of fittings?

Taxes

The government raises the money it needs in many different ways.

You pay tax to the government on the money you earn, the amount depends
on your annual income. In general, the more you earn the more you pay. Apart
from deductions from your gross pay for National Insurance Scheme (NIS) 2%,
National Housing Trust (NHT) 2.5% and Education Tax 2%, you pay *income
tax*. A person's taxable income is what remains after a fixed amount determined
by the government, and the above percentages, have been deducted from your
gross income. At present your taxable income is taxed at 25%.

There is also a tax on almost everything you buy. This is called General Consumption Tax (GCT) and is a fixed percentage of the selling price. We will assume a rate of $17\frac{1}{2}\%$ for GCT though some goods and services are charged at a higher rate.

For imported goods Import Duty at 25% is added to the total of the basic cost, insurance and freight.

All these tax rates are set by the government and can change at any time.

Exercise 9c

1 Jane Axe has a taxable income of $20 000 a week. She pays income tax on this at 20%.
 a How much tax must she pay? b Find her net income.

2 Freddy Davis has a taxable income of $520 000 a year. He pays income tax at 25%.
 a How much tax does he pay?
 b Work out his net income i a year ii a week.

3 Complete the table

Name	Taxable weekly pay	Tax rate	Tax due	Net weekly pay
M. Davis	$8200	20%		
P. Evans	$12 300	25%		
G. Brown	$17 650	25%		
A. Khan	$21 820	25%		

4 Deductions for the National Insurance Scheme (NIS) at 2%, National Housing Trust (NHT) at 2.5% and Education Tax at 2% are made from a person's total income. Copy and complete the following table:

	Weekly Income $'000s	NIS at 2%	NHT at 5%	Education Tax at 2%	Total of these deductions	Remaining income before income tax is deducted
Mrs Peacock	20					
Mr Walters	26					
Ms Morgan	32					
Mr Davis	38					
Mrs Evans	48					
Ms Bennett	120					

5 Mr Bolt is paid $70 000 a week gross. The table below shows how his net income is calculated. Copy and complete this table.

Weekly Income	$70 000
NIS at 2%	
NHT at 5%	
Education Tax at 2%	
Tax Free Allowance	$5000
Total of these deductions	
Remaining income before income tax is deducted	
Income tax at 25%	
Net income	

6 A CD costs $1200 plus general consumption tax (GCT) at $17\frac{1}{2}\%$. Find

  a the GCT to be added   **b** the price I must pay for the CD.

In questions **7** to **9** find the total purchase price of the item. Take the rate of GCT as 17.5%.

7 An electric cooker marked $40 000 + GCT.

8 A calculator costing $720 + GCT.

9 A van marked $1 450 000 + GCT.

10 The price tag on a television gives $29 000 plus GCT at 15%. What does the customer have to pay?

11 In March, Nicki looked at a camera costing $4800 plus GCT. The GCT rate at that time was $17\frac{1}{2}\%$. How much would the camera have cost in March? Nicki decided to wait until June to buy the camera but by then the GCT had been raised to 22%. How much did she have to pay?

12 An electric cooker was priced in a showroom at $45 000 plus GCT at $17\frac{1}{2}\%$.

  a What was the price to the customer?

  Later in the year GCT was increased to 20%. The showroom manager placed a notice on the cooker that read:

  Due to the increase in GCT this cooker will now cost you $54 196.88

  b Was the manager correct?

  c If your answer is 'Yes', state how the manager calculated the new price. If your answer is 'No', give your reason and find the correct price.

Banking

Most people who work will need a bank account as many employers will pay wages and salaries directly into a bank account. There are different kinds of account. Some are for everyday banking, some are for savings, others are for loans.

A simple account enables you to receive money owed to you, get money out and pay bills. Banks also offer credit cards and many other services.

Use a search engine to investigate accounts on offer from different banks.

Finding simple interest

Everybody wishes to borrow something at one time or another. Perhaps you want to borrow a video camera to record a wedding, a dress to wear to an important event or even a bicycle for a few minutes. In the same way, the time will come when you will wish to borrow money to buy a motorcycle, a car, furniture or even a house.

The cost of hiring or borrowing money is called the *interest*. The sum of money borrowed (or lent) is called the *principal* and the interest is usually an agreed *percentage* of the sum borrowed.

For example, if $10 000 is borrowed for a year at an interest rate of 12% per year, then the interest due is $\frac{12}{100}$ of $10 000, i.e. $1200.

The interest due on $20 000 for one year at 12% would be $20\,000 \times \frac{12}{100} = \$2400$, and on $P for one year at 12% would be $\$P \times \frac{12}{100}$.

If we double the period of the loan, we double the interest due, and so on. The interest on $P invested for T years at 12% would therefore be $\$P \times \frac{12}{100} \times T$.

If the interest rate was $R\%$ instead of the given 12%, the interest, I, would be $\$P \times \frac{R}{100} \times T$.

When interest is calculated this way it is called *simple interest*.

Therefore
$$I = \frac{PRT}{100}$$

where
I is the simple interest in $
P is the principal in $
R is the rate per cent per year
T is the time in years

Unless stated otherwise, $R\%$ will always be taken to mean $R\%$ each year or per annum.

Exercise 9d

Find the simple interest on:

1 $10 000 for 2 years at 10%

$1000 is the principal, 2 years is the time and 10% is the rate, so $I = \frac{10\,000 \times 10 \times 2}{100}$.

2 $10 000 for 2 years at 12%

3 $10 000 for 3 years at 8%

4 $10 000 for 4 years at 13%

5 $10 000 for 7 years at 11%

6 $20 000 for 2 years at 10%

7 $20 000 for 5 years at 8%

8 $30 000 for 4 years at 12%

9 $40 000 for 6 years at 9%

10 $60 000 for 7 years at 11%

11 $35 000 for 5 years at 7%

12 $12 500 for 4 years at 12%

13 $64 200 for 7 years at 11%

14 $174 000 for 8 years at 8%.

15 $72 400 for 3 years at 6%

16 $48 400 for 3 years at 7%

17 $37 200 for 7 years at 14%

18 $9400 for 6 years at 9%

19 $64 800 for 5 years at 13%

20 $92 600 for 9 years at 14%.

Find the simple interest on $13 466 for 5 years at 12%, giving your answer correct to the nearest dollar.

$$I = \frac{PRT}{100} \text{ where } P = 13\,466, R = 12 \text{ and } T = 5$$

$$\therefore \quad \text{simple interst} = \$\frac{13\,466 \times 12 \times 5}{100}$$

$$= \$8079.6$$

$$= \$8080 \quad \text{correct to the nearest dollar}$$

Find, giving your answers correct to the nearest dollar, the simple interest on:

21 $52 652 for 2 years at 12%

22 $9 456 for 4 years at 8%

23 $14 216 for 5 years at 11%

24 $81 340 for 4 years at 13%

25 $62 783 for 3 years at 14%

26 $55 545 for 5 years at 9%

27 $12 372 for 4 years at 2%

28 $54 389 for 7 years at 0.5%

29 $82 692 for 6 years at 7%

30 $71 747 for 4 years at 17%.

Find, giving your answers correct to the nearest dollar, the simple interest on:

31 $15 440 for 4 years at $8\frac{1}{2}$%

32 $27 380 for $4\frac{1}{2}$ years at 9%

Write $8\frac{1}{2}$ as 8.5

33 $52 749 for 3 years at $12\frac{3}{4}$%

34 $43 615 for $7\frac{1}{2}$ years at $11\frac{1}{4}$%

35 $8472 for $4\frac{1}{4}$ years at $13\frac{1}{2}$%

38 $203 448 for $1\frac{1}{2}$ years at $7\frac{1}{5}$%

36 $7358 for $5\frac{3}{4}$ years at $9\frac{3}{4}$%

39 $61 327 for $3\frac{1}{4}$ years at $15\frac{1}{2}$%

37 $36 488 for $2\frac{3}{4}$ years at $8\frac{1}{4}$%

40 $45 492 for $6\frac{1}{4}$ years at $18\frac{1}{4}$%.

Find, giving your answers correct to the nearest dollar, the simple interest on:

41 $32 000 for 100 days at 12%

42 $41 300 for 150 days at 8%

43 $100 000 for 300 days at 9%

44 $28 250 for 214 days at 16%

45 $61 394 for 98 days at $14\frac{1}{2}$%

46 $72 932 for 22 days at 11%.

> T must be in years so change the number of days to a fraction of a year. Use 365 as the number of days in a year.
> 100 days $= \frac{100}{365}$ yrs

Amount

If I borrow $25 000 for 3 years at 11% simple interest, the *sum of the interest* and *principal* is the total I must repay to clear the debt. This sum is called the *amount* and is denoted by A,

i.e. $$A = P + I$$

In this case $$I = \frac{25\,000 \times 11 \times 3}{100}$$

$$= 8250$$

So the simple interest $= 8250

$$\therefore \text{amount} = \$25\,000 + \$8250$$

$$= \$33\,250$$

Exercise 9e

Find the amount of:

1 $35 000 for 5 years at 10%

9 $73 800 for $3\frac{1}{2}$ years at 9%

2 $42 000 for 2 years at 8%

10 $18 600 for $4\frac{1}{4}$ years at 12%

3 $65 000 for 4 years at 12%

11 $28 500 for 9 years at 6%

4 $51 300 for 4 years at $13\frac{1}{2}$%

12 $82 650 for 6 years at 8%

5 $82 000 for 8 years at 14%

13 $19 263 for 5 years at 11%

6 $97 000 for 7 years at 9%

14 $56 427 for $6\frac{1}{2}$ years at 12%

7 $49 200 for 5 years at $8\frac{1}{2}$%

15 $71 855 for $4\frac{1}{4}$ years at $13\frac{1}{2}$%

8 $65 420 for 4 years at 9%

16 $31 800 for $5\frac{3}{4}$ years at $11\frac{1}{2}$%.

Inverse questions on simple interest

Consider the question of finding the rate of interest if the cost of borrowing $50 000 for 3 years is $18 000.

If we substitute these values in the formula $I = \frac{PRT}{100}$ we have

$$18\,000 = \frac{\overset{500}{\cancel{50\,000}} \times R \times 3}{\underset{1}{\cancel{100}}} \text{ where } R \text{ is the rate \%,}$$

i.e. $18\,000 = 1500R$

$\therefore \quad R = \frac{18\,000}{1500} = 12$

i.e. the rate of interest is 12%.

Similar reasoning is required if either P or T is the unknown quantity.

Exercise 9f

Find the principal that will earn:

1 $20 000 simple interest in 5 years at 8%

2 $43 200 simple interest in 6 years at 12%

3 $19 600 simple interest in 8 years at 7%

4 $8085 simple interest in $3\frac{1}{2}$ years at $5\frac{1}{2}$%

5 $39 690 simple interest in $4\frac{1}{2}$ years at $10\frac{1}{2}$%.

$I = 20\,000\ T = 5, R = 8$, giving $20\,000 = \frac{P \times 8 \times 5}{100}$.

What is the rate per cent if the cost of borrowing:

6 $50 000 for 5 years is $25 000

7 $45 000 for 4 years is $27 000

8 $70 000 for 8 years is $67 200

9 $85 000 for 3 years is $20 400

10 $34 000 for 6 years is $14 280?

$P = 50\,000, T = 5$, $I = 25\,000$.

What is the rate per cent if:

11 $25 000 will earn $7500 simple interest in 3 years

12 $37 000 will earn $29 600 simple interest in 5 years

13 $64 000 will earn $20 480 simple interest in 4 years

14 $87 000 will earn $62 640 simple interest in 6 years

15 $43 500 will earn $15 660 simple interest in 4 years?

Find the number of years in which:

16 $30 000 invested at 8% simple interest will earn $4800

17 $53 500 invested at 12% simple interest will earn $32 100

18 $47 000 invested at 14% simple interest will earn $19 740

19 $61 700 invested at 9% simple interest will earn $22 212

20 $82 400 invested at 11% simple interest will earn $27 192.

What sum of money will amount to $48 960 if invested for 3 years at 12% simple interest?

$$I = \frac{PRT}{100}$$

$$= \frac{P \times 12 \times 3}{100}$$

But
$$A = P + I$$

$$\therefore \quad 48 960 = P + \frac{36}{100}P$$

$$= \frac{136}{100}P$$

i.e.
$$4 896 000 = 136 P$$

or
$$P = \frac{4 896 000}{136}$$

$$= 36 000$$

Therefore $36 000 will amount to $48 960 if invested for 3 years at 12% simple interest.

What sum of money will amount to:

21 $34 800 if invested for 2 years at 8%

22 $84 100 if invested for 3 years at 15%

23 $68 800 if invested for 5 years at 12%

24 $124 920 if invested for 7 years at $10\frac{1}{2}$%

25 $142 750 if invested for $4\frac{1}{2}$ years at $9\frac{1}{2}$%?

! Investigation

Vivian invested $100 000 for 3 years. The interest rate when he invested the money was 4% per annum.

Vivian calculated that his investment should be worth $112 000 at the end of the three years but when he checked he found that it was worth $113 550.

Investigate possible reasons for the extra $1550.

Problems

Exercise 9g

Copy and complete the following table:

	Principal in $	Rate %	Time	Simple interest in $	Amount in $
1	23 000	10	2 years		
2	18 000		8 years	23 040	
3	95 000	9			129 200
4		18		75 240	151 240
5	63 700		6 years		121 030
6		14		4452	46 852
7	82 800		5 months	2587.5	
8	55 500	$12\frac{1}{2}$	144 days		

9 Mr Sadler's bank pays interest at 3% p.a. on money he has on deposit. How much is in his account if the interest for 7 months is $2506?

10 Miss Zeraschi pays $8487 when she borrows a sum of money from the bank for 9 months at 12% p.a. simple interest. How much does she borrow?

11 The interest I receive on $150 000 invested for a certain time would increase by $22 500 if the interest rate rose by 3%. For how long is the sum invested?

12 The interest I receive on $84 600 would decrease by $11 844 if the interest rate dropped by 2%. For how long is the sum invested?

13 A factory owner borrows $1 860 000 from his bank to buy machinery. If the annual rate of interest is $12\frac{1}{2}$%, how much must he repay the bank after 9 months to clear the debt?

14 Jane Peters borrowed $21 900 at 14% p.a. When she repaid the debt the interest due was $1050. For how many days did she borrow the money?

15 Find the sum to which $171 600 will amount to in 10 months when invested at $8\frac{1}{2}$% per annum.

Compound interest

Sometimes, when the interest is due, the whole amount remains invested or is not paid back.

Suppose $200 000 is invested for 1 year at 8%.

The interest earned is $200 000 × 0.08 = $16 000

so the amount at the end of the year is $216 000.

If this amount is invested for a year at 8% the interest due is
$216 000 × 0.08 = $17 280

The amount now is $216 000 + $17 280 = $233 280

If you invest money, do not spend the interest, and the annual rate stays the same, your money will increase by larger and larger amounts each year. The total amount by which it grows is called the *compound interest*.

Exercise 9h

Find the compound interest on $55 000 invested for 2 years at 6%.

Interest for first year at 6% is 6% of the original principal.

New principal at end of first year = 100% of original principal
                                    +6% of original principal

                                  = 106% of original principal

                                  = 1.06 × original principal

∴ principal at end of first year = 1.06 × $55 000 = $58 300

Similarly, new principal at end of second year = 106% of principal at
                                                 beginning of second year

                              = 1.06 × $58 300 = $61 798

Compound interest = principal at end of second year − original principal

           = $61 798 − $55 000 = $6798

Find the compound interest on

1   $20 000 for 2 years at 10% p.a.   3 $40 000 for 3 years at 8% p.a.

2   $30 000 for 2 years at 12% p.a.   4 $65 000 for 3 years at 9% p.a.

5 $52 000 for 2 years at 13% p.a.   **8** $500 000 for 2 years at 1.4%

6 $62 400 for 3 years at 11% p.a.   **9** $1 000 000 for 2 years at 0.8%

7 $20 000 for 2 years at 1%   **10** $400 000 for 2 years at 3.5%.

11 $4 000 000 is invested at compound interest of 10% each year.
What will it be worth in 2 years' time?

12   Brian Barnes borrows $500 000 at 12% compound interest. He agrees
to clear the debt at the end of 2 years. How much must he pay?

13   An old postage stamp increases in value by 15% each year.
If it is bought for $5000, what will it be worth in 3 years'
time?

14   A motorcycle bought for
$150 000 depreciates in
value by 10% each year.
Find its value after
3 years.

Find 10% of the purchase price and subtract this
from the purchase price to give the value after
1 year. Now find 10% of the new value. Deduct this
from the value at the end of the first year to give the
value at the end of the second year and so on.

15   A motor car bought for £2 000 000 depreciates in any one year by
20% of its value at the beginning of that year. Find its value after
2 years.

Telephone bills

There are fixed line telephones and cell phones.

The cost of a telephone call from a fixed line normally depends on three
factors:

   i   the distance between the caller and the person being called
   ii   the time of day and/or the day of the week on which the call is being
          made
  iii   the length of the call.

These three factors are put together in various ways to give metered units
of time, each unit being charged at a fixed rate.

There is also a fixed quarterly charge.

The cost of a telephone call on a cell phone depends on the package
bought. You can buy a prepaid sim card or a variety of monthly packages.
These vary depending how much you spend.

Exercise 9i

Chris Reynolds' telephone account for the last quarter showed that his telephone had been used for 546 metered units. If the standing charge was $3100 and each unit cost $9, work out how much he must pay for the quarter.

Cost of 546 units at $9 per unit $= 546 \times \$9 = \$4914$

Standing charge $= \$3100$

Total cost $=$ Cost of units $+$ Standing charge

$= \$4914 + \$3100 = \$8014$

∴ the telephone bill for the quarter was $8014.

Find the quarterly telephone bill for each of the following households.

	Name	Number of units used	Standing charge	Cost per unit
1	Mrs Keeling	750	$2800	$5
2	Mr Hodge	872	$3200	$6
3	Miss Hutton	1040	$3300	$7
4	Miss Jacob	1134	$3760	$8.5
5	Mrs Buckley	1590	$3680	$8.3
6	Mr Leeson	765	$4200	$7.68
7	Mrs Solly	965	$5100	$10.5

Calculate the monthly telephone bill for each of the following people.

	Name	up to 3 mins	Each additional min.	Number of mins	Service charge
8	Singh	$390	$130	70	$2540
9	Bird	$645	$215	28	$3000
10	Lee	$915	$305	105	$2540

Electricity: kilowatt-hours

A kilowatt-hour is electric power of one kilowatt used for one hour. The electric company charges one rate for the first number of kilowatt-hours and a higher rate for additional usage.

We all use electricity in some form and we know that some appliances cost more to run than others. For example, an electric fire costs much more to run than a light bulb. Electricity is sold in units called kilowatt-hours (kWh) and each appliance has a rating that tells us how many kilowatt-hours it uses each hour.

A typical rating for an electric fire is 2 kW. This tells us that it will use 2 kWh each hour, i.e. 2 units per hour. On the other hand, a light bulb can have a rating of 100 W. Because 1 kilowatt = 1000 watts (kilo means 'thousand' as we have already seen in kilometre and kilogram), the light bulb uses $\frac{1}{10}$ kWh each hour, or $\frac{1}{10}$ of a unit.

Exercise 9j

How many units (i.e. kilowatt-hours) will each of the given appliances use in 1 hour?

1	a 3 kW electric fire	**5**	a 60 W video recorder
2	a 100 W bulb	**6**	a 20 W radio
3	a $1\frac{1}{2}$ kW fire	**7**	an 8 kW stove
4	a 1200 W hair dryer	**8**	a 2 kW dishwasher

With the help of an adult, find the rating of any of the following appliances that you might have at home. The easiest place to find this information is probably from the instructions.

9	an electric kettle	**12**	the television set
10	the refrigerator	**13**	a bedside lamp
11	the washing machine	**14**	the electric cooker

How many units of electricity would

15	a 2 kW fire use in 8 hours	**19**	a 150 W refrigerator use in 12 hours
16	a 100 W bulb use in 10 hours	**20**	a 12 W radio use in 12 hours
17	an 8 kW cooker use in $1\frac{1}{2}$ hours	**21**	an 8 W night bulb use in a week at 10 hours per night
18	a 60 W bulb use in 50 hours	**22**	a 5 W clock use in 1 week?

For how long could the following appliances be run on one unit of electricity?

23 a 250 W bulb

24 a 2 kW electric fire

25 a 100 W television set

26 a 360 W electric drill

In the following questions assume that 1 unit of electricity costs $6.
How much does it cost to run

27 a 100 W bulb for 5 hours

28 a 250 W television set for 8 hours

29 a 3 W clock for 1 week

30 a 3 kW kettle for 5 min?

Electricity bills

It is clear from the questions in the previous exercise that lighting from electricity is cheap but heating is expensive.

Domestic electricity bills are calculated by charging every household a fixed amount, together with a charge for each unit used. The amount used is recorded on a meter, the difference between the readings at the beginning and end of a month showing how much has been used.

The following table shows a simplified electricity bill.

current	previous			usage	rate (c)	charge ($)
14 261	13 978		Energy first	100	675	675
			Energy next	183	1200	2196
			Cost charge			120
			Sub-total			2991
			Fuel & IPP charge	283	$18	5094
					Total	8085

This bill shows the number of kWh registered on the meter at the beginning and end of the charging period (usually one month). The number of units used is the difference between these two values.

The cost of the first 100 kWh is 675 c per unit and the cost of the remaining units used is 1200 c per unit. In addition there is a fixed or Cost Charge. Working these amounts out and adding them gives the Sub-total.

However there is another Fuel and IPP charge for every unit. This covers the cost, which varies from month to month, of the fuel to generate the electricity and the cost of the electricity supplied by Independent Power Producers (IPP).

In the table Cost of Energy First units $= 100 \times 675\,c = \$675$

Cost of Energy Next units $= 183 \times 1200\,c = \$2196$

Fuel & IPP charge $= 283 \times \$18 = \$5094$

It is also possible that GCT may be added to the total to give the amount the householder must pay. You can find the current charges at: **www.myjpsco.com/residential/understanding_your_bill.php**

Exercise 9k

Mrs Comerford used 196 units of electricity last month. Apart from a Cost Charge of $110 she is charged 570 c per unit for the first 100 units and 1100 c per unit for the remainder.

How much does her electricity cost for the month?

Cost of 100 units at 570 c per unit $= 100 \times 570\,c = \$570$

Cost of remaining 96 units at 1100 c per unit $= 96 \times 1100\,c = \$1056$

Cost Charge $= \$110$

Total cost $= \$570 + \$1056 + \$110 = \$1736$

1   For each householder find the total cost of their electricity for the month. For these questions we have neglected the Fuel and IPP charge.

Name	No. units used	Cost per unit for first 100 units (cents)	Cost per unit for remainder (cents)	Cost charge ($)	Total
Mr George	250	550	1120	120	
Miss Newton	320	560	1200	130	
Mr Khan	225	620	1250	105	
Mrs Wilton	174	720	1420	135	
Mr Barnes	385	645	1250	128	

2   The government introduced a General Consumption Tax of 10% for all electricity bills. For each person listed in question **1** how much extra will they have to pay?

3   For each householder find the total cost of their electricity for the month. Note that for these householders the Fuel & IPP charge is included.

Name	Number of units used	Cost per unit for first 100 units (cents)	Cost per unit for remainder (cents)	Cost charge ($)	Fuel & IPP charge per unit ($)	Total
Mrs Wan	240	510	1000	100	16	
Mr Davis	270	620	1260	130	18	
Mr Deats	166	575	1300	100	15	
Mrs Beale	342	535	1170	125	17	

4 The government introduced a General Consumption Tax of 10% for all electricity bills. For each person listed in question **3** work out the total they will now have to pay.

In questions **5** to **7** copy the table and fill in the blanks.

5

current	previous		usage	rate (c)	charge ($)
9421	9175	Energy first	100	600	600
		Energy next	146	1150	
		Cost charge			110
		Sub-total			
		Fuel & IPP charge	246	$18	
				Total	

6

current	previous		usage	rate (c)	charge ($)
8432	8156	Energy first	100	650	650
		Energy next		1250	
		Cost charge			130
		Sub-total			
		Fuel & IPP charge	276	$17	4692
				Total	

7

current	previous		usage	rate (c)	charge ($)
	10762	Energy first	120	590	
		Energy next	159	1300	
		Cost charge			140
		Sub-total			
		Fuel & IPP charge	279	$19	
				Total	

Water rates

In most countries water is supplied by the state as a public service. Businesses and householders pay for this service depending on how much water they consume.

Water usage is measured with a meter. The newest meters show consumption in cubic metres, but the bills show it in litres. One cubic metre is equal to 1000 litres.

The charges for water and sewage depend on the amount used; this charge increases the more water is used. There is a service charge that depends on the diameter of the pipe that supplies the water. There is also an adjustment to the total bill to reflect costs incurred by the water supplier. This adjustment is a percentage of the total and changes from month to month.

You can find out about these different charges from **www.nwcjamaica.com**

Payment is usually monthly.

The table shows the charges for a typical domestic customer.

Water charge	First 14 000 litres	$49.63 per 1000 litres
	Next 13 000 litres	$87.51 per 1000 litres
	Next 14 000 litres	$94.50 per 1000 litres
	....	....
	Sewerage charge	same as the water charge
	Service charge	$380.00
	Adjustments	11%

In January Mr Smith used 18 cubic metres (18 000 litres) of water. We can use the table to work out his bill.

Charge for the first 14 000 litres is $49.63 \times 14 = $694.82

Charge for remaining 4000 litres is $87.51 \times 4 = $350.04

Water charge is $\qquad$ $694.82 + $350.04 = $1044.86

The charge for sewerage is the same as the water usage.

Therefore the charges before the adjustment are

$$\text{water use} + \text{sewage} + \text{service charge} = \$1044.86 + \$1044.86 + \$380$$
$$= \$2469.72$$

$$\text{Adjustment} = 11\% \text{ of } \$2469.72 = \$2469.72 \times 0.11$$
$$= \$271.67 \text{ (to the nearest cent)}$$

Mr Smith's bill $= \$2469.72 + \$271.67 = \$2741.39$

Exercise 9I

Use the table above to calculate the charges. Give answers that are not exact to the nearest cent.

1   Mr Barnes used 10 000 litres of water in June.
    Calculate Mr Barn's water charge for June.

2   Mrs Khan used 5000 litres of water in April.
    Calculate Mrs Khan's water charge for April.

3   Mrs Weeks used 20 000 litres of water in February.
    Calculate Mrs Weeks's water charge for February.

4   Mr Anthony used 30 000 litres of water in August.
    Calculate Mr Anthony's water charge for August.

5   Mr Laynes used 10 000 litres of water in December.
    a   Calculate Mr Laynes' charges before the adjustment.
    b   Calculate Mr Laynes' bill.

6   Mrs Arnold used 18 000 litres of water in September.
    a   Calculate Mrs Arnold's charges before the adjustment.
    b   Calculate Mrs Arnold's bill.

7   Mr Amish used 28 000 litres of water in May.
    a   Calculate Mr Arnish's charges before the adjustment.
    b   Calculate Mr Amish's bill.

8   At the end of June the meter reading on Mrs Wright's water meter was 38 cubic metres. At the end of July the reading was 51 cubic metres.
    Calculate Mrs Wright's bill for July.

? Puzzle

I have two old coins. One is marked Elizabeth I and the other George VI. A knowledgeable friend told me that one was probably genuine but that the other was definitely a fake. How could she be so certain?

In this chapter you have seen...

✔ how to work out and check shopping bills

✔ how to work out commission and bonus payments

✔ that there are deductions made from earnings for tax and other things

✔ that you can find the simple interest on a sum of money by using the fomula $I = \frac{PRT}{100}$

✔ that the 'amount' is the sum of the principal and the interest

✔ that you can use the formula $I = \frac{PRT}{100}$ to find the principal or the time or the rate when you know the interest and the other two quantities

✔ that compound interest means that interest is paid on the sum of the pricipal and interest from the previous year.

✔ that a kWh is 1 kilowatt used for 1 hour

✔ that bills for domestic utilities usually include a fixed charge and a charge for the number of units used.

10 Transformations

Line symmetry

 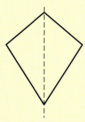 

As we saw in Grade 7, shapes like these are *symmetrical*. They have line symmetry (or bilateral symmetry); the dotted line is the *axis of symmetry* because if the shape were folded along the dotted line, one half of the drawing would fit exactly over the other half.

Exercise 10a

1   Which of the following shapes have an axis of symmetry?

a    b   c

Copy the following drawings on squared paper and complete them so that the dotted line is the axis of symmetry.

2 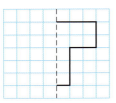   4    6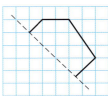

3    5 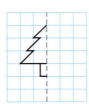   7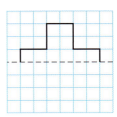

Two or more axes of symmetry

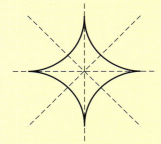   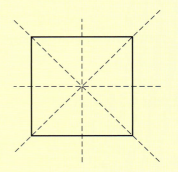

Shapes can have more than one axis of symmetry. In the drawings above, the axes are shown by dotted lines and it is clear that the first shape has four axes of symmetry, the second has three and the third has four.

Exercise **10b**

Sketch or trace the shapes in questions **1** to **12**. Mark in the axes of symmetry and say how many there are. (Some shapes may have no axis of symmetry.)

1

5

9

2

6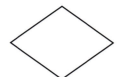

10

3

7

11

4

8

12

Copy and complete the following drawings on squared paper. The dotted lines are the axes of symmetry.

13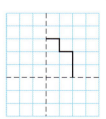

15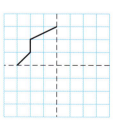

17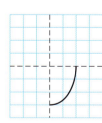

14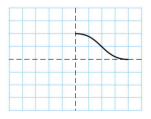

16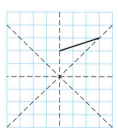

18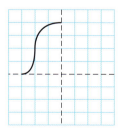

19   Draw, on squared paper or on plain paper, shapes of your own with more than one axis of symmetry.

? Puzzle

Show how sixteen counters can be arranged in ten rows with exactly four counters in each row.

Reflections

This section revises the work on reflections in Grade 7.
When an object is reflected in a line, the object and its image form a symmetrical shape.
The line is called the mirror line.

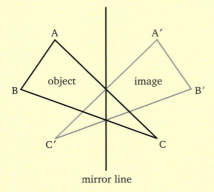

A and A′ are corresponding vertices, so are B and B′ and C and C′.

The mirror line is the perpendicular bisector of the line segment joining a pair of corresponding vertices.
You can use this fact to find a mirror line.

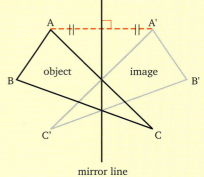

Exercise 10c

In each of questions **1** to **4**, copy the diagram. Draw the image when the object is reflected in the mirror line and label the corresponding vertices A′, B′, C′, ..., etc.

1

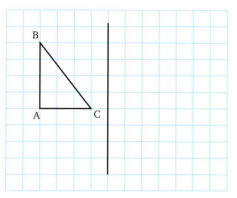

2

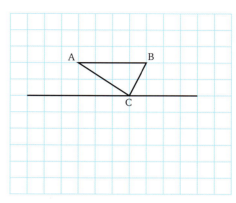

3

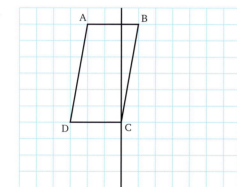

4

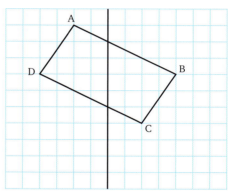

In questions 5 and 6, copy the diagram and draw in the mirror line.

5

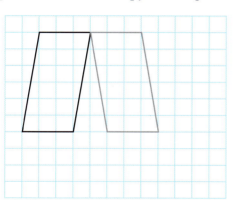

6

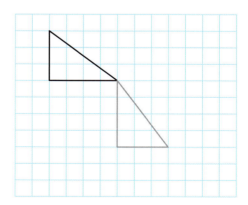

Transformations

Suppose the triangle ABC is cut out of card and is placed in the position shown.

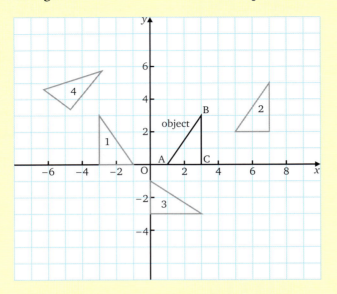

Triangle ABC can be moved to the positions shown by picking it up, turning it over and putting it down in position 1 or sliding it across the surface to positions 2, 3, or 4. All of these movements are called transformations.

Translations

All the movements in this diagram are of the same type.

Triangle ABC is slid across the paper without being turned or changed in shape or size. This type of movement is called a translation.

Although not a reflection, we still use the word object to describe triangle ABC and the word image to describe the new positions.

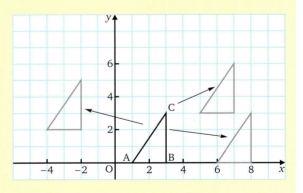

Exercise 10d

1 In the following diagram, which images of △ABC are given by translations?

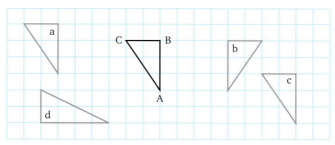

2 In the following diagram, which images of △ABC are given by translation, which by a reflection and which by neither?

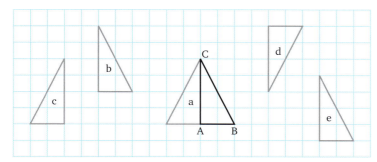

3 Is the transformation a translation? Give a reason for your answer.

a
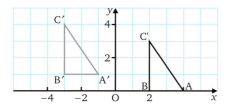

b
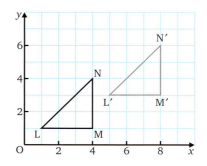

Describing a translation

We can describe a translation by giving details of the movement required to get to the new position. (This is called the displacement.)
In question 1 above, the translation of the object to the image **a** can be described as five squares to the left and one square up.

Exercise 10e

1 Describe the translation that maps △ABC to △A′B′C′.

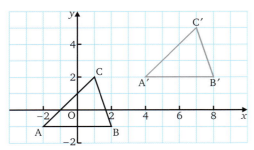

2 Describe the translations that map
   a △ABC to △PQR
   b △PQR to △ABC.

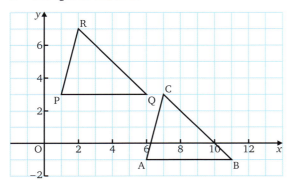

3 Describe the translations that map
   a △ABC to △PQR
   b △ABC to △LMN
   c △XYZ to △ABC
   d △ABC to △ABC.

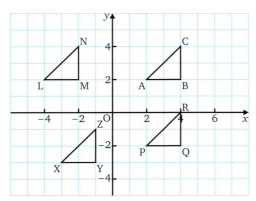

4 Draw axes for x and y from −4 to 5.
  Draw the following triangles:

  △ABC with A(2, 2), B(4, 2), C(2, 5);
  △PQR with P(1, −2), Q(3, −2), R(1, 1);
  △XYZ with X(−3, 1), Y(−1, 1), Z(−3, 4).

  Describe the translations that map
  a △ABC to △PQR
  b △PQR to △ABC
  c △PQR to △XYZ
  d △ABC to △ABC.

5 Draw axes for x and y from 0 to 9. Draw △ABC with A(3, 0), B(3, 3), C(0, 3) and △A′B′C′ with A′(8, 2), B′(8, 5), C′(5, 5).

  Is △A′B′C′ the image of △ABC under a translation? If so, describe the translation. If not, give a reason.

  Join AA′, BB′ and CC′. What type of quadrilateral is AA′B′B? Give reasons. Name other quadrilaterals of the same type in the figure.

<u>6</u>

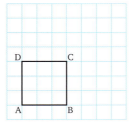

  a Square ABCD is translated parallel to AB a distance equal to AB. Sketch the diagram and draw the image of ABCD.
  b Square ABCD is translated parallel to AC a distance equal to AC. Sketch the diagram and draw the image of ABCD.

7   Draw axes for x and y from -2 to 7. Draw $\triangle$ABC with A(-2, 5), B(1, 3), C(1, 5).

Translate $\triangle$ABC 5 units to the right and 1 unit up. Label this image
$A_1B_1C_1$. Then translate $\triangle A_1B_1C_1$ 1 unit in the direction of the negative
x-axis and 3 units down the y-axis. Label this new image $A_2B_2C_2$.

Describe the translations that map
   a $\triangle$ABC to $\triangle A_2B_2C_2$ **b** $\triangle A_2B_2C_2$ to $\triangle$ABC   **c** $\triangle A_2B_2C_2$ to $\triangle A_1B_1C_1$

Rotational symmetry

Some shapes have a type of symmetry different from line symmetry.

a 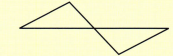   **b**   **c**

These shapes do not have an axis of symmetry but can be turned or rotated
about a centre point and still look the same. Such shapes are said to have
rotational symmetry. The centre point is called the *centre of rotation*.

Exercise 10f

1 **a**    **b** 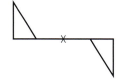   **c**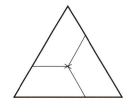

   Trace each of the shapes above, then turn the tracing paper about the centre
of rotation (put a compass point or a pencil point in the centre). Turn until the
traced shape fits over the original shape again. In each case state through what
fraction of a complete turn the shape has been rotated.

2 Which of the following shapes have rotational symmetry?

   a   **b**

   c   **d**

Order of rotational symmetry

If the smallest angle that a shape needs to be turned through is a third of a complete turn to fit, then it will need two more such turns to return it to its original position. So, starting from its original position, it takes three turns, each one-third of a revolution, to return it to its starting position.

It has *rotational symmetry of order 3*.

Exercise 10g

Give the order of rotational symmetry of the following shape.

You need to decide whether turning this shape about its centre through some angle leaves it looking unchanged to the eye. If so, how big is this angle? How many times can you do this before getting back to the starting position?

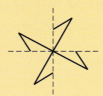

The smallest angle turned through is a right angle or one-quarter of a complete turn.

The shape has rotational symmetry of order 4.

1 Give the orders of rotational symmetry of the shapes in Exercise 10f, question **1**.
2 Give the orders of rotational symmetry, if any, of the shapes in Exercise 10f, question **2**.

Copy and complete the diagram, given that there is rotational symmetry of order 4.

If the order of rotational symmetry is to be 4 you must turn the given shape about the cross (×) through one-quarter of a turn, i.e. through 90°. Repeat this until you get back to the starting position.

Each of the diagrams in questions **3** to **8** has rotational symmetry of the order given and × marks the centre of rotation.

Copy and complete the diagrams. (Tracing paper may be helpful.)

3

Rotational symmetry of order 4

4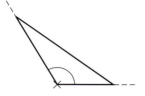

Rotational symmetry of order 3

5

Rotational symmetry of order 2

6

Rotational symmetry of order 4

7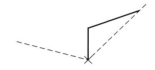

Rotational symmetry of order 3

8

Rotational symmetry of order 2

<u>**9**</u> In questions **3** to **8**, give the size of the angle, in degrees, through which each shape is turned.

Exercise 10h

Some shapes have both line symmetry and rotational symmetry:

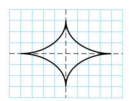

Two axes of symmetry
Rotational symmetry of order 2

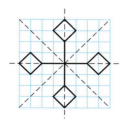

Four axes of symmetry
Rotational symmetry of order 4

Which of the following shapes have

a rotational symmetry only
b line symmetry only
c both?

You can spot line symmetry by seeing if any line acts as a mirror; you can spot rotational symmetry if you can rotate the shape about a point that leaves it looking unchanged.

1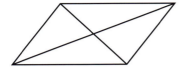

4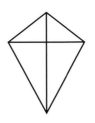

7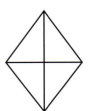

2

5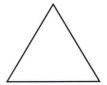

8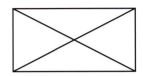

3

6

9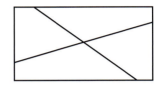

10 Make up three shapes that have rotational symmetry only. Give the order of symmetry and the angle of turn, in degrees.

11 Make up three shapes with line symmetry only. Give the number of axes of symmetry.

12 Make up three shapes that have both line symmetry and rotational symmetry.

13 The capital letter × has line symmetry (two axes) and rotational symmetry (of order 2). Investigate the other letters of the alphabet.

? Puzzle

A dried-out well is 12 metres deep. A snail starts from the bottom and tries to climb out. It climbs up 3 m every night and falls back 2 m every day. How long will it take the snail to climb out?

Rotations

a 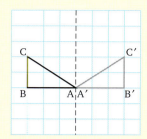   b 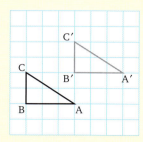   c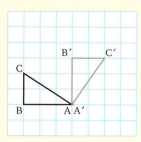

So far, in transforming an object we have used reflections, as in **a**, and translations, as in **b**, but for **c** we need a rotation.

In this case we are rotating △ABC about A through 90° clockwise (⌒). A is called the centre of rotation. We could also say △ABC was rotated through 270° anticlockwise (⌒). Clockwise rotation is sometimes referred to as positive rotation or rotating through a positive angle. Anticlockwise rotation is rotating through a negative angle.

For a rotation of 180° we do not need to say whether it is clockwise or anticlockwise.

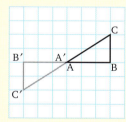

Exercise 10i

Give the angle of rotation when △ABC is mapped to △A′B′C′.

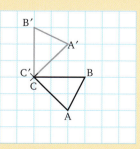

If you rotate △ABC anticlockwise about C it can be turned to position A′B′C. You must now measure the angle it has turned through. Compare the position of one side of the object triangle with the corresponding side of the image triangle, e.g. CB with C′B′.

The angle of rotation is 90° anticlockwise.

In questions **1** to **4**, give the angle of rotation when △ABC is mapped to △A'B'C'.

1

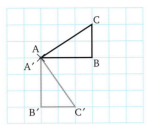

3

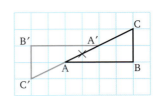

Don't forget to say, where appropriate, whether the rotation is clockwise or anticlockwise.

2

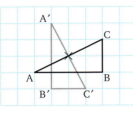

4

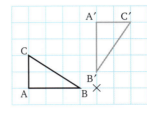

In questions **5** to **10**, state the centre of rotation and the angle of rotation. △ABC is the object in each case.

5

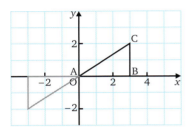

8

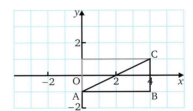

6

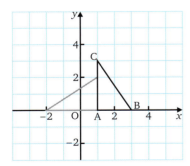

9

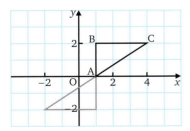

7

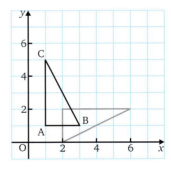

10

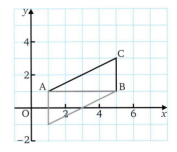

Copy the diagrams in questions **11** to **18**, using 1 cm to 1 unit.

Find the images of the given objects under the rotations described.

Be careful to do the rotation in the right direction.

11

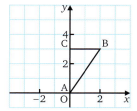

Centre of rotation (0, 0)
Angle of rotation 90° anticlockwise

14

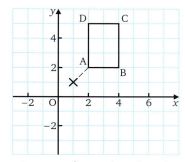

Centre of rotation (1, 1)
Angle of rotation 180°
(As the centre of rotation is not a point on the object, join it to A first.)

12

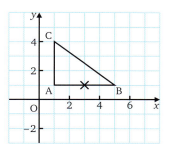

Centre of rotation (3, 1)
Angle of rotation 180°

15

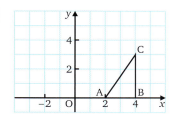

Centre of rotation (0, 0)
A negative angle of rotation of 90°

A negative angle of rotation means anticlockwise.

13

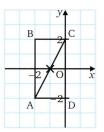

Centre of rotation (−1, 0)
Angle of rotation 180°

16

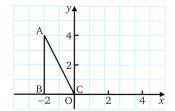

Centre of rotation (2, 0)
A positive angle of rotation of 90°

17

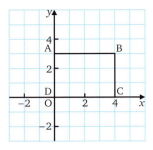

Centre of rotation (2, 0)
Angle of rotation of 90° anticlockwise

18

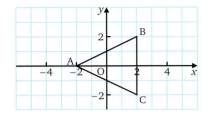

Centre of rotation (0, 0)
Angle of rotation of 180°

19 △ABC is rotated about O
through 180° to give the image,
△A′B′C′. Copy and complete the diagram,
using 1 cm to 1 unit.

  a What is the shape of the
path traced out by C as it
moves to C′?

  b Measure OC and OC′.
How do they compare?

  Repeat with OB and OB′.

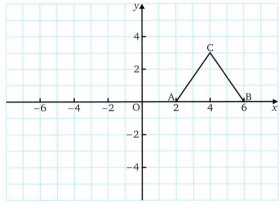

20 Why is the direction of rotation sometimes not given?

21 Draw the diagram accurately. Then draw accurately,
using a protractor, the image of △ABC under a
rotation of 60° anticlockwise about O.

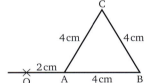

Finding the centre of rotation by drawing

As we have seen we can often spot the centre of rotation just
by looking at the diagram but sometimes it is not obvious.

In such cases we can use the fact that an object point A and
its image point A′ are the same distance from the centre.

So the centre lies on the perpendicular bisector of AA′.

It also lies on the perpendicular bisector of BB′.

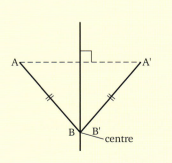

Therefore the point P, where these two bisectors meet, is the centre.

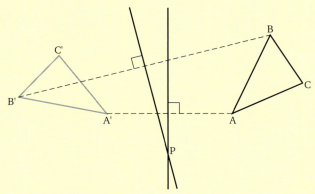

(The perpendicular bisector of CC′ will also go through P.)

Exercise 10j

1

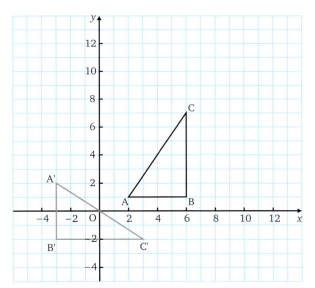

a   Copy the diagram, drawing axes for x
and y from −5 to 12. Use 1 cm to 1 unit.
△A′B′C′ is the image of △ABC under a rotation.

b   Draw the perpendicular bisectors of AA′ and BB′.

c   Mark the centre of rotation, P (that is, the point where the two
perpendicular bisectors meet).

d   Check that it is the centre by using tracing paper and the point of
your pencil.

e   Join BP and B′P. Measure BP̂B′. What is the angle of rotation?

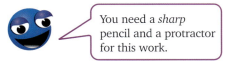

You need a *sharp*
pencil and a protractor
for this work.

2 Draw axes for *x* and *y* from −5 to 10, using 1 cm to 1 unit. Draw △ABC with A(−1, 8), B(5, 4), C(−1, 1) and △A′B′C′ with A′(4, 1), B′(0, −5), C′(−3, 1).

Repeat **b** to **e** in question 1.

3 Draw axes for *x* and *y* from −5 to 10, using 1 cm to 1 unit. Draw △ABC with A(−4, −2), B(2, −2), C(−4, 4) and △A′B′C′ with A′(4, 0), B′(4, 6), C′(−2, 0).

Repeat **b** to **e** in question 1.

Finding the angle of rotation

Having found the centre of rotation, the angle of rotation can be found by joining both an object point and its image to the centre.

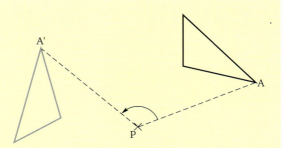

In the diagram above, A′ is the image of A and P is the centre of rotation.

Join both A and A′ to P. AP̂A′ is the angle of rotation.

In this case the angle of rotation is 120° anticlockwise.

Exercise 10k

Trace each of the diagrams and, by drawing in the necessary lines, find the angle of rotation when △ABC is rotated about the centre P to give △A′B′C′.

1

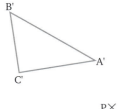

2

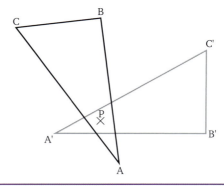

Remember that when you measure an angle, put the 0° line on the protractor over one arm of the angle, then count the number of degrees to turn to the other arm.

Investigation

Create three shapes, none of which has appeared in this chapter, such that

- one has both line symmetry and rotational symmetry
- another has line symmetry but not rotational symmetry
- and a third has rotational symmetry but not line symmetry.

Mixed questions on reflections, translations and rotations

Exercise 10I

Name the transformation, describing it fully, if the grey triangle is the image of the black one.

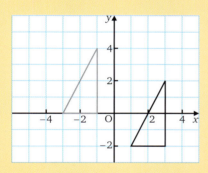

The transformation is a translation given by moving 4 units in the direction of the negative x-axis and 2 units in the direction of the positive y-axis.

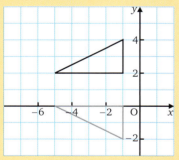

The transformation is a reflection in the line $y = 1$.

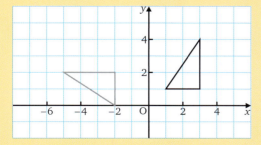

The transformation is a rotation about $(0, -1)$ through an angle of $90°$ anticlockwise.

Name the transformations in questions **1** to **10**, describing them fully. The black shape is the object, the grey shape is the image. Describe a translation by the moves left or right and up or down, for a reflection give the mirror line and for a rotation give the centre of rotation and the angle turned through.

1

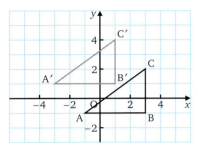

2

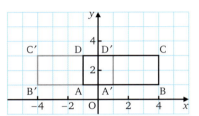

3

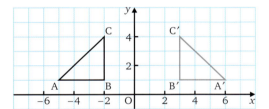

4

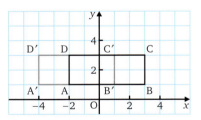

5

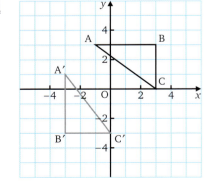

6

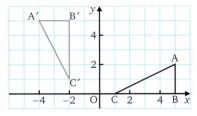

7

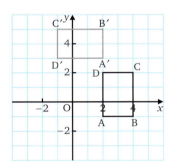

8

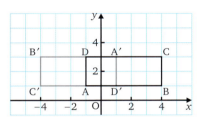

9

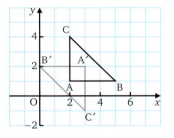

10

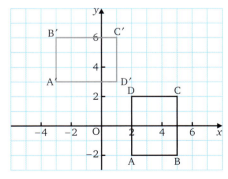

Sometimes we do not know which point is the image of a particular object point. In such cases there could be more than one possible transformation.

(Remember that a rotation of 90° anticlockwise is the same as a rotation of 270° clockwise. Do not give these as two independent transformations.)

11

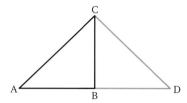

Name and describe two possible transformations that will map the object △ABC to the image △BCD.

12

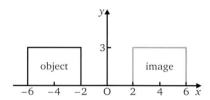

Name and describe three possible transformations that will map the object to the image.

13

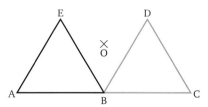

Name and describe four possible transformations that will map the left-hand triangle to the right-hand triangle.

14

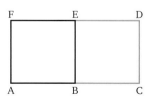

Name and describe five possible transformations that will map the left-hand square to the right-hand square.

<u>**15**</u>   A car is turning a corner and two of its positions are shown. Trace the drawing, allowing plenty of space above and below, and find the centre of the turning circle.

<u>**16**</u>   Draw axes for x and y from −5 to 5, using 1 cm to 1 unit. Draw lines AB and BC with A (2, 2), B (5, 2) and C (3, 0). Draw the images of ABC under the reflections in the four lines $y = 0$, $x = y$, $x = 0$ and $y = -x$. Draw the images of ABC under the three rotations about O through angles of 90°, 180° and 270° anticlockwise. (The seven images of $\triangle$ABC, together with $\triangle$ABC itself, form an eight-pointed star.)

<u>**17**</u>   Draw axes for x and y from −5 to 5, using 1 cm to 1 unit. Draw $\triangle$ABC with A (2, 1), B (4, 1) and C (4, 2).

    a   Reflect $\triangle$ABC in the line $y = x$ to produce the image $\triangle A_1B_1C_1$
Then rotate $\triangle A_1B_1C_1$ through 180° about O to produce $\triangle A_2B_2C_2$.
What single transformation will map $\triangle$ABC to $\triangle A_2B_2C_2$?

    b   Rotate $\triangle$ABC through 180° about O then reflect the image in the line $y = x$. Is the final image the same as $\triangle A_2B_2C_2$?

    c   Try other pairs of reflections and rotations, starting a fresh diagram where necessary. In each case find the single transformation that is equivalent to the pair. Does the order in which you do the transformations matter? Are the single transformations themselves all reflections or rotations?

In this chapter you have seen that...

✔ a shape can have one or more lines of symmetry and/or rotational symmetry

✔ when an object is reflected in a mirror line, the object and the image are symmetrical about the mirror line

✔ the mirror line is the perpendicular bisector of the line joining of a pair corresponding points on the object and the image

✔ a translation moves an object without turning it, reflecting it or changing its size or shape

✔ a translation can be described by giving the movements needed to change the position.

11 Circles: circumference and area

At the end of this chapter you should be able to...

1 Use a pair of compasses to draw circles and accurate line segments.
2 State the relationship between the circumference and diameter of a circle.
3 Calculate the circumference of a circle given its diameter or radius.
4 Solve problems involving the calculation of circumferences of circles.
5 Calculate the radius of a circle of given circumference.
6 Calculate the area of a circle of given radius.
7 Calculate the area of a sector of a circle of given angle.

Did you know?

Circumference means 'the perimeter of a circle'. The word comes from the Latin *circumferre* – 'to carry around'. The symbol π is the first letter of the Greek word for circumference – *perimetron*. In Germany, π is identified as the ludolphine number, because of the work of Ludolph van Ceulen, who tried to find a better estimate of the number.

You need to know...

✔ the units of area
✔ how to estimate an answer
✔ how to correct a number to a given place value

Key words

annulus, arc, circle, circumference, cone, cylinder, diameter, irrational number, perimeter, pi (π), quadrant, radius, rational number, rectangle, revolution, sector, segment, semicircle, square

Using a pair of compasses

Using a pair of compasses is not easy: it needs practice. Draw several circles. Make some of them small and some large. You should not be able to see the place at which you start and finish.

Now try drawing the daisy pattern below.

Draw a circle of radius 5 cm. Keeping the radius the same, put the point of the compasses at A and draw an arc to cut the circle in two places, one of which is B. Move the point to B and repeat. Carry on moving the point of your compasses round the circle until the pattern is complete.

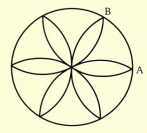

Repeat the daisy pattern but this time draw complete circles instead of arcs.

There are some more patterns using compasses in the practical work on page **175**.

Drawing straight lines of a given length

To draw a straight line that is 5 cm long, start by using your ruler to draw a line that is *longer* than 5 cm.

Then mark a point on this line near one end as shown.
Label it A.

Next use your compasses to measure 5 cm on your ruler.

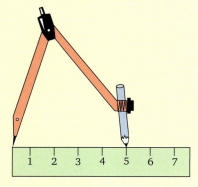

Then put the point of the compasses on the line at A and draw an arc to cut the line as shown.

The length of line between A and B should be 5 cm. Measure it with your ruler.

Exercise 11a

Draw, as accurately as you can, straight lines of the following lengths:

1	6 cm	**3**	12 cm	**5**	8.5 cm	**7**	4.5 cm
2	2 cm	**4**	9 cm	**6**	3.5 cm	**8**	6.8 cm

The patterns below are made using a pair of compasses. Try copying them. Some instructions are given, which should help.

9

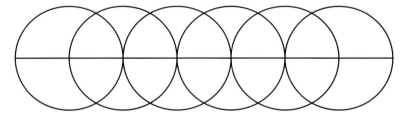

Draw a straight line. Open your compasses to a radius of 3 cm and draw a circle with its centre on the line. Move the point of the compasses 3 cm along the line and draw another circle. Repeat as often as you can.

10 Draw a square of side 4 cm. Open your compasses to a radius of 4 cm and with the point on one corner of the square draw an arc across the square. Repeat on the other three corners.

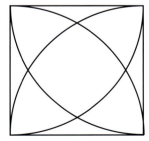

Try the same pattern, but leave out the sides of the square; just mark the corners. A block of four of these looks good.

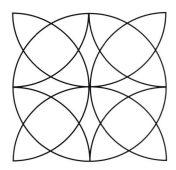

11   Draw a square of side 8 cm. Mark the midpoint of each side. Open
your compasses to a radius of 4 cm, and with the point on the middle
of one side of the square, draw an arc. Repeat at the other three
midpoints.

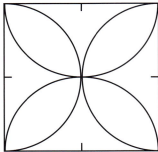

Diameter, radius and circumference

When you use a pair of compasses to draw a circle, the place where you
put the point is the *centre* of the circle. The length of the line that the pencil
draws is the *circumference* of the circle.

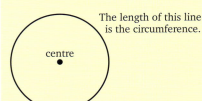

The length of this line
is the circumference.

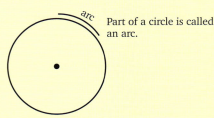

Part of a circle is called
an arc.

Any straight line joining the centre to a point on the
circumference is a *radius*.

A straight line across the full width of a circle
(i.e. going through the centre) is a *diameter*.

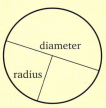

The diameter is twice as long as the radius. If *d* stands for the length of a
diameter and *r* stands for the length of a radius, we can write this as a formula:

$$d = 2r$$

A line joining two points on a circle
is called a chord.

A chord divides a circle into two segments.

A section of a circle enclosed by two radii and the arc
between them is called a sector.

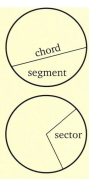

Exercise 11b

In questions **1** to **6**, write down the length of the diameter of the circle
whose radius is given

1

2

3 15 mm

4 3.5 cm

5 1 km

6 4.6 cm

7 For this question you will need some thread and a cylinder (e.g. a tin
of soup, a soft drink can, the cardboard tube from a roll of kitchen
paper).

Measure across the top of the cylinder to get a value for the diameter.
Wind the thread 10 times round the can. Measure the length of thread
needed to do this and then divide your answer by 10 to get a value
for the circumference. If C stands for the circumference and d for the
length of the diameter, find, approximately, the value of $C \div d$.

(Note that you can also use the label from the cylindrical tin. If you
are careful you can reshape it and measure the diameter and then
unroll it to measure the circumference.)

8 Compare the results from the whole class for the value of $C \div d$.

Introducing π

From the last exercise you will see that, for any circle,

circumference ≈ 3 × diameter

The number that you have to multiply the diameter by to get the
circumference is slightly larger than 3.

This number is unlike any number that you have met so far. It cannot be written down exactly, either as a fraction or as a decimal: as a fraction it is approximately, but *not* exactly, $\frac{22}{7}$; as a decimal it is approximately 3.142, correct to 3 decimal places.

Now with a computer to do the arithmetic we can find its value to as many decimal places as we choose: it is a never-ending, never-repeating decimal fraction. To sixty decimal places, the value of this number is

3.141592653589793238462643383279502884197169399375105820974944...

Because we cannot write it down exactly we use the Greek letter π (pi) to stand for this number. Then we can write a formula connecting the circumference and diameter of a circle in the form $C = \pi d$. But $d = 2r$ so we can rewrite this formula as

$$C = 2\pi r$$

where C = circumference and r = radius

The symbol π was first used by an English writer, William Jones, in 1706. It was later adopted in 1737 by Euler.

Did you know?

Numbers that can be written exactly as a fraction or a terminating decimal are called rational numbers. Numbers like π are called irrational numbers.

Calculating the circumference

Exercise 11c

Using 3.142 as an approximate value for π, find the circumference of a circle of radius 3.8 m.

Using $C = 2\pi r$
with $\pi = 3.142$ and $r = 3.8$
gives $\qquad C = 2 \times 3.142 \times 3.8$
$\qquad\qquad = 23.9$ to 1 d.p.
Circumference $= 23.9$ m to 1 d.p.

3.8 m

Use 3.142 as an approximate value for π to find the circumference of a circle of radius. (Give your answers correct to the accuracy given in brackets.)

1	2.3 m (1 d.p.)	**6**	250 mm (10)	**11**	7 cm (1 d.p.)
2	4.6 cm (1 d.p.)	**7**	36 cm (unit)	**12**	28 mm (unit)
3	2.9 cm (1 d.p.)	**8**	4.8 m (1 d.p.)	**13**	1.4 m (2 d.p.)
4	53 mm (unit)	**9**	1.8 m (1 d.p.)	**14**	35 mm (unit)
5	8.7 m (1 d.p.)	**10**	0.014 km (4 d.p.)	**15**	5.6 cm (1 d.p.)

For questions **16** to **23** you can use $C = 2\pi r$ or $C = \pi d$.

Read the question carefully before you decide which one to use.

Using $\pi \approx 3.14$ and giving your answer correct to the place value given, find the circumference of a circle of:

16	radius 154 mm (10)	**20**	radius 34.6 cm (10)
17	diameter 28 cm (unit)	**21**	diameter 511 mm (100)
18	diameter 7.7 m (unit)	**22**	diameter 630 cm (100)
19	radius 210 mm (100)	**23**	diameter 9.1 m (unit)

! Investigation

Count Buffon's experiment

Count Buffon was an 18th-century scientist who carried out many probability experiments. The most famous of these is his 'Needle Problem'. He dropped needles on to a surface ruled with parallel lines and considered the drop successful when a needle fell between two lines. His amazing discovery was that the number of successful drops divided by the number of unsuccessful drops was an expression involving π.

You can repeat his experiment and get a good approximation for the value of π from it.

Take a matchstick or a similar small stick and measure its length. Now take a sheet of paper measuring about $\frac{1}{2}$ m each way and fill the sheet with a set of parallel lines whose distance apart is equal to the length of the stick. With the sheet on the floor, drop the stick on to it from a height of about 1 m. Repeat this about a hundred times and keep a tally of the number of times the stick touches or crosses a line and of the number of times it is dropped. Then find the value of

$$\frac{2 \times \text{number of times it is dropped}}{\text{number of times it crosses or touches a line}}$$

Problems

Exercise 11d

Use the value of π on your calculator and give your answers correct to 1 d.p.

Find the perimeter of the given semicircle.

(The prefix 'semi' means half.)

Remember that perimeter means distance all round, so you need to find half the circumference then add the length of the straight edge.

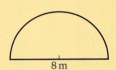

8 m

The complete circumference of the circle is $2\pi r$.

The curved part of the semicircle is $\frac{1}{2} \times 2\pi r$

$$= \frac{1}{2} \times 2 \times \pi \times 4\,\text{m}$$

$$= 12.57\,\text{m to 2 d.p.}$$

The perimeter = curved part + straight edge

$$= (12.57 + 8)\,\text{m}$$

$$= 20.57\,\text{m}$$

$$= 20.6\,\text{m to 1 d.p.}$$

Find the perimeter of each of the following shapes.

1
4 cm

2
3 cm

(This is called a *quadrant*: it is one quarter of a circle.)

3
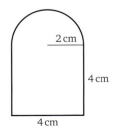
2 cm
4 cm
4 cm

4
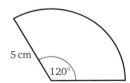
5 cm
120°

This is one-third of a circle because 120° is $\frac{1}{3}$ of 360°.

5

10 cm
45°

A 'slice' of a circle is called a *sector*. $\frac{45}{360} = \frac{1}{8}$ so this sector is $\frac{1}{8}$ of a circle.

6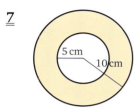
5 cm

8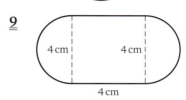
10 mm
10 mm   10 mm
10 mm

7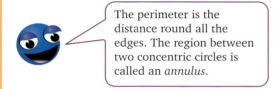
5 cm
10 cm

9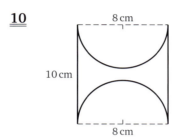
4 cm   4 cm
4 cm

The perimeter is the distance round all the edges. The region between two concentric circles is called an *annulus*.

10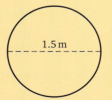
8 cm
10 cm
8 cm

Exercise 11e

Use the value of π on your calculator and give your answers correct to 1 d.p.

A circular flower bed has a diameter of 1.5 m. A metal edging is to be placed round it. Find the length of edging needed and the cost of the edging if it is sold by the metre (i.e. you can only buy a whole number of metres) and costs $60 a metre.

First find the circumference of the circle, then how many metres you need.

Using $C = \pi d$,

$$C = \pi \times 1.5$$
$$= 4.712...$$

1.5 m

Length of edging needed $= 4.71$ m to 3 s.f.

(Note that if you use $C = 2\pi r$, you must remember to halve the diameter.)

As the length is 4.71 m we have to buy 5 m of edging.

$$\text{Cost} = 5 \times \$60$$
$$= \$300$$

1 Measure the diameter, in millimetres, of a 25 c coin. Use your measurement to find the circumference of a 25 c coin.

2 Repeat question **1** with a 10 c coin and a 1 c coin.

3 A circular tablecloth has a diameter of 1.4 m. How long is the hem of the cloth?

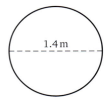

4 A rectangular sheet of metal measuring 50 cm by 30 cm has a semicircle of radius 15 cm cut from each short side as shown. Find the perimeter of the shape that is left.

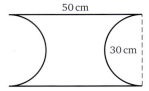

5 A bicycle wheel has a radius of 28 cm. What is the circumference of the wheel?

6 How far does a bicycle wheel of radius 28 cm travel in one complete revolution? How many times will the wheel turn when the bicycle travels a distance of 352 m?

7 A cylindrical tin has a radius of 2 cm. What length of paper is needed to put a label on the tin if the edges just meet?

8 A square sheet of metal has sides of length 30 cm. A quadrant (one quarter of a circle) of radius 15 cm is cut from each of the four corners. Sketch the shape that is left and find its perimeter.

9 A boy flies a model aeroplane on the end of a wire 10 m long. If he keeps the wire horizontal, how far does his aeroplane fly in one revolution?

10 If the aeroplane described in question 9 takes 1 second to fly 10 m, how long does it take to make one complete revolution? If the aeroplane has enough power to fly for 1 minute, how many turns can it make?

11 A cotton reel has a diameter of 2 cm. There are 500 turns of thread on the reel. How long is the thread?

12 A bucket is lowered into a well by unwinding rope from a cylindrical drum. The drum has a radius of 20 cm and with the bucket out of the well there are 10 complete turns of the rope on the drum. When the rope is fully unwound the bucket is at the bottom of the well. How deep is the well?

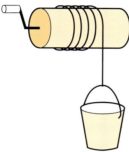

13 A garden hose is 100 m long. For storage it is wound on a circular hose reel of diameter 45 cm. How many turns of the reel are needed to wind up the hose?

14 The cage that takes miners up and down the shaft of a coal mine is raised and lowered by a rope wound round a circular drum of diameter 3 m. It takes 10 revolutions of the drum to lower the cage from ground level to the bottom of the shaft. How deep is the shaft?

⚠ Investigation

Ken entered a 50 km sponsored cycle ride. He wondered how many pedal strokes he made. The diameter of each wheel is 70 cm.

a Investigate this problem if one pedal stroke gives one complete turn of the wheels.

b What happens if Ken uses a gear that gives two turns of the wheel for each pedal stroke?

c Find out how the gears on a racing bike affect the ratio of the number of pedal strokes to the number of turns of the wheels.

Discuss the assumptions made in order to answer parts **a** and **b**.

Write a short report on how these assumptions affect the reasonableness of your answers.

Finding the radius of a circle given the circumference

If a circle has a circumference of 24 cm, we can find its radius from the formula $C = 2\pi r$

i.e. $24 = 2 \times 3.142 \times r$

and solving this equation for r.

Exercise 11f

Use the value of π on your calculator and give your answers correct to 1 d.p.

The circumference of a circle is 36 m. Find the radius of this circle.

$$\text{Using } C = 2\pi r \text{ gives}$$
$$36 = 2 \times \pi \times r$$
$$36 = 6.283 \times r$$

(Writing down the first 4 digits in the calculator display)

$$\frac{36}{6.283} = r \text{ (dividing both sides by 6.283)}$$
$$= 5.729...$$
$$r = 5.7 \text{ to } 1 \text{ d.p.}$$

Therefore the radius is 5.7 cm to 1 d.p.

Find the radius of the circle whose circumference is:

1	44 cm	**3**	550 m	**5**	462 mm	**7**	36.2 mm	**9**	582 cm
2	121 mm	**4**	275 cm	**6**	831 cm	**8**	391 m	**10**	87.4 m

11   Find the diameter of a circle whose circumference is 52 m.

12   A roundabout at a major road junction is to be built. It has to have a minimum circumference of 188 m. What is the corresponding minimum diameter?

13   A bicycle wheel has a circumference of 200 cm. What is the radius of the wheel?

14   A car has a turning circle whose circumference is 63 m. What is the narrowest road that the car can turn round in without going on the pavement?

15 When the label is taken off a tin of soup it is found to be 32 cm long. If there was an overlap of 1 cm when the label was on the tin, what is the radius of the tin?

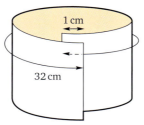

16 The diagram shows a quadrant of a circle. If the curved edge is 15 cm long, what is the length of a straight edge?

17 A tea cup has a circumference of 24 cm. What is the radius of the cup? Six of these cups are stored edge to edge in a straight line on a shelf. What length of shelf do they occupy?

18 Make a cone from a sector of a circle as follows:

On a sheet of paper draw a circle of radius 8 cm. Draw two radii at an angle of 90°. Make a tab on one radius as shown. Cut out the larger sector and stick the straight edges together. What is the circumference of the circle at the bottom of the cone?

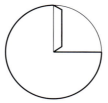

19 A cone is made by sticking together the straight edges of the sector of a circle, as shown in the diagram.

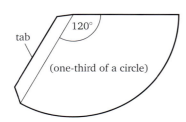

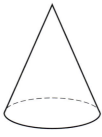

The circumference of the circle at the bottom of the finished cone is 10 cm. What is the radius of the circle from which the sector was cut?

20 The shape in the diagram is made up of a semicircle and a square. Find the length of a side of this square.

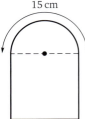

15 cm

21 The curved edge of a sector of angle 60° is 10 cm. Find the radius and the perimeter of the sector.

The area of a circle

The formula for finding the area of a circle is

$$A = \pi r^2$$

You can see this if you cut a circle up into sectors and place the pieces together as shown to get a shape which is roughly rectangular. Consider a circle of radius r whose circumference is $2\pi r$.

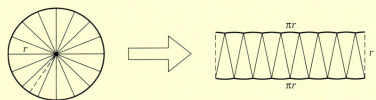

Area of circle = area of 'rectangle'
= length × width
= $\pi r \times r = \pi r^2$

Exercise 11g

Use the value of π on your calculator and give your answers correct to 1 d.p.

Find the area of a circle of radius 2.5 cm.

Using $A = \pi r^2$
With $r = 2.5$
gives $A = \pi \times (2.5)^2$
$= 19.63...$
$= 19.6$ to 1 d.p.
Area is 19.6 cm² to 1 d.p.

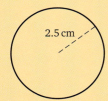

2.5 cm

Find the areas of the following circles

1

4

7   Be careful!

2

5

8 <u>8</u>

3

6

9 <u>9</u>

This is a *sector* of a circle. Find its area.

If we find 45° as a fraction of 360° this will tell us what fraction this sector is of a circle.

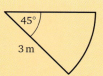

$$\frac{\overset{9}{\cancel{45}}}{\underset{72}{\cancel{360}}} = \frac{\overset{1}{\cancel{9}}}{\underset{8}{\cancel{72}}} = \frac{1}{8}$$

∴ area of sector = $\frac{1}{8}$ of area of circle of radius 3 m

Area of sector = $\frac{1}{8}$ of πr^2

$= \frac{1}{8} \times \pi \times 9 \, \text{m}^2$

$= 3.534 \, \text{m}^2$

$= 3.53 \, \text{m}^2$ to 2 d.p.

Find the areas of the following shapes:

10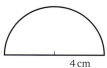

12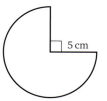

11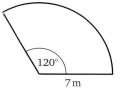

13 <u>13</u>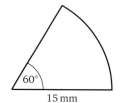

14

18

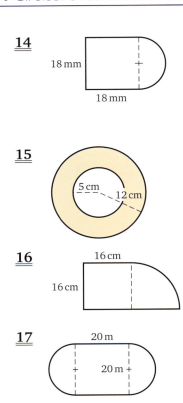

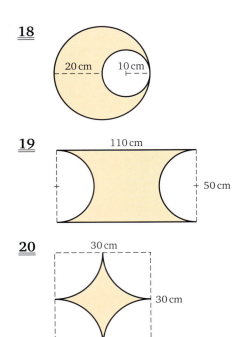

15

19

16

20

17

Problems

Exercise 11h

Use the value of π on your calculator and make a rough sketch to illustrate each problem.

A circular table has a radius of 75 cm. Find the area of the table top. The top of the table is to be varnished. One tin of varnish covers 4 m². Will one tin be enough to give the table top three coats of varnish?

The area to varnish is three times the area of the top of the table.

Area of table top is πr^2

$$= \pi \times 75 \times 75 \text{ cm}^2$$
$$= 17\,671.4...$$
$$= 17\,670... \text{ cm}^2$$
$$= \frac{17\,670}{100^2} \text{ m}^2$$
$$= 1.767... \text{ m}^2$$

For three coats, enough varnish is needed to cover

$$3 \times 1.767... \text{ m}^2 = 5.30 \text{ m}^2 \text{ to 2 d.p.}$$

So one tin of varnish is not enough.

1   The minute hand on a clock is 15 cm long. What area, to the nearest unit, does it pass over in 1 hour?

2   What area does the minute hand described in question 1 cover in 20 minutes to the nearest unit?

3   The diameter of a 25 c coin is 25 mm. Find the area of one of its flat faces to the nearest mm².

4   The hour hand of a clock is 10 cm long. What area does it pass over in 1 hour to 1 d.p.?

5   A circular lawn has a radius of 5 m. A bottle of lawn weedkiller says that the contents are sufficient to cover 50 m². Is one bottle enough to treat the whole lawn?

6   The largest possible circle is cut from a square of paper 10 cm by 10 cm. What area of paper is left, to 1 d.p.?

7   Circular place mats of diameter 8 cm are made by stamping as many circles as possible from a rectangular strip of card measuring 8 cm by 64 cm. How many mats can be made from the strip of card and what area of card is wasted to the nearest 10 cm²?

8   A wooden counter top is a rectangle measuring 280 cm by 45 cm. There are three circular holes in the counter, each of radius 10 cm. Find the area of the wooden top to the nearest 100 cm².

9   The surface of the counter top described in question 8 is to be given four coats of varnish. If one tin covers 3.5 m², how many tins will be needed?

10 Take a cylindrical tin of food with a paper label:

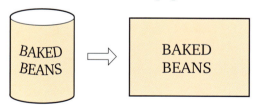

Measure the diameter of the tin and use it to find the length of the label. Find the area of the label. Now find the total surface area of the tin (two circular ends and the curved surface).

Investigation

In addition to using the formula $A = \pi r^2$ to calculate the area of a circle, there are other methods that provide reasonably good results. One such method is by weighing.

This method requires floor tiles, linoleum, or other material of measurable weight that can be easily cut. Trace the circle whose area is required on the tile and cut out the resulting circular region.

Use the remaining tiles to cut out three 10 cm squares, five 10 cm by 1 cm rectangles, and ten 1 cm squares. You will need more cutouts if the circle is large. These rectangular cutouts will be used as weights.

Place the circular cutout in the scale pan and use the rectangular pieces as weights. When the scale balances, remove the rectangular pieces and find their total area. This area will be the area of the circular cutout.

Why does this method of weighing to find area make sense?

Mixed exercises

Use the value of π on your calculator. Give your answers to a reasonable degree of accuracy.
(A reasonable degree of accuracy should match the accuracy of the measurements.)

Exercise 11i

1   Find the circumference of a circle of radius 2.8 mm.

2   Find the radius of a circle of circumference 60 m.

3   Find the circumference of a circle of diameter 12 cm.

4   Find the area of a circle of radius 2.9 m.

5   Find the area of a circle of diameter 25 cm.

6   Find the perimeter of the quadrant in the diagram.

8 mm

7   Find the area of the sector in the diagram.

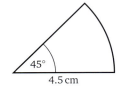
45°
4.5 cm

Exercise 11j

1 Find the circumference of a circle of diameter 20 m.

2 Find the area of a circle of radius 12 cm.

3 Find the radius of a circle of circumference 360 cm.

4 Find the area of a circle of diameter 8 m.

5 Find the diameter of a circle of circumference 280 mm.

6 Find the perimeter of the sector in the diagram.

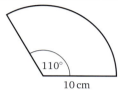

7 Find the area of the shaded part of the diagram.

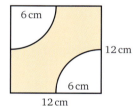

Exercise 11k

1 Find the area of a circle of radius 2 km.

2 Find the circumference of a circle of radius 49 mm.

3 Find the radius of a circle of circumference 88 m.

4 Find the area of a circle of diameter 14 cm.

5 Find the area of a circle of radius 3.2 cm.

6 An ornamental pond in a garden is a rectangle with a semicircle on each short end. The rectangle measures 5 m by 3 m and the radius of each semicircle is 1 m. Find the area of the pond.

Did you know?

Over the centuries mathematicians have spent a lot of time trying to find the true value of π. The ancient Chinese used 3. Three is also the value given in the Old Testament (1 Kings 7:23). The Egyptians (c. 1600 BCE) used $4 \times \left(\frac{8}{9}\right)^2$. Archimedes (c. 225 BCE) was the first person to use a sound method for finding its value and a mathematician called Van Ceulen (1540–1610) spent most of his life finding it to 35 decimal places!

In this chapter you have seen that...

✔ for any circle the circumference divided by the diameter gives a fixed value; this value is denoted by π and its approximate value is 3.142

✔ you can find the circumference of a circle using either the formula $C = 2\pi r$ or the formula $C = \pi d$ when you know the radius or diameter of the circle

✔ you can use the formula $A = \pi r^2$ to find the area of a circle

✔ a sector of a circle is shaped like a slice of cake; the fraction that the angle at the point is of 360° gives the fraction that its area is of the area of the circle.

12 Polygons

Polygons

A polygon is a plane (flat) figure formed by three or more points joined by line segments. The points are called vertices (singular – 'vertex'). The line segments are called sides.

 This is a nine-sided polygon.

Some polygons have names that you already know:

a three-sided polygon
is a triangle

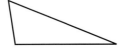

a four-sided polygon
is a quadrilateral

a five-sided polygon
is a pentagon

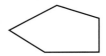

a six-sided polygon
is a hexagon

an eight-sided polygon
is an octagon

Regular polygons

A polygon is called regular when all its sides are the same length *and* all its angles are the same size. The polygons below are all regular:

Exercise 12a

State which of the following figures are regular polygons. Give a brief reason for your answer:

1	Rhombus	**4**	Parallelogram	**7**	Equilateral triangle
2	Square	**5**	Isosceles triangle	**8**	Circle
3	Rectangle	**6**	Right-angled triangle		

Make a rough sketch of each of the following polygons. (Unless you are told that a polygon is regular, you must assume that it is *not* regular.)

9	A regular quadrilateral	**12**	A regular triangle	**15**	A quadrilateral
10	A hexagon	**13**	A regular hexagon	**16**	A ten-sided polygon
11	A triangle	**14**	A pentagon		

When the vertices of a polygon all point outwards, the polygon is convex.

Sometimes one or more of the vertices point inwards, in which case the polygon is concave.

convex polygon

concave polygon

In this chapter we consider only convex polygons.

Interior angles

The angles enclosed by the sides of a polygon are the *interior angles*.
For example

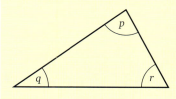

p, *q* and *r* are the interior angles of the triangle

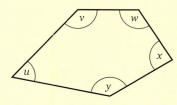

u, *v*, *w*, *x* and *y* are the interior angles of the pentagon.

Exterior angles

If we produce (extend) one side of a polygon, an angle is formed outside the polygon. It is called an *exterior angle*.

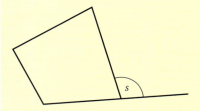

For example, *s* is an exterior angle of the quadrilateral.

If we produce all the sides in order we have all the exterior angles.

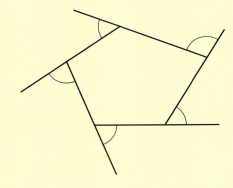

Exercise 12b

1 What is the sum of the interior angles of any triangle?

2 What is the sum of the interior angles of any quadrilateral?

3 In triangle ABC, find
   a the size of each marked angle
   b the sum of the exterior angles.

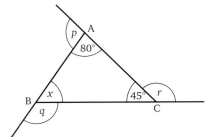

4 ABCD is a parallelogram. Find
   a the size of each marked angle
   b the sum of the exterior angles.

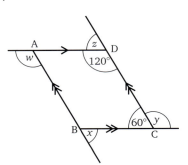

5   In a triangle ABC, write down the value of

   a $x+q$

   b   the sum of all six marked angles

   c   the sum of the interior angles

   d   the sum of the exterior angles.

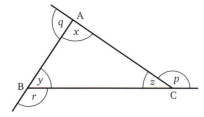

6   Draw a pentagon. Produce the sides in order to form the five exterior angles. Measure each exterior angle and then find their sum.

7   Construct a regular hexagon of side 5 cm. (Start with a circle of radius 5 cm and then with your compasses still open to a radius of 5 cm, mark off points on the circumference in turn.) Produce each side of the hexagon in turn to form the six exterior angles.

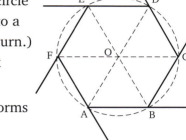

   If O is the centre of the circle, joining O to each vertex forms six triangles:

   a   What kind of triangle is each of these triangles?

   b   What is the size of each interior angle in these triangles?

   c   Write down the value of AB̂C.

   d   Write down the value of CB̂G.

   e   Write down the value of the sum of the six exterior angles of the hexagon.

The sum of the exterior angles of a polygon

In the last exercise, we found that the sum of the exterior angles is 360° in each case. This is true of any polygon, whatever its shape or size.

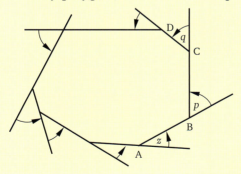

Consider walking round this polygon. Start at A and walk along AB. When you get to B you have to turn through angle p to walk along BC. When you get to C you have to turn through angle q to walk along CD, ... and so on until you return to A. If you then turn through angle z you are facing in the direction AB again. You have now turned through each exterior angle and have made just one complete turn, i.e.

   the sum of the exterior angles of a polygon is 360°.

Exercise 12c

Find the size of the angle marked p.

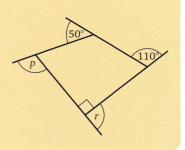

$p + r + 110° + 50° = 360°$ (sum of exterior angles of a polygon)

but $r = 90°$ (angles on a straight line)

∴ $p = 360° - 90° - 110° - 50°$

                  $p = 110°$

In each case find the size of the angle marked p:

1

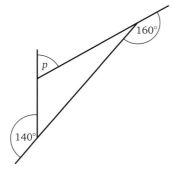

4

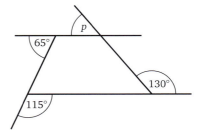

2

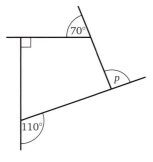

5

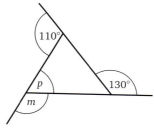

3

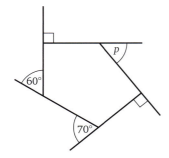

6

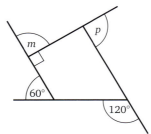

7

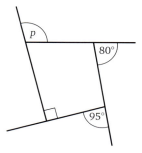

9

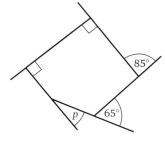

8

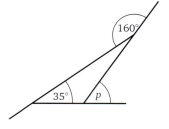

10

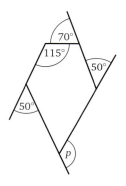

In questions **11** and **12** find the value of x.

11

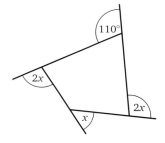

12

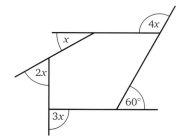

13 The exterior angles of a hexagon are x, $2x$, $3x$, $4x$, $3x$ and $2x$. Find the value of x.

14 Find the number of sides of a polygon if each exterior angle is
   a 72° **b** 45°.

The exterior angle of a regular polygon

If a polygon is regular, all its exterior angles are the same size. We know that the sum of the exterior angles is 360°, so the size of one exterior angle is easily found; we just divide 360° by the number of sides of the polygon, i.e.

in a *regular* polygon with n sides, the size of an exterior angle is $\frac{360°}{n}$

Find the size of each exterior angle of a 24-sided regular polygon.

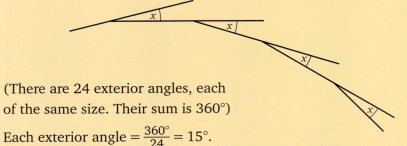

(There are 24 exterior angles, each
of the same size. Their sum is 360°)

Each exterior angle $= \frac{360°}{24} = 15°$.

Find the size of each exterior angle of a regular polygon with:

1	10 sides	**4**	6 sides	**7**	9 sides
2	8 sides	**5**	15 sides	**8**	16 sides
3	12 sides	**6**	18 sides	**9**	20 sides.

The sum of the interior angles of a polygon

Consider an octagon:

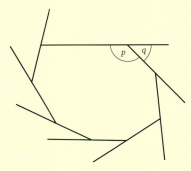

At each vertex there is an interior angle and an exterior angle and the sum of
these two angles is 180° (angles on a straight line), i.e. $p + q = 180°$ at each
one of the eight vertices.

Therefore, the sum of the interior angles and exterior angles together is

$$8 \times 180° = 1440°$$

The sum of the eight exterior angles is 360°.
Therefore, the sum of the interior angles is

$$1440° - 360° = 1080°$$

Exercise 12e

Find the sum of the interior angles of a 14-sided polygon.

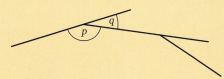

At each vertex $p + q = 180°$

If there are 14 sides there are 14 pairs of exterior and interior angles.

∴ sum of interior angles and exterior angles is
$$14 \times 180° = 2520°$$
∴ sum of interior angles $= 2520° - 360°$
$$= 2160°$$

Find the sum of the interior angles of a polygon with:

1	6 sides	**4**	4 sides	**7**	18 sides
2	5 sides	**5**	7 sides	**8**	9 sides
3	10 sides	**6**	12 sides	**9**	15 sides.

Formula for the sum of the interior angles

If a polygon has n sides, then the sum of the interior and exterior angles together is $n \times 180° = 180n°$ so the sum of the interior angles only is $180n° - 360°$ which, as $360° = 180° \times 2$, can be written as $180°(n-2)$,

i.e. in a polygon with n sides, the sum of the interior angles is
$$(180n - 360)° \quad \text{or} \quad 180°\,(n-2)$$

We can also show this by taking a point inside a polygon and joining it to each vertex.

This polygon has six sides and is divided into six triangles.

This divides the polygon into a number of triangles.
If the polygon has n sides, this gives n triangles.

The sum of the angles in all n triangles is $n \times 180° = 180n°$
If we subtract the sum of the angles round the point (360°), this leaves the sum of the interior angles of the polygon,
i.e. for a polygon with n sides, the sum of the interior angles $= (180n - 360)°$

Exercise 12f

1 Find the sum of the interior angles of a polygon with
    a 20 sides   **b** 16 sides   **c** 11 sides.

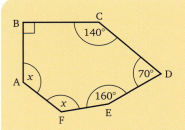

In the hexagon ABCDEF, the angles marked x are equal. Find the value of x.

The sum of the interior angles is $180° \times 6 - 360° = 1080° - 360° = 720°$

$$\therefore 90° + 140° + 70° + 160° + 2x = 720°$$
$$460° + 2x = 720°$$
$$2x = 260°$$
$$x = 130°$$

In each of the following questions find the size of the angle(s) marked x:

2

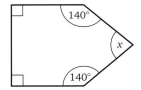

5

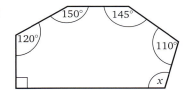

3

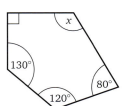

6

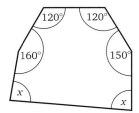

4

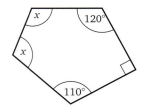

7

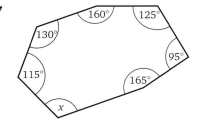

Find the size of each interior angle of a regular nine-sided polygon.

(As the polygon is regular, all the exterior angles are equal and all the interior angles are equal.)

Method 1   Sum of exterior angles $= 360°$

$\therefore$ each exterior angle $= 360° \div 9 = 40°$

$\therefore$ each interior angle $= 180° - 40° = 140°$

Method 2   Sum of interior angles $= 180° \times 9 - 360° = 1260°$

$\therefore$ each interior angle $= 1260° \div 9 = 140°$

Find the size of each interior angle of a:

8   regular pentagon   **11**   regular ten-sided polygon

9   regular hexagon   **12**   regular 12-sided polygon

10   regular octagon   **13**   regular 20-sided polygon.

14   How many sides has a regular polygon if each exterior angle is
     a $20°$ **b** $15°$?

15   How many sides has a regular polygon if each interior angle is
     a $150°$ **b** $162°$?

Find the exterior angle first.

16   Is it possible for each exterior angle of a regular polygon to be
     a $30°$ **b** $40°$ **c** $50°$ **d** $60°$ **e** $70°$ **f** $90°$?
     In those cases where it is possible, give the number of sides.

17   Is it possible for each interior angle of a regular polygon to be
     a $90°$ **b** $120°$ **c** $180°$ **d** $175°$ **e** $170°$ **f** $135°$?
     In those cases where it is possible, give the number of sides.

18   Construct a regular pentagon with sides 5 cm long.

19   Construct a regular octagon of side 5 cm.

Find the size of each interior angle, then use your protractor.

? **Puzzle**

Arrange ten counters in such a way as to form five rows with four counters only in each row.

Mixed problems

Exercise 12g

ABCDE is a pentagon, in which the interior angles
at A and D are each $3x°$ and the interior angles at B, C and E
are each $4x°$. AB and DC are produced
until they meet at F.

Find $B\hat{F}C$.

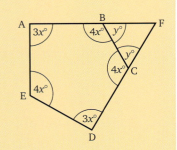

Sum of the interior angles of a pentagon $= 180 × 5360°$
$$= 540°$$
$$3x + 4x + 3x + 4x + 4x = 540$$
$$18x = 540$$
$$x = 30$$

∴ $A\hat{B}C = 120°$ and $B\hat{C}D = 120°$

∴ $y = 60°$

∴ $B\hat{F}C = 180° − 2 × 60°$ (angle sum of △BFC)

$$= 60°$$

In questions **1** to **10** find the value of x:

1
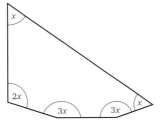

3
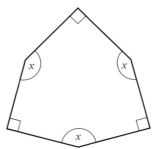

5
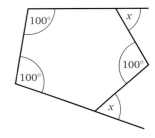

2
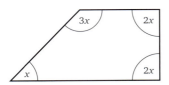

4
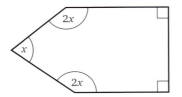

6
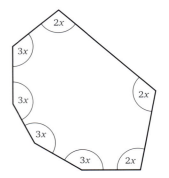

7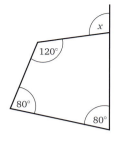

9

11 ABCDE is a regular pentagon.
OA = OB = OC = OD = OE.
Find the size of each angle at O.

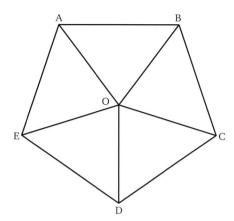

8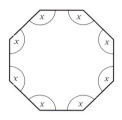

10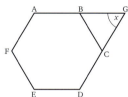

12 ABCDEFGH is a regular octagon. O is a point in the middle of the octagon such that O is the same distance from each vertex. Find AÔB.

Draw a diagram.

13 ABCDEF is a regular hexagon.
AB and DC are produced until they meet at G. Find BĜC.

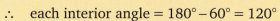

14 ABCDE is a regular pentagon.
AB and DC are produced until they meet at F. Find BF̂C.

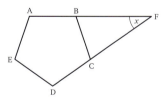

ABCDEF is a regular hexagon.
Find AD̂B.

As ABCDEF is regular, the exterior angles are all equal.

Each exterior angle = 360° ÷ 6 = 60°

∴ each interior angle = 180° − 60° = 120°

△BCD is isosceles (BC = DC).

∴ CB̂D = BD̂C = 30°   (angle sum of △BCD)

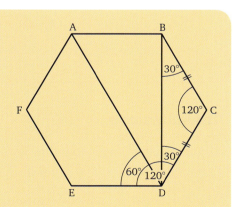

AD is a line of symmetry for the hexagon.

∴   ED̂A = CD̂A = 60°

∴   AD̂B = 60° − 30°

                     = 30°

In questions **15** to **20**, each polygon is regular. Give answers correct to one decimal place where necessary:

15

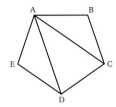

Find   **a**   AĈB
        b   DÂC.

16

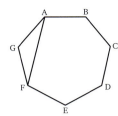

Find   **a**   AĜF
        b   GÂF.

17

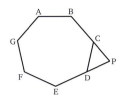

Find CP̂D.

18

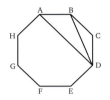

Find   **a**   CB̂D
        b   BD̂A.

19

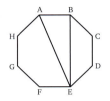

Find AÊB.

20

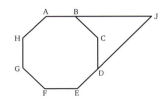

Find BĴD.

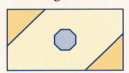

Puzzle

This flag is to be coloured red, white and blue.

Adjoining regions must have different colours.

How many different flags are possible?

Tessellations

Regular hexagons fit together without leaving gaps, to form a flat surface. We say that they *tessellate*.

The hexagons tessellate because each interior angle of a regular hexagon is 120°, so three vertices fit together to make 360°.

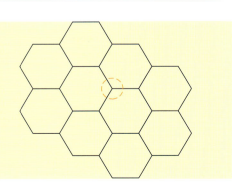

Exercise 12h

1   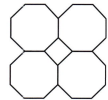   This is a pattern using regular octagons.

They do not tessellate:

a   Explain why they do not tessellate.

b   What shape is left between the four octagons?

c   Continue the pattern. (Trace one of the shapes above, cut it out and use it as a template.)

2   Trace this regular pentagon and use it to cut out a template:

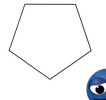

a   Will regular pentagons tessellate?

Make sure that you have a sharp pencil.

b

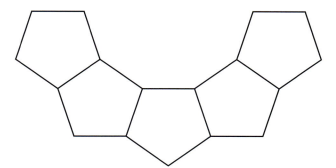

Use your template to copy and continue this pattern until you have a complete circle of pentagons. What shape is left in the middle?

c   Make up a pattern using pentagons.

3

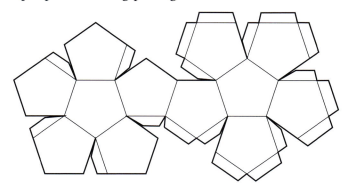

Use your template from question 2 to copy this net on to thick paper. Cut it out and fold along the lines. Stick the edges together using the flaps. You have made a regular dodecahedron.

4   Apart from the hexagon, there are two other regular polygons that tessellate. Which are they, and why?

5   Regular hexagons, squares and equilateral triangles can be combined to make interesting patterns. Some examples are given below:

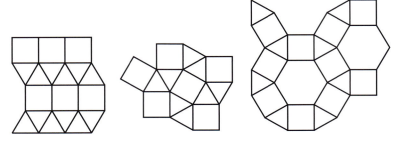

Copy these patterns and extend them. (If you make templates to help you, make each shape of side 2 cm.)

6   Make some patterns of your own using the shapes in question **5**.

Did you know?

There are only five regular convex polyhedra that can be made from regular shapes.

They are:

*   a tetrahedron which uses 4 equilateral triangles
*   an octahedron which uses 8 equilateral triangles
*   an icosahedron which uses 20 equilateral triangles
*   a cube which uses 6 squares
*   a dodecahedron which uses 12 regular pentagons.

These five solids are called the Platonic solids.

In this chapter you have seen that...

✔ the sum of the exterior angles of any polygon is 360°

✔ if a polygon is regular (i.e. equal sides) the exterior angles are equal and the size of each one is 360° ÷ the number of sides

✔ for a polygon with n sides the sum of the interior angles is $(180n - 360)°$ or, in a slightly more useful form, $180°(n - 2)$

✔ some regular polygons tessellate, i.e. they fit together without leaving gaps.

13 Relations

At the end of this chapter you should be able to...

1 Recognise a relation.

2 Describe a relation in words.

3 Use an arrow diagram to show a relation.

4 Use an ordered pair to show a relation.

5 Find the domain and range of a relation.

6 Identify the different types of relation.

7 Use tables, equations and graphs to represent relations.

8 Substitute values into an equation

You need to know...

✔ about sets

✔ how to plot points

✔ what x^2 and x^3 means

Key words

domain, mapping, ordered pair, range, relation, table of values

Relations

Look at the pairs of numbers in the set {(1, 2), (2, 4), (6, 12)}.

There is the same relation between the numbers in each pair: the second number in each pair is twice the first number.

The pairs (1, 2), (2, 4) and (6, 12) are called ordered pairs because the order of the numbers in them is important. This is because if, for example, we change (1, 2) to (2, 1), it is no longer true that the second number is twice the first.

The set {(1, 2), (2, 4), (6, 12)} is an example of a relation.

A relation is a set of ordered pairs with a rule that connects the two objects in each pair.

The objects do not have to be numbers, and the relation does not have to be mathematical.

For example, John, David and Mary are friends.

John is taller than David and David is taller than Mary; this is a relation between two children.

We can write this information as a set of pairs: (David, John), (Mary, David).

John must also be taller than Mary, so we can add another pair with the same relation: (Mary, John).

Again, the order of the two names in each pair is important. For example, for the pair (John, David), the relation is not true because David is not taller than John.

We can describe the relation as 'the second child in each pair is taller than the first child'.

We can write this relation as the set of ordered pairs

{(David, John), (Mary, David), (Mary, John)}

Exercise 13a

1   Describe the relation between the second and the first number in each pair in this set.

{(1, 2), (2, 3), (5, 6), (10, 11)}

2   Describe the relation between the second and the first number in each pair in this relation.

{(1, 3), (2, 4), (6, 8), (10, 12)}

3 Describe the relation between the second and the first number in each pair in this relation.

$$\{(2, 4), (3, 9), (4, 16), (5, 25)\}$$

4 This table shows the subject and number of pages in three school books.

Title	Number of pages
Maths	160
Spanish	210
Science	140

Write down the set of ordered pairs in the relation described as 'The second book in each pair has more pages than the first book.'

5 This is a set of shapes {□, ⌂, ▲}.

Write down the relation described as the set of ordered pairs where the first object is the number of sides and the second object is the shape.

Remember that a relation is set of ordered pairs

6 This table lists some countries and their populations.

Country	Population
Jamaica	2 500 000
Trinidad	1 300 000
Barbados	300 000
St Lucia	150 000

Give the relation described as 'the second country has a larger population than the first country'

7 The second number in each pair in this relation is the square of the first number. Fill in the missing numbers.

$$\{(2, 4), (5, \quad), (\quad, 64)\}$$

8 The second number in each pair in this relation is the next prime number that is larger than the first number. Fill in the missing numbers.

$$\{(8, 11), (6, \quad), (1, \quad), (14, \quad)\}$$

Domain and range

The domain of a relation is the set of the first objects in the ordered pairs.

For example, the domain of the relation {(1, 2), (2, 4), (6, 12)} is the set {1, 2, 6} and the domain of the relation {(David, John), (Mary, David), (Mary, John)} is the set {David, Mary}.

Notice that we do not include Mary twice as it is the same person.

The range of a relation is the set of the second objects in each ordered pair.

For example, the range of the relation {(1, 2), (2, 4), (6, 12)} is the set {2, 4, 12} and the range of the relation {(David, John), (Mary, David), (Mary, John)} is the set {John, David}.

Exercise 13b

1   Write down the domain and the range of the relation
    {(1, 2), (2, 3), (5, 6), (10, 11)}.

2   Write down the domain and range of each relation.
    a {(a, b), (a, c), (b, c)}   **b** {(▭, ▢). (△, △). (▱, ▢)}

3   The set {2, 4, 6} is the domain of a relation. The second number in
    each ordered pair is the square of the first number. What is the range?

4   Fred, Dwayne and Scott are three boys. Fred is older than Dwayne and
    Dwayne is older than Scott.
    a Write down the relation described as 'the second boy in each pair
          is older than the first boy.'
    b Give the domain and range.

5   Write down the domain and range of each relation.
    a {(10°, acute), (150°, obtuse), (45°, acute), (175°, obtuse)}
    b {(t, t), (t, u), (s, t), (s, w)}

Mapping diagrams

We can represent a relation with a mapping diagram.

This mapping diagram represents the relation {(1, 2), (2, 4), (6, 12)}

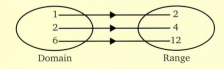

The members of the domain are placed in one oval and the members in the range are placed in a second oval. The arrows show the association between the members in the domain and the members in the range.

We say that 1 maps to 2, 2 maps to 4 and 6 maps to 12.

This mapping diagram represents the relation
{(David, David), (Mary, David), (Mary, John)}

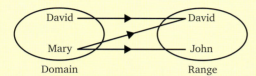

This shows clearly that David maps to David and that Mary maps to John and David. This mapping diagram represents another relation.

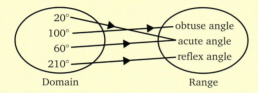

We can use this diagram to write down the relation as a set of ordered pairs:

{(20°, acute angle), (100°, obtuse angle), (60°, acute angle), (210°, reflex angle)}

Exercise 13c

1 Draw a mapping diagram to represent these relations.
   a {(1, 2), (2, 3), (5, 6), (10, 11)}
   b {(a, b), (a, c), (b, c)}
   c {(2, 4), (3, 9), (4, 16), (5, 25)}
   d {(a, 2a), (b, 2b), (c, 2c)}

> Start by writing down the domain and the range. Remember that these are sets so only list the different members of the set.

2 Each diagram represents a relation. Write down the relation as a set of ordered pairs.

a
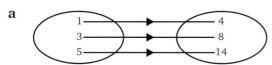

b
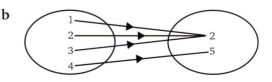

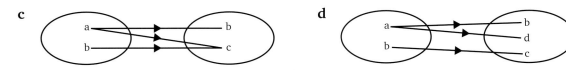

3 A relation is described as 0 maps to 0, 90 maps to 1 and 180 maps to 0. Draw a diagram to represent this relation.

Types of relation

Look again at the relation {(1, 2), (2, 4), (6, 12)}

No two ordered pairs have the same first number, and no two ordered pairs have the same second number.

We can see this clearly from the mapping diagram:

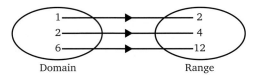

There is only one arrow from every member of the domain and there is only one arrow to every member of the range, so every member of the domain maps to only one member of the range, and every member of the range comes from only one member of the domain.

Any relation where this is true is called a 'one to one' relation. This is written as $1:1$ or $1–1$.

There are other types of relation.
One type is where more than one member of the domain maps to one member of the range. This is the case with this relation.

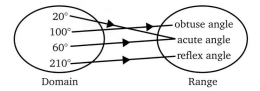

This type is called a 'many to one' relation, which we write as $n:1$

Another type is where a member of the domain maps to more than one member of the range. This is the case with this relation.

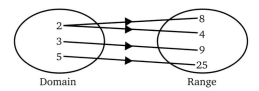

This is an example of a 'one to many' relation, written as $1:n$.

The last type of relation is where a member of the domain maps to more than one member of the domain *and* more than one member of the domain maps to one member of the domain.

This is the case with the relation {(David, John), (Mary, David), (Mary, John)}

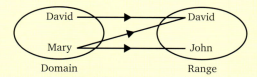

This type is called a 'many to many' relation. We write this as $n:n$. This diagram summarises the different types of relation.

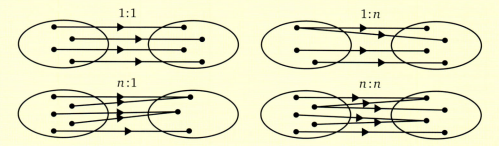

 Exercise 13d

Describe the type of relation in each question in exercise **13c**.

Investigation

Kim, David, Jenny, and Clare are Emma's family. They are Emma's mother, father, younger brother, and younger sister.
1   David is older than Emma.
2   Clare is not Emma's younger brother.
3   Jenny is not Emma's father. She is also not Emma's younger sister.

Who is the mother, father, younger brother and younger sister?

Using tables and graphs

When the ordered pairs in a relation are numbers, as in {(1, 2), (2, 4), (6, 12)}, we can think of them as sets of coordinates.

This means we can represent them in a table and plot them as points, {x, y}, on a plane.

We use x to stand for the values of the first number in each pair and y to stand for the second number in each pair.

So the relation {(1, 2), (2, 4), (6, 12)} can be represented by the table:

x	1	2	6
y	2	4	12

The values of x give the members of the domain and the values of y give the members of the range.

We can then represent these ordered pairs as points on a graph.

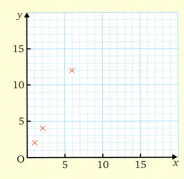

Exercise 13e

1 Represent each relation as a table of values of x and y and illustrate them on a graph.

a {(1, 2), (2, 4), (3, 6), (4, 8)}   b {(2, 1), (4, 2), (6, 3), (9, 4.5)}
c {(10, 0), (7, 3), (5, 5), (0, 10)}   d {(0, 0), (1, 4), (2, 6), (3, 8), (4, 6)}

2 A relation is represented by this table.

x	2	6	10
y	1	3	5

Illustrate the relation on a graph.

3 This graph illustrates a relation.
  a Represent the relation as a table.
  b Give the relation as a set of ordered pairs.
  c What type of relation is this?

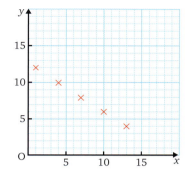

4 This graph represents a relation.
  a Represent this relation as a table.
  b Draw an arrow diagram to show the relation.
  c What type of relation is this?

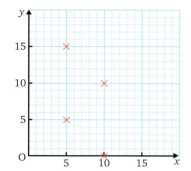

5 The points A, B, C, D and E illustrate a relation.
  a Represent these points as a table.
  b How is the y-coordinate of each point related to its x-coordinate?
  c The points A, B, C, D and E all lie on the same straight line.
  G is another point on this line. Its x-coordinate is 8; what is its y-coordinate?
  d F, H and I are also points on this line. Find the missing coordinates.
  (5,), (16,), (a,)

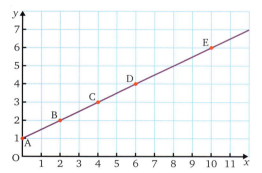

6 The points A, B, C and D illustrate a relation.
  a Represent these points as a table.
  b How is the y-coordinate of each point related to its x-coordinate?
  c The points A, B, C, and D all lie on the same straight line.
  E is another point on this line. Its x-coordinate is 5; what is its y-coordinate?
  d F, H and I are also points on this line. Find the missing coordinates.
  (1,), (12,), (a,)

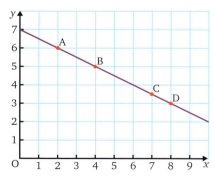

7 The table represents a relation.

x	1	4	4	8	8
y	1	0	2	0	6

a Plot the points on a graph.

b What type of relation is this? Give a reason for your answer.

Equations

We have already seen that we can describe the connection between the two numbers in each ordered pair in the relation {(1, 2), (2, 4), (6, 12)} as 'the second number in each pair is twice the first number'.

Using x to stand for the first number and y to stand for the second number, we can describe the connection more briefly as the equation $y = 2x$.

If we also give the values that x can have, we can use the equation to define the relation as

$$\{(x, y)\} \quad \text{where} \quad y = 2x \text{ for } x = 1, 2, 6.$$

Now consider the relation $\{(x, y)\}$ where $y = 2x - 1$ for $x = 2, 4, 6, 8$.

This means that the value of x in the first ordered pair is 2 and we use the equation $y = 2x - 1$ to find the value of y by substituting 2 for x.

Remember that $2x$ means $2 \times x$, so when $x = 2$,

$$y = 2 \times 2 - 1$$
$$= 4 - 1 \text{ (do multiplication before}$$
$$\text{subtraction)}$$
$$= 3$$

We can find the other ordered pairs in the same way and represent the relation in a table.

x	2	4	6	8
y	3	7	11	15

Exercise 13f

1 A relation is given by $\{(x, y)\}$ where $y = 3x$ for $x = 1, 2, 3$.

Complete this table of values.

x	1	2	3
y		6	

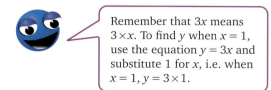

Remember that $3x$ means $3 \times x$. To find y when $x = 1$, use the equation $y = 3x$ and substitute 1 for x, i.e. when $x = 1$, $y = 3 \times 1$.

2 A relation is given by $\{(x, y)\}$ where $y = 4x - 1$ for $x = 1, 2, 3$.

Complete this table of values.

x	1	2	3
y		7	

Remember that multiplication and division must be done before addition and subtraction.

3 A relation is given by $\{(x, y)\}$ where $y = 10 - x$ for $x = 2, 4, 6, 8$.

Complete this table of values.

x	2	4	6	8
y		6		

4 A relation is given by $\{(x, y)\}$ where $y = 12 - 2x$ for $x = 1, 2, 5, 6$.

Complete this table of values.

x	1	2	5	6
y			2	

5 A relation is given by $\{(x, y)\}$ where $y = x^2 + 1$ for $x = 1, 2, 3, 4$.

Complete this table of values.

x	1	2	3	4
y			10	

Remember that x^2 means $x \times x$, so when $x = 2$, $x^2 + 1 = 2 \times 2 + 1$.

6 A relation is given by $\{(x, y)\}$ where $y = 2x^2 - 1$ for $x = 1, 2, 3, 4$.

Complete this table of values.

x	1	2	3	4
y			17	

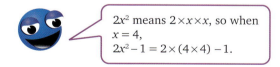

$2x^2$ means $2 \times x \times x$, so when $x = 4$, $2x^2 - 1 = 2 \times (4 \times 4) - 1$.

7 A relation is given by $\{(x, y)\}$ where $y = x^2 - 3x + 4$ for $x = 1, 2, 3, 4$.

Complete this table of values.

x	1	2	3	4
y		2		

8 A relation is given by $\{(x, y)\}$ where
$y = x^2 - x$ for $x = 1, 2, 3, 4$.

a Complete this table of values.

x	1	2	3	4
y		2		

b Write down the domain and range.

c Represent the relation with an arrow
diagram.

This means, is it a 1:1
relation or is it one of the
other types?

d What type of relation is this?

9 A relation is given by $\{(x, y)\}$ where $y = 3x + \frac{1}{2}$ for $x = 1, 1\frac{1}{2}, 2$.

a Complete this table of values.

x	1	$1\frac{1}{2}$	2
y	$3\frac{1}{2}$		

b Write down the domain and range.

c Represent the relation with an arrow diagram.

d What type of relation is this?

10 A relation is given by $\{(x, y)\}$ where $y = x^2 - 5x + 6$ for $x = 1, 2, 3, 4$.

a Complete this table of values.

x	1	2	3	4
y		0	0	

b Write down the domain and range.

c Represent the relation with an arrow diagram.

d What type of relation is this?

e Draw a graph to represent this relation.

11 A relation is given by $\{(x, y)\}$ where $y = x^3 - 8x^2 + 15x$ for $x = 0, 2, 3, 5$.

a Complete this table of values.

x	0	2	3	5
y			0	

b Write down the domain and range.

c Represent the relation with an arrow diagram.

d What type of relation is this?

e Draw a graph to represent this relation.

12 This graph represents a relation.

   a Write the relation as a table of values of x and y.

   b Work out how the y-coordinate of each point is related to its x-coordinate.

   c Hence give the relation in the form $\{(x, y)\}$ where $y = \ldots$ for $x = \ldots$

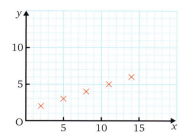

? Puzzle

Complete these calculations:

$132 \div 6 =$ $102 + 65 \div 5 =$ $42 - 2 \times 8 =$

$24 \times 8 =$ $14 \times 3 - 25 =$ $186 \div 31 - 6 =$

Now do a number search to find the complete calculations.

```
4  2  −  2  ×  8  =  2  6  7  −  9  5
+  1  3  1  4  ×  3  −  2  5  =  1  7
5  3  0  2  4  ×  9  =  2  1  5  4  9
=  2  ×  2  0  1  8  ×  2  4  =  4  7
2  ÷  7  0  +  3  6  =  2  4  +  1  2
9  6  +  1  8  6  ÷  3  1  −  6  =  0
7  =  6  0  3  =  5  7  =  9  +  1  2
×  2  =  2  +  8  6  ÷  6  ÷  2  3  1
0  2  2  +  6  ×  7  =  5  2  +  5  6
4  =  8  5  −  4  2  0  0  =  2  8  +
=  7  1  =  5  2  −  3  ×  4  1  =  7
3  −  2  1  +  3  1  −  6  5  =  3  =
÷  6  =  7  1  8  6  ÷  3  1  −  6  5
```

In this chapter you have seen that...

✔ a relation is a set of ordered pairs

✔ the set of the first objects in each pair is the domain and the set of the second objects in each pair is the range

✔ a relation can be represented by a mapping diagram, e.g.

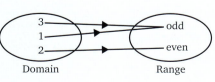

✔ a relation where a number maps to a number can be represented by an equation, e.g. $y = 2x$ together with the values that x can take

✔ there are four types of relation: $1:1$, $n:1$, $1:n$ and $n:n$.

14 Similar figures

At the end of this chapter you should be able to...

1 State whether or not given figures are similar.

2 Determine if two triangles are similar or not, by comparing the sizes of their angles or their sides.

3 Write down equal ratios in two similar triangles.

4 Calculate the length of one side of a triangle from necessary data on a pair of similar triangles.

5 Use the scale factor for enlarging one triangle into another, to find a missing length.

You need to know...

✔ the angle sum of a triangle

✔ how to find one quantity as a fraction of another

✔ how to solve equations involving fractions

✔ the meaning of ratio

✔ how to find a missing quantity in a ratio

Key words

alternate angles, enlargement, ratio, scale factor, similar, vertically opposite angles

Similar figures

Two figures are similar if they are the same shape but not necessarily the same size. One figure is an enlargement of the other.

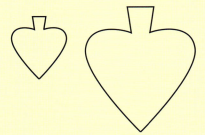

One may be turned round compared with the other.

One figure may be turned over compared with the other.

The following figures are not similar although their angles are equal.

Exercise 14a

State whether or not the pairs of figures in questions **1** to **10** are similar.

1

2

3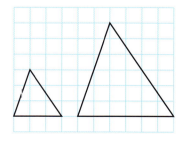

4

5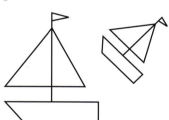

6

7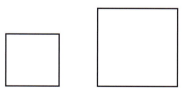

8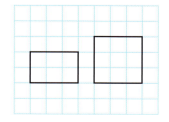

9 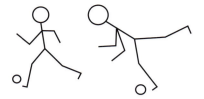 10

11 Which two rectangles are similar?

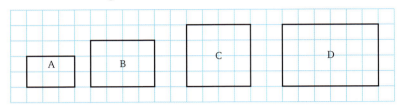

Similar triangles

Some of the easiest similar figures to deal with are triangles.
This is because only a small amount of information is needed
to prove them to be similar.

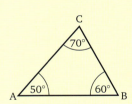 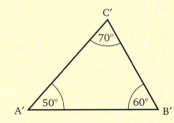

In these triangles the corresponding angles are equal and so the
triangles are the same shape. One triangle is an enlargement of
the other. These triangles are *similar*.

Exercise 14b

1 Draw the following triangles accurately:
   a Are the triangles similar?
   b Measure the remaining sides.
   c Find $\dfrac{A'B'}{AB}$, $\dfrac{B'C'}{BC}$ and $\dfrac{C'A'}{CA}$
      (as decimals if necessary)
   d What do you notice about the
      answers to part **c**?

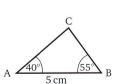 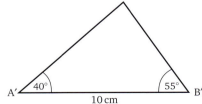

Repeat question **1** for the pairs of triangles in questions **2** to **5**.

2

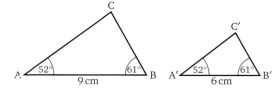

4

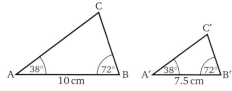

3

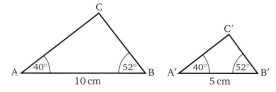

5

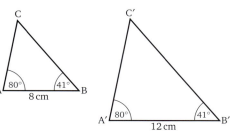

Sketch the following pairs of triangles and find the sizes of the missing angles. In each question state whether the two triangles are similar. (One triangle may be turned round or over compared with the other.)

6

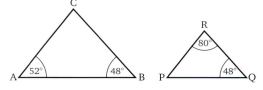

8

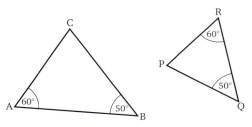

7

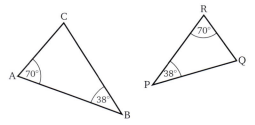

9

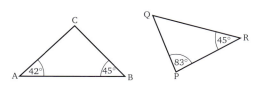

Corresponding vertices

These two triangles are similar and we can see that X corresponds to A, Y to B and Z to C.

We can write $\Delta s \dfrac{ABC}{XYZ}$ are similar

Make sure that X is written below A, Y below B and Z below C.

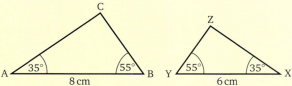

The pairs of corresponding sides are in the same ratio,

that is $\dfrac{AB}{XY} = \dfrac{BC}{YZ} = \dfrac{CA}{ZX}$

Exercise 14c

State whether triangles ABC and PQR are similar and, if they are, give the ratios of the sides.

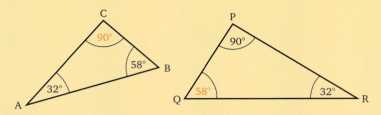

First find the third angle in each triangle

$$\widehat{C} = 90° \quad \text{(angles of a triangle)}$$

and $$\widehat{Q} = 58°$$

Now you can see that the triangles are similar and that A corresponds to R, and so on,

so $$\triangle s \begin{matrix} ABC \\ RQP \end{matrix} \text{ are similar}$$

and $$\frac{AB}{RQ} = \frac{BC}{QP} = \frac{CA}{PR}$$

In questions **1** to **8**, state whether the two triangles are similar and, if they are, give the ratios of the sides.

1
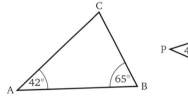

3
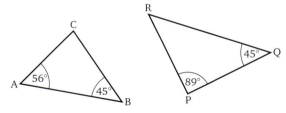

2
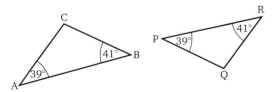

4
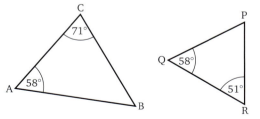

5 Use the triangles given in question **6** of Exercise 14b.

6 Use the triangles given in question **7** of Exercise 14b.

7 Use the triangles given in question **8** of Exercise 14b.

8 Use the triangles given in question **9** of Exercise 14b.

Puzzle

There are 13 stations on a railway line. All tickets are printed with the name of the station you board the train and the station you leave the train. How many different tickets are needed?

Finding a missing length

Exercise 14d

State whether the two triangles are similar. If they are, find AB.

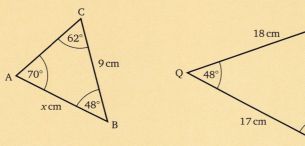

First find the third angle in each triangle.

$$\widehat{C} = 62° \quad \text{and} \quad \widehat{Q} = 48° \quad \text{(angles of a triangle)}$$

so $\Delta s \begin{smallmatrix} ABC \\ PQR \end{smallmatrix}$ are similar and $\dfrac{AB}{PQ} = \dfrac{BC}{QR} = \dfrac{CA}{RP}$

Ring the side you want and the sides you know: $\dfrac{\textcircled{AB}}{\textcircled{PQ}} = \dfrac{\textcircled{BC}}{\textcircled{QR}} = \dfrac{CA}{RP}$

Now substitute the values $\quad \dfrac{x}{17} = \dfrac{9}{18}$

$$17 \times \dfrac{x}{17} = \dfrac{9}{18} \times 17$$

$$x = \dfrac{17}{2} = 8.5$$

$$AB = 8.5\,\text{cm}$$

In questions **1** to **4**, state whether the pairs of triangles are similar. If they are, find the required side.

1   Find PR.

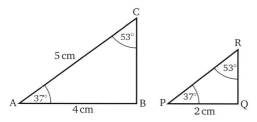

2   Find QR.

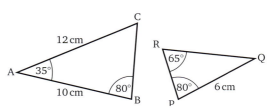

3 Find BC.

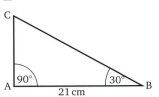

4 Find PR.

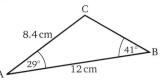

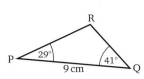

In some cases we do not need to know the sizes of the angles as long as we know that pairs of angles are equal. (Two pairs only are needed as the third pair must then be equal.)

In △s ABC and DEF, Â = Ê and B̂ = D̂. AB = 4 cm, DE = 3 cm and AC = 6 cm. Find EF.

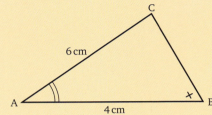

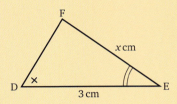

△s $_{ABC}^{EDF}$ are similar because they are equiangular

(we put the triangle with the unknown side on top)

$$\frac{FE}{CA} = \frac{ED}{AB} = \frac{DF}{BC}$$

Ringing the sides known and wanted gives $\dfrac{\boxed{FE}}{\boxed{CA}} = \dfrac{\boxed{ED}}{\boxed{AB}} = \dfrac{DF}{BC}$

Substituting values gives

$$\frac{x}{6} = \frac{3}{4}$$

$$\overset{1}{\cancel{6}} \times \frac{x}{\cancel{6}_1} = \frac{3}{\cancel{4}_2} \times \cancel{6}^{3}$$

$$x = \frac{9}{2} = 4.5$$

so EF = 4.5 cm

5 In △s ABC and XYZ, Â = X̂ and B̂ = Ŷ.
AB = 6 cm, BC = 5 cm and XY = 9 cm.
Find YZ.

6 In △s ABC and PQR, Â = P̂ and Ĉ = R̂.
AB = 10 cm, PQ = 12 cm and QR = 9 cm. Find BC.

7 In △s ABC and DEF, Â = Ê and B̂ = F̂. AB = 3 cm, EF = 5 cm and AC = 5 cm. Find DE.

8 In △s ABC and PQR, Â = Q̂ and Ĉ = R̂. AC = 8 cm, BC = 4 cm and QR = 9 cm. Find PR.

Draw the triangles and mark the equal angles and the lengths of the sides that are given. Label the side you have to find x cm.

a Show that triangles ABC and CDE are similar.

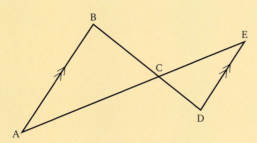

b Given that AC = 15 cm, CE = 9 cm and DE = 8 cm, find AB.

a $\hat{A} = \hat{E}$ (alternate angles, AB || DE)

$\hat{B} = \hat{D}$ (alternate angles, AB || DE)

(Or we could use $B\hat{C}A = E\hat{C}D$ as these are vertically opposite angles.)

so $\quad \Delta s \begin{smallmatrix} ABC \\ EDC \end{smallmatrix}$ are similar.

b Sketch the diagram and mark the equal angles and the lengths of the sides that are given, and the side you need to find.

$$\frac{AB}{ED} = \frac{BC}{DC} = \frac{CA}{CE}$$

$$\frac{x}{8} = \frac{15}{9}$$

$$\cancel{8} \times \frac{x}{\cancel{8}_1} = \frac{\cancelto{5}{15}}{\cancelto{3}{9}} \times 8$$

$$x = \frac{40}{3}$$

$$= 13\tfrac{1}{3}$$

$$AB = 13\tfrac{1}{3} \text{ cm}, \quad \text{or} \quad 13.3 \text{ cm correct to 1 d.p.}$$

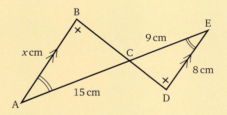

9 a Show that △s ABC and BDE are similar.
b If AB = 6 cm, BD = 3 cm and DE = 2 cm, find BC.

Draw your own diagram.

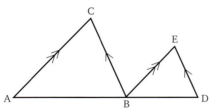

10 a Show that △s ABC and CDE are similar.
b If AB = 7 cm, BC = 6 cm, AC = 4 cm and CE = 6 cm, find CD and DE.

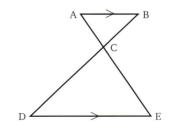

11 **a** ABCD is a square. EF is at right angles to BD.
Show that △s ABD and DEF are similar.

b If AB = 10 cm, DB = 14.1 cm and DF = 7.1 cm,
find EF.

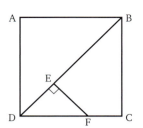

12 **a** Show that △s ABC and ADE are similar.
(Notice that Â is *common* to both triangles.)

b If AB = 10 cm, AD = 15 cm, BC = 12 cm and AC = 9 cm,
find DE, AE and CE.

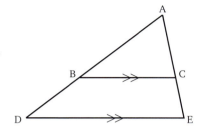

 Puzzle

Arrange four 9s to make 100.

Scale factor

Similar triangles are the same shape but one
triangle is larger than the other. In the diagram,
AB and PQ are corresponding sides and
AB = 3 × PQ. This means that the lengths of the
other sides of triangle ABC are three times the
lengths of the corresponding sides in triangle PQR.

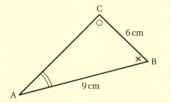

Therefore CB = 3 × QR, so 6 = 3x, i.e. $x = 2$

For any pair of similar triangles, the lengths of all the sides of one triangle
will be the same multiple of the lengths of the corresponding sides of the
other triangle. This multiple is called the scale factor.

If we want to find a length in the first
triangle, we can see that the scale factor
is 4

so $x = 4 \times 2\frac{1}{2} = 10$

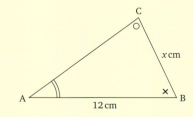
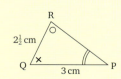

Exercise 14e

Find QR.

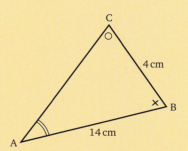

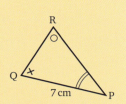

$\triangle s \begin{matrix} PQR \\ ABC \end{matrix}$ are similar.

QP and BA are corresponding sides and
$QP = \frac{1}{2} AB$

So the scale factor is $\frac{1}{2}$

QR and BC are corresponding sides

$\therefore \qquad QR = \frac{1}{2} \times 4\,cm = 2\,cm$

1 Find BC.

Identify the corresponding sides and use them to find the scale factor.

4 Find XY.

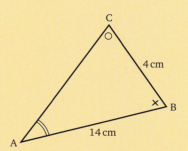

2 Find PR.

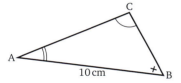

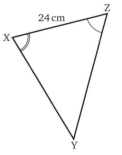

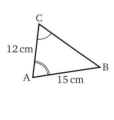

3 Find PR.

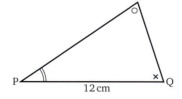

5 Find LN.

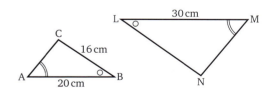

Corresponding sides

If the three pairs of sides of two triangles are in the same ratio, then the triangles are similar and their corresponding angles are equal.

When finding the ratio of three sides give the ratio as a whole number or as a fraction in its lowest terms.

Exercise 14f

State whether triangles ABC and PQR are similar.
Say which angle, if any, is equal to $\widehat{A}$.

Start with the shortest side of each
triangle: $\dfrac{PR}{AC} = \dfrac{9}{3} = 3$

Now the longest sides: $\dfrac{PQ}{AB} = \dfrac{13\frac{1}{2}}{4\frac{1}{2}} = \dfrac{27}{9} = 3$

Lastly the third sides $\dfrac{QR}{BC} = \dfrac{12}{4} = 3$

so $\qquad\qquad \dfrac{PR}{AC} = \dfrac{PQ}{AB} = \dfrac{QR}{BC}$

$\therefore \qquad\qquad \triangle s \; \overset{PQR}{\underset{ABC}{}} \;$ are similar

and $\qquad\qquad \widehat{P} = \widehat{A}$

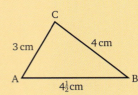

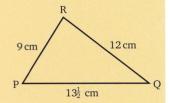

State whether the following pairs of triangles are similar. In each case say
which angle, if any, is equal to $\widehat{A}$.

1
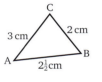

4
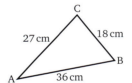

2
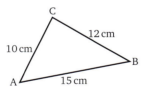

5
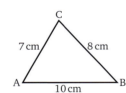

3
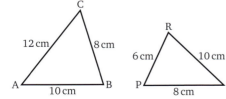

6
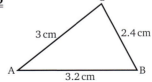

7 Are the triangles ABC and ADE similar?
Which angles are equal?
Are BC and DE parallel?
Give a reason for your answer.

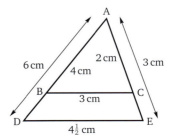

One pair of equal angles and two pairs of sides

The third possible set of information about similar triangles concerns a pair of angles and the sides containing them.

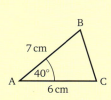

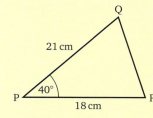

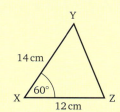

$$\frac{PR}{AC} = \frac{18}{6} = 3 \quad \text{and} \quad \frac{PQ}{AB} = \frac{21}{7} = 3$$

i.e.
$$\frac{PR}{AC} = \frac{PQ}{AB}$$

and
$$\hat{A} = \hat{P}$$

so $\Delta s \; \begin{matrix} ABC \\ PQR \end{matrix}$ are similar.

We can see that △PQR is an enlargement of △ABC and that the scale factor is 3.
$\left(\text{It is given by } \frac{PQ}{AB}.\right)$

On the other hand, △XYZ is a different shape from the other two and is not similar to either of them, even though two pairs of sides are in the same ratio, because the angles between the pairs of sides are not the same.

Exercise 14g

State whether triangles ABC and PQR are similar. If they are, find PQ.

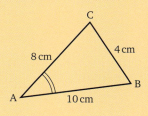

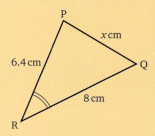

$\hat{A} = \hat{R}$ so compare the ratios of the arms containing these angles.

$$\frac{RP}{AC} = \frac{6.4}{8} = 0.8 \quad \text{(comparing the two shorter sides)}$$

$$\frac{RQ}{AB} = \frac{8}{10} = 0.8 \quad \text{(comparing the other two arms)}$$

∴
$$\frac{RP}{AC} = \frac{RQ}{AB} \quad \text{and} \quad \hat{A} = \hat{R}$$

so $\Delta s \; \begin{matrix} RQP \\ ABC \end{matrix}$ are similar.

Now, $\dfrac{PQ}{CB} = \dfrac{RQ}{AB}$

∴ PQ = 3.2 cm

$\dfrac{x}{4} = \dfrac{8}{10}$

$4 \times \dfrac{x}{4} = \dfrac{8}{10} \times 4$

$x = 3.2$

PQ = 3.2 cm

or: BC is half AC so PQ is half PR

State whether the following pairs of triangles are similar. If they are,
find the missing lengths.

1
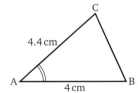

4
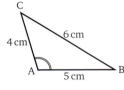

2
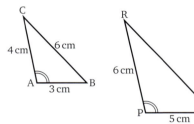

5
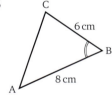

3
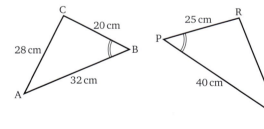

6 In △s ABC and PQR, $\widehat{A} = \widehat{P}$, AB = 8 cm, BC = 8.5 cm, CA = 6.5 cm,
PQ = 4.8 cm and PR = 3.9 cm. Find QR.

7 In △s PQR and XYZ, $\widehat{P} = \widehat{X}$, PQ = 4 cm, PR = 3 cm, QR = $2\frac{1}{4}$ cm,
XY = $5\frac{1}{3}$ cm and XZ = 4 cm. Find ZY.

Summary: similar triangles

If two triangles are the same shape (but not necessarily the same size)
they are said to be *similar*. This word, when used in mathematics, means
that the triangles are *exactly* the same shape and not vaguely alike, as two
sisters may be.

One triangle may be turned over or round compared with the other.

Pairs of corresponding sides are in the same ratio. This ratio is the
scale factor for the enlargement of one triangle into the other.

To check that two triangles are similar we need to show *one* of the three
following sets of facts:

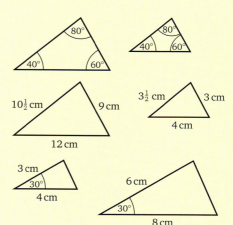

a The angles of one triangle are equal to the
angles of the other (as in Exercise 14c).

b The three pairs of corresponding sides are
in the same ratio (as in Exercise 14f).

c There is one pair of equal angles and the
sides containing the known angles are in
the same ratio (as in Exercise 14g).

Mixed exercises

Exercise 14h

State whether or not the pairs of triangles in questions **1** to **10** are similar,
giving your reasons. If they are similar, find the required side or angle.

1 Find BC.

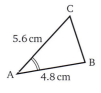
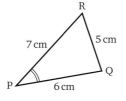

2 Find QR.

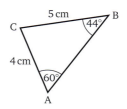
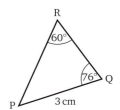

<u>3</u> Find $\widehat{Q}$.

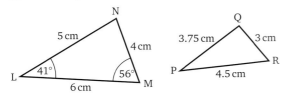

<u>6</u> Find $\widehat{Q}$.

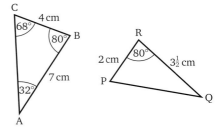

<u>4</u> Find FE.

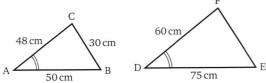

<u>7</u> Find YZ.

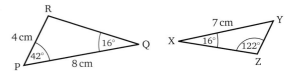

<u>8</u> Find AC.

<u>5</u> Find $\widehat{P}$.

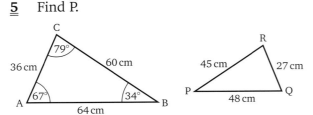

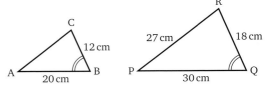

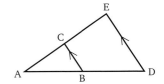

<u>9</u> **a** Show that △s ABC and ADE are similar.

   b AB = 3.6 cm, AD = 4.8 cm and AE = 4.2 cm.
      Find AC and CE.

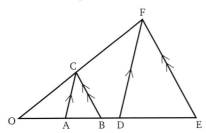

<u>10</u> **a** Show that △s ABC and DEF are similar.

   b AB = 40 cm, BC = 52 cm and DE = 110 cm. Find EF.

<u>11</u> In the figure below there are three overlapping triangles.

   a Show that △s ABC and ABD are similar.

   b Show that △s ABC and BDC are similar.

   c Are △s ABD and BDC similar?

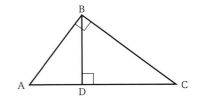

12   A pole, AB, 2 m high, casts a shadow, AC, that is 3 m long.
Another pole, PQ, casts a shadow 15 m long.
How high is the second pole?

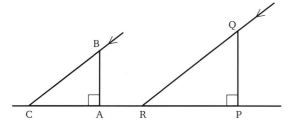

13   The shadow of a 1 m stick held upright on the ground is 2.4 m long.
How long a shadow would be cast by an 8 m telegraph pole?

14   A slide measures 1.8 cm by 2.4 cm. A picture 90 cm by 120 cm is cast
on the screen. On the slide, a house is 1.2 cm high. How high is the
house in the picture on the screen?

! Investigation

Draw a quadrilateral ABCD with four unequal sides. Mark the midpoints of
the sides AB, BC, CD and DA with the letters P, Q, R and S in that order.

Join P, Q, R and S to give a new quadrilateral. Are the two quadrilaterals
similar?

Investigate what happens if you repeat this with other quadrilaterals,
including the special quadrilaterals this time.

In this chapter you have seen that...

✔ two shapes are similar if one is an enlargement of the other

✔ to prove that two triangle are similar you need to show that

- the triangles have the same angles **or**

- all three sides of each triangle are in the same ratio **or**

- one pair of angles are equal and the sides round those angles
  are in the same ratio.

REVIEW TEST 2: CHAPTERS 8–14

In questions 1 to 10 choose the letter for the correct answer.

1   The simple interest on $2000 for 3 years at 5% is

   A $100   B $150   C $300   D $600

2   An item bought for $1200 is sold to make a profit of 20%. The selling price is

   A $960   B $1250   C $1400   D $1440

3   In an exam marked out of 120 marks, a student requires a minimum of 80% to attain a grade 'A'. The minimum mark required for a grade 'A' is

   A 80   B 96   C 100   D 120

4   Each interior angle of a regular 5-sided polygon measures

   A 144°   B 108°   C 72°   D 36°

5   The exterior angle of a regular polygon is 40°. The number of side the polygon has is

   A 6   B 8   C 9   D 12

6   The image of (2, 2) under a clockwise rotation of 90 degrees about the origin is

   A (−2, 2)   B (2, −2)   C (−2, 0)   D (2, 0)

7   The circumference of a circle is 88 cm. What is its radius? $\left(\text{Take } \pi = \frac{22}{7}\right)$

   A 11 cm   B 14 cm   C 22 cm   D 44 cm

8   In the diagram LM =

   A 14 cm
   B 15 cm
   C 16 cm
   D 20 cm

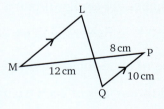

9   A relation is given by {x, y} where y = 2x − 5. For x = 5, y =

   A 15   B 5   C 1   D 0

10   The number of hours I can run a 100 W bulb on 1 unit of electricity, i.e. 1 kWh, is

   A 4   B 5   C 10   D 20

11 Find **a** the area of a circle of radius 7 cm $\left(\text{Take } \pi = \frac{22}{7}\right)$

         b the circumference of a circle of diameter 21 mm

         c the radius of a circle of circumference 44 cm.

12 Find the compound interest on $35 000 invested for 2 years at 5%.

13 The water rates due on my house this year are 8% more than they were last year. Last year I paid $36 500. What must I pay this year?

14 Miss White earns $25 000 a week. Deductions from her gross pay are NIS 2%, NHT 2.5% and Education tax 2.5%. After these deductions she also pays income tax at 25% on her remaining pay over $5000. Work out
a her total deductions   **b** her net weekly pay.

15 State whether or not triangles ABC and PQR are similar. If they are, find the length of PR.

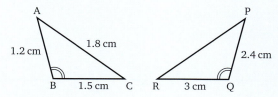

16 Draw a triangle PQR with vertices at P(2, 5), Q(3, 8) and R(7, 1). Draw the images of PQR after reflections in
a the x-axis   **b** the y-axis   **c** the line $y = x$.
Name the coordinates of the image in each case.

17 Describe the transformation that maps the rectangle ABCD to
a A′B′C′D′   **b** A″B″C″D″.

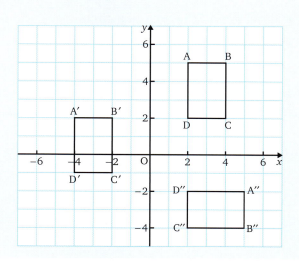

15 Area and volume

At the end of this chapter you should be able to...

1 Find the area of a rectangle and a triangle.

2 Recognise volume as a measure of space.

3 Measure volume using standard units.

4 Convert from one standard metric unit to another.

5 Calculate the volumes of solids with uniform cross-sections.

6 Calculate the volume of a cylinder, given its radius and height.

7 Calculate the volumes of compound solids.

8 Use flat paper to construct nets of:

   a   a cube

   b   a tetrahedron

   c   a octahedron

   d   a square-based pyramid

   e   a prism with triangular cross-section

   f   a prism with hexagonal cross-section

   g   an eight-pointed star.

You need to know...

✔ how to change from one metric unit of length to another

✔ how to multiply fractions and decimals

✔ how to find the area of a circle

✔ what a right angle is

✔ how to measure length and draw angles accurately using a protractor

✔ how to use a pair of compasses

Key words

capacity, cube, cuboid, cross-section, cylinder, diagonal, diameter, dimension, hexagon, kite, litre, millilitre, net, octahedron, pi(π), prism, radius, rectangle, rhombus, square, square-based pyramid, symmetry, tetrahedron, triangle, triangular prism, uniform, volume

Area of a triangle

We know that the area of a rectangle is found by multiplying its length by its breadth. We can use this to find the area of a triangle.

If we enclose a triangle in a rectangle as shown below, we can see that the area of the triangle is half the area of the rectangle.

Area $= \frac{1}{2}$ (base $\times$ height)

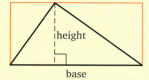

These diagrams can be drawn on squared paper and then cut out to show how the pieces fit.

Height of a triangle

When we talk about the height of a triangle we mean its perpendicular height and not its slant height.

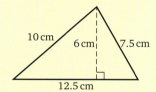

If we draw the given triangle accurately on squared paper, we can see that the height of the triangle is not 10 cm or 7.5 cm but 6 cm. (We can also see that the foot of the perpendicular is *not* the midpoint of the base.)

Finding areas of triangles

Exercise 15a

Find the area of a triangle with base 7 cm and height 6 cm.

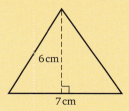

$$\text{Area} = \frac{1}{2} (\text{base} \times \text{height})$$
$$= \frac{1}{2} \times 7 \times 6 \, \text{cm}^2$$
$$= 21 \, \text{cm}^2$$

Find the areas of the following triangles.

1

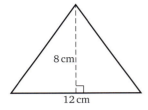

8 cm

12 cm

4

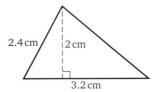

2.4 cm

2 cm

3.2 cm

2

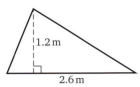

1.2 m

2.6 m

5

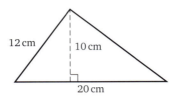

12 cm

10 cm

20 cm

3

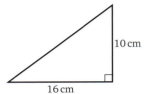

10 cm

16 cm

6

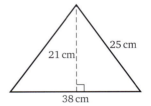

21 cm

25 cm

38 cm

7 In questions **4**, **5** and **6**, one of the given measurements is redundant. Which one is it?

8

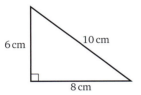

6 cm

10 cm

8 cm

12

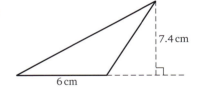

7.4 cm

6 cm

9

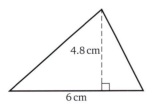

4.8 cm

6 cm

13

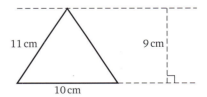

11 cm

9 cm

10 cm

10

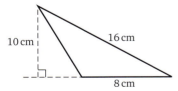

10 cm

16 cm

8 cm

Remember that you want the perpendicular height, that is the perpendicular distance from the base to the opposite vertex.

11

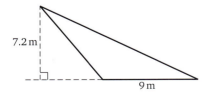

7.2 m

9 m

Find the area of the triangle.

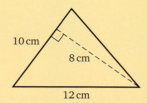

10 cm

8 cm

12 cm

(Look at this diagram from the direction of the arrow.)

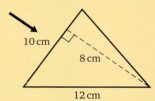

10 cm

8 cm

12 cm

$$\text{Area} = \tfrac{1}{2}(\text{base} \times \text{height})$$
$$= \tfrac{1}{2} \times 10 \times \overset{4}{8} \, \text{cm}^2$$
$$= 40 \, \text{cm}^2$$

If necessary turn the page round and look at the triangle from a different direction.

14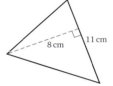
8 cm 11 cm

18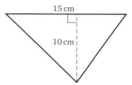
15 cm 10 cm

22
12 cm 5 cm 6 cm

15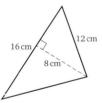
16 cm 12 cm 8 cm

19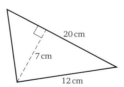
20 cm 7 cm 12 cm

23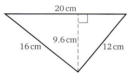
20 cm 16 cm 9.6 cm 12 cm

16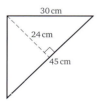
30 cm 24 cm 45 cm

20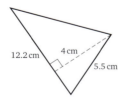
12.2 cm 4 cm 5.5 cm

24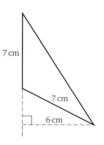
7 cm 7 cm 6 cm

17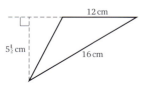
12 cm $5\tfrac{1}{2}$ cm 16 cm

21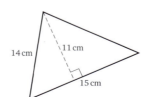
14 cm 11 cm 15 cm

25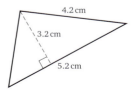
4.2 cm 3.2 cm 5.2 cm

For questions **26** to **31**, use squared paper to draw axes for x and y from 0 to 6 using 1 square to 1 unit. Find the area of each triangle.

26 △ABC with A(1, 0), B(6, 0) and C(4, 4)

27 △PQR with P(0, 2), Q(6, 0) and R(6, 4)

28 △DEF with D(1, 1), E(1, 5) and F(6, 0)

29 △LMN with L(5, 0), M(0, 6) and N(5, 6)

30 △ABC with A(0, 5), B(5, 5) and C(4, 1)

31 △PQR with P(2, 1), Q(2, 6) and R(5, 3)

Finding missing measurements

Exercise 15b

The area of a triangle is $20\,\text{cm}^2$. The height is $8\,\text{cm}$. Find the length of the base.

Let the base be $b\,\text{cm}$ long.

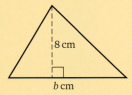

Using the formula for the area of a triangle, we substitute the values we know and b for the base.

$$\text{Area} = \tfrac{1}{2}\,(\text{base} \times \text{height})$$

$$20 = \tfrac{1}{2} \times b \times 8$$

Now we can solve this equation for b

$$20 = 4b$$

$$b = 5$$

The base is $5\,\text{cm}$ long.

Find the missing measurements of the following triangles.

	Area	Base	Width
1	24 cm²	6 cm	
2	30 cm²		10 cm
3	48 cm²		16 cm
4	10 cm²	10 cm	
5	36 cm²	24 cm	
6	108 cm²		6 cm

	Area	Base	Width
7	96 cm²		64 cm
8	4 cm²		3 cm
9	2 cm²	10 cm	
10	1.2 cm²	0.4 cm	
11	72 cm²		18 cm
12	1.28 cm²	0.64 cm	

2D compound shapes

Exercise 15c

ABCE is a square of side 8 cm. The total height of the shape is 12 cm.
Find the area of ABCDE.

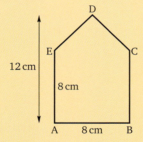

We can divide this shape into a square and a triangle. From the measurements given we see that the height of the triangle is 4 cm and the length of the base is 8 cm.

Area of △ECD = $\frac{1}{2}$ (base × height)

$\qquad = \frac{1}{2} \times 8 \times 4$ cm²

$\qquad = 16$ cm²

Area of ABCE = 8 × 8 cm²

$\qquad = 64$ cm²

Total area = 80 cm²

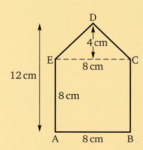

Find the areas of the following shapes.

1

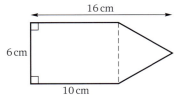

Remember to draw a diagram for each question and mark in all the measurements.

2

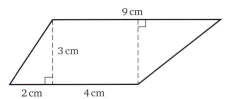

3

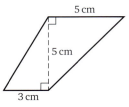

4

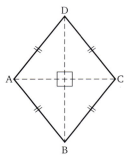

ABCD is a rhombus.

AC = 9 cm

BD = 12 cm

5

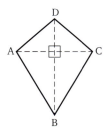

ABCD is a kite.

(A kite has two pairs of equal in length. BD is the axis of symmetry. The diagonals cut at right angles.)

AC = 10 cm   BD = 12 cm

6

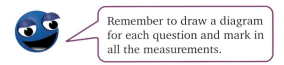

A square ABCD, of side 9 cm, has a triangle EAF cut off it.

7

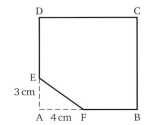

8

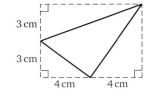

9 ABCD is a rhombus whose diagonals measure 7 cm and 11 cm.

10 ABCD is a kite whose diagonals measure 12 cm and 8 cm. (There are several possible kites you can draw with these measurements but their areas are all the same. See question **5** for the properties of a kite.)

Cubic units

The volume of a solid is the amount of space it occupies.

As with area, we need a convenient unit for measuring volume. The most suitable unit is a cube. A cube has six faces. Each face is a square.

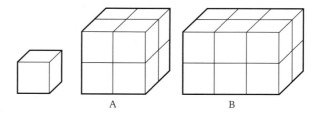

How many of the smallest cubes are needed to fill the same space as each of the solids A and B? Careful counting will show that 8 small cubes fill the same space as solid A and 12 small cubes fill the same space as solid B.

A cube with a side of 1 cm has a volume of one cubic centimetre, which is written 1 cm^3.

Similarly a cube with a side of 1 mm has a volume of 1 mm^3 and a cube with a side of 1 m has a volume of 1 m^3.

Volume of a cuboid

A cuboid is the mathematical name for a rectangular block. Each face of a cuboid is a rectangle.

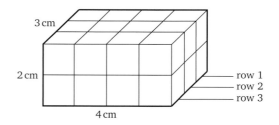

The diagram shows a rectangular block or cuboid measuring 4 cm by 3 cm by 2 cm. To cover the area on which the block stands we need three rows of cubes measuring 1 cm by 1 cm by 1 cm, with four cubes in each row, i.e. 12 cubes. A second layer of 12 cubes is needed to give the volume shown, so the volume of the block is 24 cubes.

But the volume of one cube is 1 cm^3.

Therefore the volume of the solid is 24 cm^3.

This is also given when we calculate length × breadth × height,

i.e.   the volume of the cuboid $= 4 \times 3 \times 2 \text{ cm}^3$

the volume of any cuboid = length × breadth × height

Find the volume of a cuboid measuring 12 cm by 10 cm by 5 cm.

$$\text{Volume of cuboid} = \text{length} \times \text{breadth} \times \text{height}$$

$$= 12 \times 10 \times 5 \text{ cm}^3$$

i.e. $$\text{Volume} = 600 \text{ cm}^3$$

Find the volume of each of the following cuboids:

	Length	Breadth	Height		Length	Breadth	Height
1	4 cm	4 cm	3 cm	**7**	4 m	3 m	2 m
2	20 mm	10 mm	8 mm	**8**	8 m	5 m	4 m
3	45 mm	20 mm	6 mm	**9**	8 cm	3 cm	$\frac{1}{2}$ cm
4	5 mm	4 mm	0.8 mm	**10**	12 cm	1.2 cm	0.5 cm
5	6.1 m	4 m	1.3 m	**11**	4.5 m	1.2 m	0.8 m
6	3.5 cm	2.5 cm	1.2 cm	**12**	1.2 m	0.9 m	0.7 m

Find the volume of a cube with the given side:

13	4 cm	**16**	$\frac{1}{2}$ cm	**19**	8 km
14	5 cm	**17**	2.5 cm	**20**	$1\frac{1}{2}$ km
15	2 m	**18**	3 km	**21**	3.4 m

The edges of a cube are all the same length so the volume is $(4 \times 4 \times 4)$ cm³.

? **Puzzle**

Which two of these shapes will fit together to form a cube?

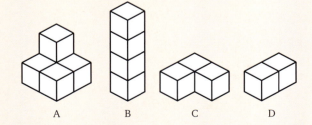

A   B   C   D

Exercise 15e

Draw a cube of side 8 cm. How many cubes of side 2 cm would be needed to fill the same space?

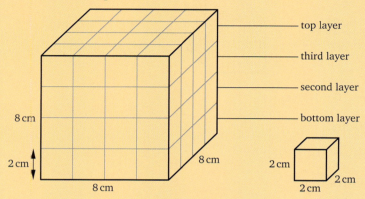

top layer

third layer

second layer

bottom layer

8 cm

2 cm

8 cm

8 cm

2 cm

2 cm

2 cm

The bottom layer requires 4×4, i.e. 16 cubes of side 2 cm, and there are four layers altogether.

Therefore 64 cubes are required.

1   Draw a cube of side 4 cm. How many cubes of side 2 cm would be needed to fill the same space?

2   Draw a cuboid measuring 6 cm by 4 cm by 2 cm. How many cubes of side 2 cm would be needed to fill the same space?

3   Draw a cube of side 6 cm. How many cubes of side 3 cm would be needed to fill the same space?

4   Draw a cuboid measuring 8 cm by 6 cm by 2 cm. How many cubes of side 2 cm would be needed to fill the same space?

? Puzzle

The large cube in the worked example is painted red.

How many of the 64 small cubes are unpainted.

Changing units of volume

Consider a cube of side 1 cm. If each edge is divided into 10 mm the cube can be divided into 10 layers, each layer with 10×10 cubes of side 1 mm.

100 cubes, each with a volume of $1\,mm^3$, in every one of these layers

i.e. $\qquad 1\,cm^3 = 10 \times 10 \times 10\,mm^3$

Similarly, since 1 m = 100 cm

1 cubic metre = $100 \times 100 \times 100\,cm^3$

i.e. $\qquad 1\,m^3 = 1\,000\,000\,cm^3$

Exercise 15f

Express $2.4\,m^3$ in   **a** cm^3 **b** mm^3.

a   Since $1\,m^3 = 100 \times 100 \times 100\,cm^3$

$\qquad 2.4\,m^3 = 2.4 \times 100 \times 100 \times 100\,cm^3$

$\qquad\qquad = 2\,400\,000\,cm^3$

b   Since $1\,m^3 = 1000 \times 1000 \times 1000\,mm^3$

$\qquad 2.4\,m^3 = 2.4 \times 1000 \times 1000 \times 1000\,mm^3$

$\qquad\qquad = 2\,400\,000\,000\,mm^3$

1   Which metric unit would you use to measure the volume of
     a   a room   **b**   a teaspoon   **c**   a can of cola?

Express in mm^3:

2 $8\,cm^3$ **3** $14\,cm^3$ **4** $6.2\,cm^3$ **5** $0.43\,cm^3$ **6** $0.092\,m^3$ **7** $0.04\,cm^3$

Express in cm^3:

8 $3\,m^3$ **10** $0.42\,m^3$ **12** $22\,mm^3$

9 $2.5\,m^3$ **11** $0.0063\,m^3$ **13** $731\,mm^3$

Capacity

When we buy a bottle of milk or a can of engine oil we are not usually interested in the external measurements or volume of the container. What really concerns us is the *capacity* of the container, i.e. how much milk can the bottle hold, or how much engine oil is inside the can.

The most common unit of capacity in the metric system is the *litre*. (A litre is usually the size of a large bottle of water.) A litre is much larger than a cubic centimetre but much smaller than a cubic metre. The relationship between these quantities is:

$$1000\,cm^3 = 1\ litre$$

i.e.   a litre is the volume of a cube of side 10 cm

and $1000\ litres = 1\,m^3$

When the amount of liquid is small, such as dosages for medicines, the millilitre (ml) is used. A millilitre is a thousandth part of a litre, i.e.

$$1000\,ml = 1\ litre \quad or \quad 1\,ml = 1\,cm^3$$

Exercise 15g

Express 5.6 litres in cm^3.

$$1\ litre = 1000\,cm^3$$

so $5.6\ litres = 5.6 \times 1000\,cm^3$

$$= 5600\,cm^3$$

Express in cm^3:

1	2.5 litres	**3**	0.54 litre	**5**	35 litres
2	1.76 litres	**4**	0.0075 litres	**6**	0.028 litres

Express in litres:

7	7000 cm^3	**8**	4000 cm^3	**9**	24 000 cm^3	**10**	600 cm^3

Express in litres:

11	5 m^3	**12**	12 m^3	**13**	4.6 m^3	**14**	0.067 m^3

Volumes involving a change of units

Sometimes the dimensions of a cuboid are given in more than one unit or the volume is asked for in a different unit from those given for the dimensions.

For example, if the length and width are given in centimetres while the height is given in millimetres we must express the height in centimetres to be able to give the volume in cubic centimetres.

Exercise 15h

Find the volume, in cm³, of a cuboid of length 9 cm, width 6 cm and height 35 mm.

Changing the height to millimetres, 35 mm = 35 ÷ 10 cm = 3.5 cm

$$\text{Volume of cuboid} = \text{length} \times \text{width} \times \text{height}$$

$$= 9 \times 6 \times 3.5 \text{ cm}^3$$

∴ volume of cuboid = 189 cm³

1 Find the volume, in cm³, of a cuboid of length 8 cm, width 50 mm and height 4 cm.

Remember that all measurements must be in the same units before they are multiplied together. Give the units in your answer.

2 Find the volume, in cm³, of a cuboid of length 5 cm, width 35 mm and height 20 mm.

3 Find the volume, in m³, of a cuboid of length 300 cm, width 1.5 m and height 0.5 m.

Find the volumes of the following cuboids, changing the units first if necessary. Do *not* draw a diagram.

	Length	Width	Height	Volume units
4	3 cm	5 mm	10 mm	mm³
5	1.4 cm	0.9 cm	0.32 cm	mm³
6	1.2 cm	0.4 cm	1 cm	mm³
7	140 cm	1 m	80 cm	m³
8	540 cm	180 cm	0.2 m	m³
9	1 m	30 cm	12 cm	cm³
10	4 cm	3.5 cm	12 cm	cm³
11	4.2 cm	30 mm	0.15 cm	cm³
12	0.48 m	3.2 m	15 cm	cm³

! Investigation

A rectangular piece of steel sheet measuring 20 cm by 14 cm is to be used to make an open rectangular box. The diagram shows one way of doing this. The four corner squares are removed and the sides folded up. The four vertical seams at the corners are then sealed.

a Copy and complete the following table, which gives the measurements of the base and the capacity of the box when squares of different sizes are removed from the corners. Continue to add numbers to the first column in the table as long as it is reasonable to do so.

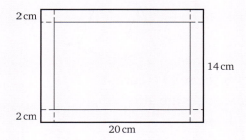

These numbers should follow the pattern indicated.

Length of edge of square (cm)	Measurements of base (cm)	Capacity of box (cm³)
0.5	13×19	$0.5 \times 13 \times 19 = 123.5$
1	12×18	$1 \times 12 \times 18 = 216$
1.5		
2		
2.5		
3		

b What is the last number you entered in the first column of the table? Justify your choice.

c What size of square should be removed to give the largest capacity recorded in the table?

d Investigate whether you can find a number that you have not already entered in the first column that gives a larger capacity than any value you have found so far.

Volumes of solids with uniform cross-sections

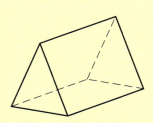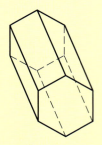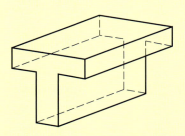

When we cut through any one of the solids above, parallel to the ends, we always get the same shape as the end. This shape is called the cross-section.

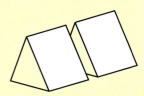

As the cross-section is the same shape and size wherever the solid is cut, the cross-section is said to be *uniform* or *constant*. These solids are also called *prisms* and we can find the volumes of some of them.

First consider a cuboid (which can also be thought of as a rectangular prism).

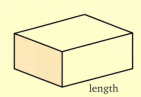

Volume = length × width × height

= (width × height) × length

= area of shaded end × length

= area of cross-section × length

Now consider a triangular prism. If we enclose it in a cuboid we can see that its volume is half the volume of the cuboid.

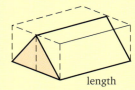

Volume = ($\frac{1}{2}$ × width × height) × length

= area of shaded triangle × length

= area of cross-section × length

This is true of any prism so that

Volume of a prism = area of cross-section × length

Exercise 15i

Find the volume of the solid below.

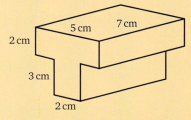

To find the volume you need first to find the area of the cross-section.

Draw the cross-section, then divide it into two rectangles.

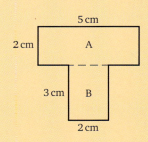

Area of A $= 2 \times 5 \, cm^2 = 10 \, cm^2$

Area of B $= 2 \times 3 \, cm^2 = 6 \, cm^2$

Area of cross-section $= 16 \, cm^2$

Volume $=$ area $\times$ length

$= 16 \times 7 \, cm^3$

$= 112 \, cm^3$

Find the volumes of the following prisms. Draw a diagram of the cross-section but do *not* draw a picture of the solid.

1

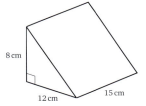

3

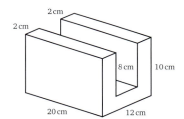

5

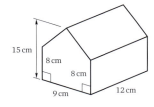

2

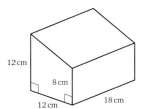

4

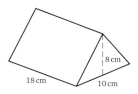

6

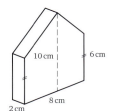

Be careful that you get the right cross-section.

The following two solids are standing on their ends so the vertical measurement is the length.

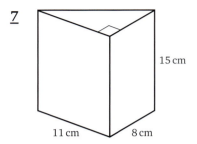

7

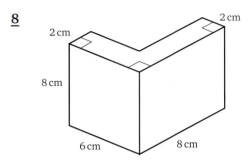

8

In questions **9** to **16**, the cross-sections of the prisms and their lengths are given. Find their volumes.

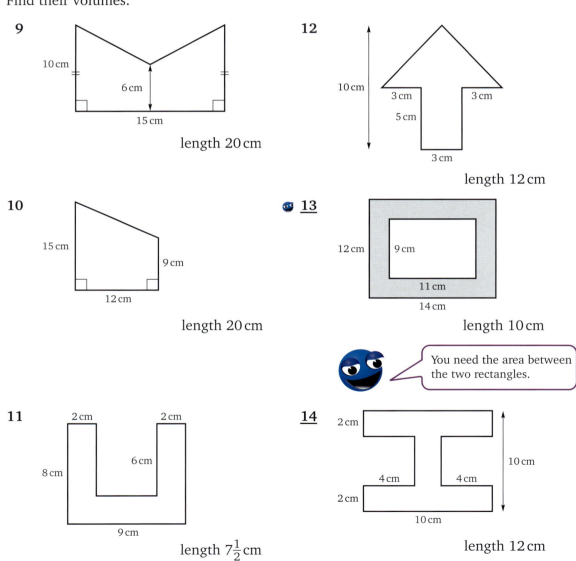

9

length 20 cm

12

length 12 cm

10

length 20 cm

13

length 10 cm

You need the area between the two rectangles.

11

length $7\frac{1}{2}$ cm

14

length 12 cm

15

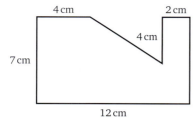

length 12 cm

16

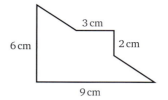

length 24 cm

17 A tent is in the shape of a triangular prism. Its length is 2.4 m, its height 1.8 m and the width of the triangular end is 2.4 m. Find the volume enclosed by the tent.

18

The area of the cross-section of the given solid is 42 cm² and the length is 32 cm.

Find its volume.

19

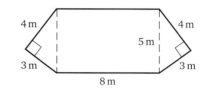

A solid of uniform cross-section is 12 m long. Its cross-section is shown in the diagram.

Find its volume.

Volume of a cylinder

Reminder: The circumference of a circle is given by $C = 2\pi r$ and the area of a circle by $A = \pi r^2$

A cylinder can be thought of as a circular prism so its volume can be found using

$$\text{volume} = \text{area of cross-section} \times \text{length}$$

$$= \text{area of circular end} \times \text{length}$$

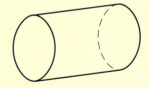

From this we can find a formula for the volume.

We usually think of a cylinder as standing upright so that its length is represented by h (for height).

If the radius of the end circle is r, then the area of the cross-section is πr^2

$\therefore$ $\text{volume} = \pi r^2 \times h$

$$= \pi r^2 h$$

Exercise 15j

Find the volume inside a cylindrical mug of internal diameter 8 cm and height 6 cm. Use the π button on your calculator.

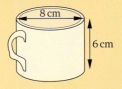

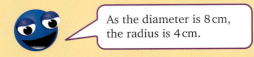

As the diameter is 8 cm, the radius is 4 cm.

The volume is given by the formula

$V = \pi r^2 h$

$= \pi \times 4 \times 4 \times 6 = 301.59\ldots$

Therefore volume of mug is 302 cm² (correct to the nearest cm²).

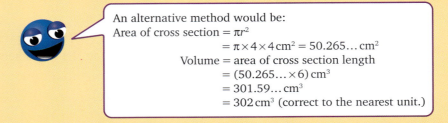

An alternative method would be:
Area of cross section $= \pi r^2$
$= \pi \times 4 \times 4\,\text{cm}^2 = 50.265\ldots\,\text{cm}^2$
Volume $=$ area of cross section length
$= (50.265\ldots \times 6)\,\text{cm}^3$
$= 301.59\ldots\,\text{cm}^3$
$= 302\,\text{cm}^3$ (correct to the nearest unit.)

Use the value of π on your calculator and give all your answers correct to the place value in brackets.

(1) means to the nearest unit.

Find the volumes of the following cylinders:

1 Radius 2 cm, height 10 cm (1)

2 Radius 3 cm, height 4 cm (1)

3 Radius 5 cm, height 4 cm (1)

4 Radius 3 cm, height 2.1 cm (1 d.p)

5 Diameter 2 cm, height 1 cm (2 d.p)

6 Radius 1 cm, height 4.8 cm (1 d.p)

7 Diameter 4 cm, height 3 cm (1 d.p)

8 Diameter 6 cm, height 1.8 cm (1 d.p)

9 Radius 12 cm, height 10 cm (10)

10 Radius 7 cm, height 9 cm (10)

11 Radius 3.2 cm, height 10 cm (1)

12 Radius 6 cm, height 3.6 cm (1)

13 Diameter 10 cm, height 4.2 cm (1)

14 Radius 7.2 cm, height 4 cm (1)

15 Diameter 64 cm, height 22 cm (100)

16 Diameter 2.4 cm, height 6.2 cm (10)

17 Radius 4.8 mm, height 13 mm (1)

18 Diameter 16.2 cm, height 4 cm (1)

19 Radius 76 cm, height 88 cm (1)

20 Diameter 0.02 m, height 0.14 m (1 d.p.)

3D compound shapes

Exercise 15k

Find the volumes of the following solids. Use the value of π on your calculator and give your answers correct to $1\,cm^3$.

Draw diagrams of the cross-sections but do *not* draw pictures of the solids.

Find the area of the cross-section first.

1

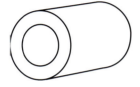

A tube of length 20 cm. The inner radius is 3 cm and the outer radius is 5 cm.

2

A half-cylinder of length 16 cm and radius 4 cm.

3

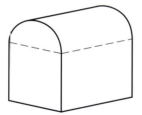

A solid of length 6.2 cm, whose cross-section consists of a square of side 2 cm surmounted by a semicircle.

4

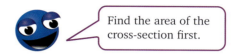

A disc of radius 9 cm and thickness 0.8 cm.

5

A solid made of two cylinders each of height 5 cm. The radius of the smaller one is 2 cm and of the larger one is 6 cm.

6

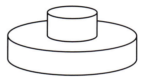

A solid made of two half-cylinders each of length 11 cm. The radius of the larger one is 10 cm and the radius of the smaller one is 5 cm.

? Puzzle

How many cubes can you see?

Making solids from nets

To make a solid object from a sheet of flat paper you need to construct a *net*: this is the shape that has to be cut out, folded and stuck together to make the solid. A net should be drawn as accurately as possible, otherwise you will find that the edges will not fit together properly.

Exercise 15I

Each solid in this exercise has flat faces (called *plane* faces) and is called a polyhedron. 'Poly' is a prefix used quite often; it means 'many'.

1 **Cube**

This net will make a cube.

a Which edge meets AB?

b Which other corners meet at H?

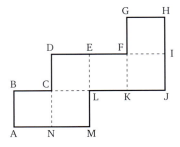

2 **Tetrahedron**

This net consists of four equilateral triangles. Draw the net accurately, making the sides of each triangle 6 cm long. Start by drawing one triangle of side 12 cm; mark the midpoints of the sides and join them up. Draw flaps on the edges shown. (These are not part of the net.)

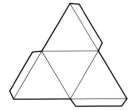

Cut out the net. Score the solid lines (use a ruler and ballpoint pen – an empty one is best) and fold the outer triangles up so that their vertices meet. Use the flaps to stick the edges together.

This solid is called a *regular* tetrahedron. A regular solid is one in which all the faces are identical. These make good Christmas tree decorations if painted or if made out of foil-covered paper.

3 **Octahedron**

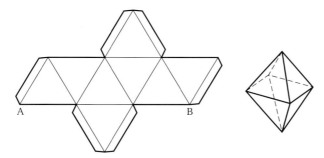

This net consists of equilateral triangles: make the sides of each triangle 4 cm long, and start by making AB 12 cm long. Is this octahedron a regular solid?

4 **Square-based pyramid**

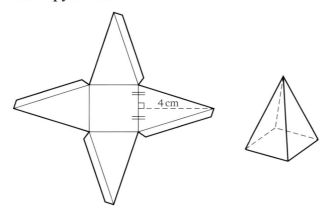

This net consists of a square with an isosceles triangle on each side of the square. Make the sides of the square 3 cm and the heights of the triangles 4 cm. Is this a regular solid?

5 **Prism with triangular cross-section**

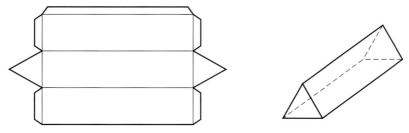

This net consists of three rectangles, each 8 cm long and 4 cm wide, and two equilateral triangles (sides 4 cm).

6 **Eight-pointed star (stella octangula)**

This model needs time and patience. If you have both it is worth the effort!

It consists of a regular octahedron (see question **3**) with a regular tetrahedron (see question **2**) stuck on each face.

You will need 8 tetrahedra. In all the nets make the triangles have sides of length 4 cm.

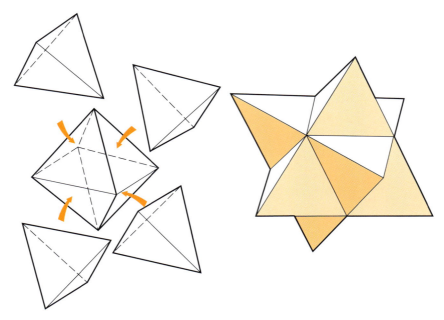

? Puzzle

A set of three encyclopaedias is placed in the normal way on a shelf.

A bookworm takes $\frac{1}{4}$ day to eat through a cover and $\frac{1}{2}$ day to eat through the pages of one book.

How long will it take this bookworm to eat its way from the first page of Volume 1 to the last page of Volume 3?

In this chapter you have seen that...

✔ the area of a triangle is equal to $\frac{1}{2}$ base × perpendicular height

✔ the volume of a cuboid is found by multiplying its length by its width by its height, i.e. $V = l \times w \times h$

✔ any solid with a uniform cross-section is called a prism and its volume is equal to the area of the cross-section multiplied by its length

✔ in particular the volume of a cylinder is given by $V = \pi r^2 h$

✔ a net is a flat shape that can be folder to made a solid.

16 Squares and square roots

At the end of this chapter you should be able to...

1 Find the square of a given number.

2 Find a rough estimate of the square of a number.

3 Find the square of a number using a calculator.

4 Complete a table of given numbers and their squares.

5 Use values in (4) to draw the graph of $y = x^2$.

6 Use the graph of $y = x^2$ to find the squares of numbers.

7 Find, by inspection, the square root of a number.

8 Find rough estimates of square roots.

9 Use a calculator to find square roots of numbers to a given degree of accuracy.

You need to know...

✔ how to multiply decimals

✔ how to draw and scale x- and y-axes and plot points

✔ how to find the area of a square

✔ the meaning of the first significant figure

Key words

significant figures, square, square root, the symbol ($\sqrt{}$)

Squares

We obtain the *square* of a number when we multiply the number by itself.

Exercise 16a

The square of 3 means 3×3.

Find the square of:

1	3	**3**	9	**5**	0.4	**7**	300
2	5	**4**	30	**6**	50	**8**	0.02

Take care with the decimal point.

9	500	**11**	0.3	**13**	0.004	**15**	0.03	
10	10	**12**	2000	**14**	1			

Write 32 correct to the first significant figure and use this to give a
rough estimate of the square of 32.

$$32 \approx 30$$
$$32^2 \approx 30 \times 30 = 900$$

In questions **16** to **27**, give each number correct to the first significant figure
then use this to give a rough estimate of the square of the number.

16	28	**19**	0.27	**22**	1212	**25**	87	
17	99	**20**	7.9	**23**	73	**26**	0.081	
18	4.2	**21**	37.2	**24**	0.0312	**27**	249	

Finding squares

Using a calculator

Enter the number to be squared and press the 'square' button, which is usually labelled x^2. If
there is no 'square' button, then multiply the number by itself.

Use your rough estimate to check your calculator answer is reasonable.

Exercise 16b

Find the square of:

1	7.8	**9**	51.3					
2	38	**10**	9.8					
3	79.2	**11**	12.1					
4	0.41	**12**	2.94	**17**	0.824	**22**	0.24	
5	0.16	**13**	1.02	**18**	0.879	**23**	0.072	
6	0.032	**14**	13.6	**19**	0.036	**24**	14	
7	48.2	**15**	17	**20**	72.4	**25**	140	
8	11.3	**16**	241	**21**	3.78	**26**	0.14	

Find a rough estimate first.

27 **a** Copy and complete the following table:

x	0	0.5	1	1.5	2	2.5	3	3.5	4
x^2	0			2.25	4				

b Draw axes for x from 0 to 4 using 2 cm to 1 unit and for y from 0 to 16 using 1 cm to 1 unit. Use the table to draw the graph of $y = x^2$.

c From the graph, find the values of y when $x = 2.2$, 1.8, 3.1 and 2.7.

d Use a calculator to find 2.2^2, 1.8^2, 3.1^2 and 2.7^2. How do these answers compare with your answers to part **c**?

e Repeat parts **c** and **d** with other values of your own choice.

28 **a** Copy and complete the following table:

x	2	4	6	8	10	12	14	15
x^2	4		36		100			225

b Draw axes for x from 0 to 15 using 1 cm ≡ 1 unit and for y from 0 to 240 using 1 cm ≡ 10 units. Use your table to draw the graph of $y = x^2$.

c From the graph, find the values of y when $x = 5.5$, 8.4, 12.8 and 13.6.

d Use a calculator to find 5.5^2, 8.4^2, 12.8^2 and 13.6^2. How do these answers compare with your answers to part **c**?

 Investigation

Is there a shortcut for finding squares of some two-digit numbers?

$$25^2 = (2 \times 3)100 + 25 = 625$$
$$35^2 = (3 \times 4)100 + 25 = 1225$$

Do you see a pattern?

Use the above method to find the following:

$$55^2, 65^2, 75^2, 85^2.$$

Areas of squares

Exercise 16c

Find the area of a square of side 7.2 m.

Area $= (7.2 \times 7.2)$ m²

$\approx (7 \times 7)$ m² $= 49$ m²

Area $= 51.84$ m²

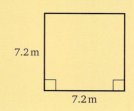

Find the areas of the squares whose sides are given in questions **1** to **9**.

1	2.4 cm	**4**	1.06 m	**7**	0.062 m
2	9.6 m	**5**	17.2 cm	**8**	320 km
3	32.4 cm	**6**	52 mm	**9**	0.31 cm

The area of a square is found by multiplying the length of a side by itself, i.e. by squaring the length of a side.

Square roots

The square root of a number is the number that, when multiplied by itself, gives the original number,

e.g. because $4^2 = 16$, the square root of 16 is 4.

The square root could also be -4 because $(-4) \times (-4) = 16$ but we will deal only with positive square roots in this chapter.

The symbol for the positive square root is $\sqrt{}$ so $\sqrt{16} = 4$

Exercise 16d

Find the square roots in questions **1** to **18** without using a calculator.

Check your answer by squaring it.

1	$\sqrt{9}$	**8**	$\sqrt{64}$
2	$\sqrt{25}$	**9**	$\sqrt{1}$
3	$\sqrt{4}$	**10**	$\sqrt{8100}$

4	$\sqrt{81}$	**11**	$\sqrt{0.81}$	**15**	$\sqrt{0.04}$
5	$\sqrt{100}$	**12**	$\sqrt{0.64}$	**16**	$\sqrt{400}$
6	$\sqrt{36}$	**13**	$\sqrt{4900}$	**17**	$\sqrt{2500}$
7	$\sqrt{49}$	**14**	$\sqrt{490\,000}$	**18**	$\sqrt{10\,000}$

Use the answers to Exercise 16a, questions **1** to **15**, to find the following square roots.

19	$\sqrt{0.09}$	**21**	$\sqrt{0.0004}$	**23**	$\sqrt{4\,000\,000}$
20	$\sqrt{0.16}$	**22**	$\sqrt{250\,000}$	**24**	$\sqrt{0.000\,016}$

Check by squaring your answer.

Rough estimates of square roots

So far, we have been able to find exact square roots of the numbers we have been given. Most numbers, however, do not have exact square roots; $\sqrt{23}$, for example, lies between 4 and 5 because $4 \times 4 = 16$, and $5 \times 5 = 25$.

$\sqrt{23}$, if given as a decimal, will start with 4.

i.e. $\qquad \sqrt{23} = 4.\text{–} \text{–} \text{–}$

Exercise 16e

Find the first significant figure of the square root of 30.

$$\sqrt{30} = 5.---$$

($\sqrt{30}$ lies between 5 and 6 because $5^2 = 25$ and $6^2 = 36$)

Find the first significant figure of the square roots of the following numbers:

1	17	4	40	7	85	10	0.05	13	14.2
2	10	5	3	8	15	11	0.20	14	0.50
3	38	6	10.2	9	4.6	12	90	15	5.7

Notice that $\sqrt{3} = 1.---$ While $\sqrt{30} = 5.---$
and that $\sqrt{300} = 1-.---$ While $\sqrt{3000} = 5-.---$

Every *pair of figures* added to the original number adds *one* figure to the approximate square root. We can pair off the figures from the decimal point, i.e. $\sqrt{3|00|00}$. Looking at the figure or figures in front of the first dividing line we can find the first significant figure of the square root.

Then $\sqrt{3|00|00}. = 1--.--$
≈ 100 (*Check*: $100 \times 100 = 10\,000$)

And $\sqrt{30|00|00}. = 5--.--$
≈ 500 (*Check*: $500 \times 500 = 250\,000$)

Exercise 16f

Find a rough value for the square root of 5280.

$7^2 = 49$ and $8^2 = 64$ so $\sqrt{52}$ is between 7 and 8

$$\sqrt{52|80}. = 7-.--$$
$$\approx 70$$

(*Check*: $70 \times 70 = 4900$)

By finding the first significant figure of the square root, give a rough value for the square root of each of the following numbers:

1	1400	5	720	9	4160	13	756	17	729.4
2	62 300	6	14 000	10	14 860	14	75 600	18	15.26
3	623	7	3260	11	396 000	15	7 560 000	19	3.698
4	7200	8	41 600	12	396	16	4128	20	39.46

Finding square roots

Using a calculator

Press the square root button, which is usually labelled $\sqrt{x}$, then enter the number, say 5280. You will usually get a number that fills the display.

$\sqrt{5280} = 72.66$ correct to 2 decimal places.

Check that it agrees with your rough estimate.

Exercise 16g

Find the square roots of the following numbers correct to 1 d.p. Give a rough estimate first in each case.

1	38.4	**5**	32	**9**	650	**13**	24	**17**	728
2	19.8	**6**	9.8	**10**	65	**14**	19	**18**	7280
3	428	**7**	67	**11**	11.2	**15**	10 300	**19**	61
4	4230	**8**	5.7	**12**	58	**16**	412 000	**20**	7 280 000

Rough estimates of square roots of numbers less than 1

$$0.2 \times 0.2 = 0.04 \qquad \text{so} \qquad \sqrt{0.04} = 0.2$$

and $\qquad \sqrt{0.05} = 0.2 ---$ also $\qquad \sqrt{0.0004} = 0.02$

So $\qquad \sqrt{0.0005} = 0.02 ---$ but $\qquad \sqrt{0.004}$ is neither 0.2 nor 0.02

It is easiest to find a rough estimate of the square root by again pairing off from the decimal point, but this time going to the right instead of to the left: $\sqrt{0.\overset{|}{\,}00\overset{|}{\,}40}$, adding a zero to complete the pair.

Now $\sqrt{40} = 6.---$ so we see that $\sqrt{0.004} = 0.06 ---$

(*Check*: $0.06 \times 0.06 = 0.0036 \approx 0.004$)

Using a calculator or tables we find
$$\sqrt{0.004} = 0.0632 \text{ correct to 4 d.p.}$$

Note that each pair of zeros after the decimal point gives one zero after the decimal point in the answer.

Exercise 16h

Find the square roots of 0.007 32 and 0.000 732 correct to 4 d.p.

$\sqrt{0.\mathbf{|}00\mathbf{|}73\mathbf{|}2} = 0.08 - - -$

$\sqrt{0.007\,32} = 0.0856$ correct to 4 d.p.

$\sqrt{0.\mathbf{|}00\mathbf{|}07\mathbf{|}32} = 0.02 - - -$

$\sqrt{0.000\,732} = 0.0271$ correct to 4 d.p.

Find a rough estimate (as far as the first significant figure) and then use your calculator to find the square root of each of the following numbers to 3 d.p.

1	0.042	8	0.278	15	0.0432
2	0.42	9	0.0278	16	0.009 61
3	0.014	10	0.002 78	17	0.832
4	0.56	11	0.3	18	0.32
5	0.000 14	12	0.173	19	0.052
6	0.5	13	0.2	20	0.75
7	0.6014	14	0.69	21	0.000 073

Exercise 16i

Find the side of the square whose area is 50 m².

Length of the side = $\sqrt{50}$ m

                    = 7.– – – m

Length of the side = 7.07 m correct to 2 d.p.

50 m²

Find the sides of the squares whose areas are given below. Give your answers correct to 2 d.p.

1	85 cm²	5	0.06 m²	9	0.0085 km²
2	120 cm²	6	15.1 cm²	10	59 cm²
3	500 m²	7	749 mm²	11	241 m²
4	32 m²	8	84 300 km²	12	61 cm²

Puzzle

An explorer leaves his tent and walks 1 km south, he then walks 2 km due east and finally 1 km north back to his tent. Where is his tent?

Did you know?

The Pythagoreans believed that everything could be explained in terms of whole numbers? When they discovered that $\sqrt{2}$ could not be written as a ratio of whole numbers (i.e. a fraction), they tried to keep it secret. $\sqrt{2}$ is an irrational number.

In this chapter you have seen that...

✔ the square of a number bigger than 1 is bigger than the original number whereas the square of a number less than one is smaller than the original number

✔ you can find the first significant figure of a square root by pairing the numbers out from the decimal point and estimating the square root of the first pair

✔ the square root of a number bigger than 1 is less than the number

✔ the square root of a number smaller than one is bigger than the number.

17 Travel graphs

At the end of this chapter you should be able to...

1 Draw distance–time graphs using suitable scales.

2 Calculate the distance travelled in a given time, by an object moving at a constant speed.

3 Calculate the time taken to travel a given distance at a constant speed.

4 Calculate the average speed of a body, given the distance travelled and time taken.

5 Calculate the average speed for a body covering different distances at different speeds.

6 Read information from a distance–time graph.

Did you know?

The measure of distance used in air and sea navigation is the nautical mile.

One nautical mile is slightly less than 2 kilometres. Speeds of nautical miles per hour are called knots.

You need to know...

✔ how to work with fractions and decimals

✔ how to read and draw graphs

✔ metric and imperial units of distance

✔ units of time, including the 12-hour and 24-hour clock, and how to convert between them

Key words

average speed, constant speed, distance, knot, speed, the symbol ≡

Finding distance from a graph

When we went on holiday we travelled by car to our holiday resort at a steady speed of 30 kilometres per hour (km/h), i.e. in each hour we covered a distance of 30 km.

This graph shows our journey. It plots distance against time and shows that

in 1 hour we travelled 30 km

in 2 hours we travelled 60 km

in 3 hours we travelled 90 km

in 4 hours we travelled 120 km

in 5 hours we travelled 150 km.

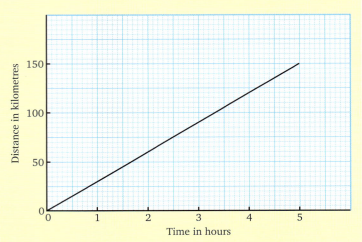

Exercise 17a

The graphs that follow show four different journeys.

For each journey find:

a   the distance travelled

b   the time taken

c   the distance travelled in 1 hour.

Make sure you understand what the subdivisions on the scales represent.

1

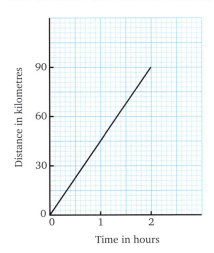

2

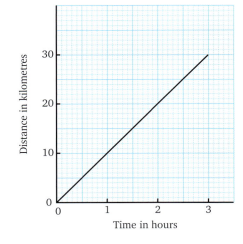

3

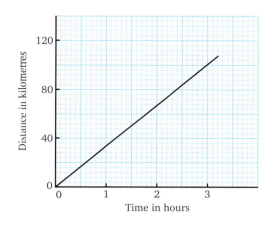

4

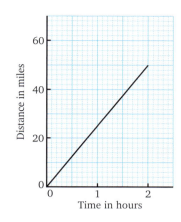

Drawing travel graphs

If Peter walks at 6 km/h, we can draw a graph to show this, using 2 cm to represent 6 km on the distance axis and 2 cm to represent 1 hour on the time axis.

Plot the point that shows in 1 hour he has travelled 6 km. Join the origin to this point and extend the straight line to give the graph shown. From this graph we can see that in 2 hours Peter travels 12 km and in 5 hours he travels 30 km.

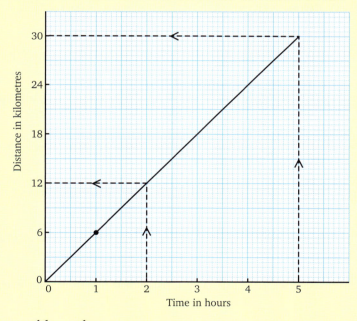

Alternatively we could say that

If he walks 6 km in 1 hour

he will walk 6×2 km $= 12$ km in 2 hours

and he will walk 6×5 km $= 30$ km in 5 hours

The distance walked is found by multiplying the speed by the time,

i.e.   distance $=$ speed $\times$ time

Exercise 17b

Draw a travel graph to show a journey of 150 km in 3 hours. Plot distance along the vertical axis and time along the horizontal axis.

Let 4 cm represent 1 hour and 2 cm represent 50 km.

Draw a line from 0 to the point above 3 on the time axis and along from 150 on the distance axis.

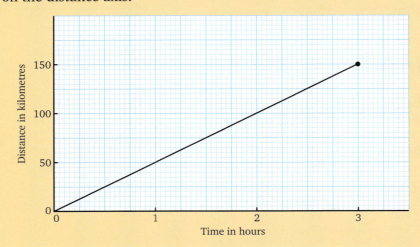

Draw travel graphs to show the following journeys. Plot distance along the vertical axis and time along the horizontal axis. Use the scales given in brackets. ('≡' means 'is equivalent to')

1 60 km in 2 hours
(4 cm ≡ 1 hour, 1 cm ≡ 10 km)

2 180 km in 3 hours
(4 cm ≡ 1 hour, 2 cm ≡ 50 km)

3 300 km in 6 hours
(1 cm ≡ 1 hour, 1 cm ≡ 50 km)

4 80 miles in 2 hours
(6 cm ≡ 1 hour, 1 cm ≡ 10 miles)

5 140 miles in 4 hours
(2 cm ≡ 1 hour, 1 cm ≡ 25 miles)

6 100 km in $2\frac{1}{2}$ hours
(2 cm ≡ 1 hour, 2 cm ≡ 25 km)

7 105 km in $3\frac{1}{2}$ hours
(2 cm ≡ 1 hour, 4 cm ≡ 50 km)

8 75 miles in $1\frac{1}{4}$ hours
(8 cm ≡ 1 hour, 2 cm ≡ 25 miles)

9 90 m in 5 sec
(2 cm ≡ 1 sec, 2 cm ≡ 10 m)

10 240 m in 12 sec
(1 cm ≡ 1 sec, 2 cm ≡ 50 m)

11 Alan walks at 5 km/h. Draw a graph to show him walking for 3 hours. Take 4 cm to represent 5 km and 4 cm to represent 1 hour. Use your graph to find how far he walks in **a** $1\frac{1}{2}$ hours **b** $2\frac{1}{4}$ hours.

12   Julie can jog at 10 km/h. Draw a graph to show her jogging for 2 hours. Take 1 cm to represent 2 km and 8 cm to represent 1 hour. Use your graph to find how far she jogs in   **a** $\frac{3}{4}$ hour   **b** $1\frac{1}{4}$ hours.

13   Jo drives at 35 m.p.h. Draw a graph to show her driving for 4 hours. Take 1 cm to represent 10 miles and 4 cm to represent 1 hour. Use your graph to find how far she drives in   **a** 3 hours   **b** $1\frac{1}{4}$ hours.

14   John walks at 4 m.p.h. Draw a graph to show him walking for 3 hours. Take 1 cm to represent 1 m.p.h. and 4 cm to represent 1 hour. Use your graph to find how far he walks in   **a** $\frac{1}{2}$ hour   **b** $3\frac{1}{2}$ hours.

The remaining questions should be solved by calculation.

15   An express train travels at 200 km/h. How far will it travel in
a 4 hours   **b** $5\frac{1}{2}$ hours?

16   Ken cycles at 24 km/h. How far will he travel in
a 2 hours   **b** $3\frac{1}{2}$ hours   **c** $2\frac{1}{4}$ hours?

17   An aeroplane flies at 300 m.p.h. How far will it travel in
a 4 hours   **b** $5\frac{1}{2}$ hours?

18   A bus travels at 60 km/h. How far will it travel in
a $1\frac{1}{2}$ hours   **b** $2\frac{1}{4}$ hours?

19   Susan can cycle at 12 m.p.h. How far will she ride in
a $\frac{3}{4}$ hour   **b** $1\frac{1}{4}$ hours?

20   An athlete can run at 10.5 m/s. How far will he travel in
a 5 sec   **b** 8.5 sec?

21   A boy cycles at 12 m.p.h. How far will he travel in
a 2 hours 40 min   **b** 3 hours 10 min?

22   Majid can walk at 8 km/h. How far will he walk in
a 30 min   **b** 20 min   **c** 1 hour 15 min?

23   A racing car travels at 111 m.p.h. How far will it travel in
a 20 min   **b** 1 hour 40 min?

24   A bullet travels at 100 m/s. How far will it travel in
a 5 sec   **b** $8\frac{1}{2}$ sec?

25   A Boeing 747 travels at 540 m.p.h. How far does it travel in
a 3 hours 15 min   **b** 7 hours 45 min?

26   A racing car travels around a 2 km circuit at 120 km/h. How many laps will it complete in
a 30 min   **b** 1 hour 12 min?

Calculating the time taken

Georgina walks at 6km/h so we can find how long it will take her to walk

a 24 km   **b** 15 km.

a If she takes 1 hour to walk 6 km, she will take $\frac{24}{6}$ hours,
i.e. 4h, to walk 24 km.

b If she takes 1 hour to walk 6 km, she will take $\frac{15}{6}$ hours,
i.e. $2\frac{1}{2}$ hours, to walk 15 km.

i.e. $$\text{time} = \frac{\text{distance}}{\text{speed}}$$

Exercise 17c

1. How long will Zena, walking at 5 km/h, take to walk
   a 10km   **b** 15 km?

2. How long will a car travelling at 80 km/h, take to travel
   a 400 km   **b** 260 km?

3. How long will it take David, running at 10 m.p.h. to run
   a 5 miles   **b** $12\frac{1}{2}$ miles?

4. How long will it take an aeroplane flying at 450 m.p.h. to fly
   a 1125 miles   **b** 2400 miles?

5. A cowboy rides at 14 km/h. How long will it take him to ride
   a 21 km   **b** 70 km?

6. A rally driver drives at 50 m.p.h. How long does it take him to cover
   a 75 miles   **b** 225 miles?

7. An athlete runs at 8m/s. How long does it take him to cover
   a 200 m   **b** 1600 m?

8. A dog runs at 20 km/h. How long will it take him to travel
   a 8 km   **b** 18 km?

9. A liner cruises at 28 nautical miles per hour. How long will it take to travel
   a 6048 nautical miles   **b** 3528 nautical miles?

10. A car travels at 56 m.p.h. How long does it take to travel
    a 70 miles   **b** 154 miles?

11. A cyclist cycles at 12 m.p.h. How long will it take him to cycle
    a 30 miles   **b** 64 miles?

12. How long will it take a car travelling at 64 km/h to travel
    a 48 km   **b** 208 km?

Average speed

Russell Compton left home at 8 a.m. to travel the 50 km to his place of work. He arrived at 9 a.m. Although he had travelled at many different speeds during his journey he covered the 50 km in exactly 1 hour. We say that his *average speed* for the journey was 50 kilometres per hour, or 50 km/h. If he had travelled at the same speed all the time, he would have travelled at 50 km/h.

Judy Smith travelled the 135 miles from her home to Georgetown in 3 hours. If she had travelled at the same speed all the time, she would have travelled at $\frac{135}{3}$ m.p.h., i.e. 45 m.p.h. We say that her average speed for the journey was 45 m.p.h.

In each case: average speed = $\dfrac{\text{total distance travelled}}{\text{total time taken}}$

This formula can also be written:

distance travelled = average speed × time taken

and   time taken = $\dfrac{\text{distance travelled}}{\text{average speed}}$

Suppose that a car travels 35 km in 30 min, and we wish to find its speed in kilometres per hour. To do this we must express the time taken in hours instead of minutes,

i.e.   time taken = 30 min = $\frac{1}{2}$ hour

Then   average speed = $\dfrac{35}{\frac{1}{2}}$ km/h = $35 \times \frac{2}{1}$ km/h

= 70 km/h

Great care must be taken with units. If we want a speed in kilometres per hour, we need the distance in kilometres and the time in hours. If we want a speed in metres per second, we need the distance in metres and the time in seconds.

Exercise 17d

Find the average speed for each of the following journeys:

1	80 km in 1 hour	**7**	150 km in 3 hours
2	120 km in 2 hours	**8**	520 km in 8 hours
3	60 miles in 1 hour	**9**	245 miles in 7 hours
4	480 miles in 4 hours	**10**	104 miles in 13 hours
5	80 m in 4 sec	**11**	252 m in 7 sec
6	135 m in 3 sec	**12**	255 m in 15 sec

Find the average speed in km/h for a journey of 39 km that takes 45 min.

To find a speed in km/h you need the distance in kilometres and the time in hours.

First, convert the time taken to hours:

$$45 \text{ min} = \frac{45}{60} \text{ hour} = \frac{3}{4} \text{ hour}$$

Then average speed $= \dfrac{\text{distance travelled}}{\text{time taken}}$

$$= \frac{39 \text{ km}}{\frac{3}{4} \text{ hour}}$$

$$= 39 \times \frac{4}{3} \text{ km/h}$$

$$= 52 \text{ km/h}$$

Find the average speed in km/h for a journey of:

13 40 km in 30 min **15** 48 km in 45 min

14 60 km in 40 min **16** 66 km in 33 min.

Find the average speed in km/h for a journey of:

17 4000 m in 20 min **19** 40 m in 8 sec

18 6000 m in 45 min **20** 175 m in 35 sec.

> Make sure that the time is in hours and the distance is in kilometres.

Find the average speed in m.p.h. for a journey of:

21 27 miles in 30 min **23** 25 miles in 25 min

22 18 miles in 20 min **24** 28 miles in 16 min.

The following table shows the distances in kilometres between various towns in the West Indies.

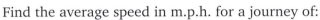

	St John's	Roseau	Castries	Basse-Terre	Kingstown	St Georges	Port of Spain
Roseau	174						
Castries	382	211					
Basse-Terre	100	478	621				
Kingstown	446	272	74	557			
St Georges	549	570	554	1040	118		
Port of Spain	723	659	534	1218	528	176	
Georgetown	1234	1224	1099	1694	1093	741	565

Use this table to find the average speeds for journeys between:

25 St John's, leaving at 1025 h, and Kingstown, arriving at 1625 h.

26 St Georges, leaving at 0330 h, and Castries, arriving at 0730 h.

27 Basse-Terre, leaving at 1914 h, and St Georges, arriving at 2044 h.

28 Port of Spain, leaving at 0620 h, and St John's, arriving at 0750 h.

> **29** Roseau, leaving at 1537 h, and St Georges, arriving at 1907 h.
>
> **30** Castries, leaving at 1204 h, and Georgetown, arriving at 1624 h.
>
> **31** Roseau, leaving at 1014 h, and Port of Spain, arriving at 1638 h.

Problems frequently occur where different parts of a journey are travelled at different speeds in different times but we wish to find the average speed for the whole journey.

Consider for example a motorist who travels the first 50 miles of a journey at an average speed of 25 m.p.h. and the next 90 miles at an average speed of 30 m.p.h.

One way to find his average speed for the whole journey is to complete the following table by using the relationship:

$$\text{time in hours} = \frac{\text{distance in miles}}{\text{speed in m.p.h.}}$$

	Speed in m.p.h.	Distance in miles	Time in hours
First part of journey	25	50	2
Second part of journey	30	90	3
Whole journey		140	5

We can add the distances to give the total length of the journey, and add the times to give the total time taken for the journey.

$$\text{average speed for whole journey} = \frac{\text{total distance}}{\text{total time}}$$

$$= \frac{140 \text{ miles}}{5 \text{ hours}}$$

$$= 28 \text{ m.p.h.}$$

Note: Never add or subtract average speeds.

We could also solve this problem, without using a table, as follows:

$$\text{time to travel 50 miles at 25 m.p.h.} = \frac{\text{distance}}{\text{speed}}$$

$$= \frac{50 \text{ miles}}{25 \text{ m.p.h.}}$$

$$= 2 \text{ hours.}$$

time to travel 90 miles at 30 m.p.h. $= \dfrac{\text{distance}}{\text{speed}}$

$$= \frac{90 \text{ miles}}{30 \text{ m.p.h}}$$

$$= 3 \text{ hour}$$

∴ total distance of 140 miles is travelled in 5 hours

i.e. average speed for whole journey $= \dfrac{\text{total distance}}{\text{total time}}$

$$= \frac{140 \text{ miles}}{5 \text{ hours}}$$

$$= 28 \text{ m.p.h.}$$

Exercise 17e

1 I walk for 24 km at 8 km/h, and then jog
   for 12 km at 12 km/h. Find my average
   speed for the whole journey.

 To find the average speed you need the total distance travelled and the *total time* taken.

2 A cyclist rides for 23 miles at an average speed of $11\frac{1}{2}$ m.p.h. before
   his cycle breaks down, forcing him to push his cycle the remaining
   distance of 2 miles at an average speed of 4 m.p.h. Find his average
   speed for the whole journey.

3 An athlete runs 6 miles at 8 m.p.h., then walks 1 mile at 4 m.p.h.
   Find his average speed for the total distance.

4 A woman walks 3 miles at an average speed of $4\frac{1}{2}$ m.p.h. and then
   runs 4 miles at 12 m.p.h. Find her average speed for the whole
   journey.

5 A motorist travels the first 30 km of a journey at an average speed of
   120 km/h, the next 60 km at 60 km/h, and the final 60 km at 80 km/h.
   Find the average speed for the whole journey.

6 Phil Sharp walks the 1 km from his home to the bus stop in 15 min,
   and catches a bus immediately which takes him the 9 km to the airport
   at an average speed of 36 km/h. He arrives at the airport in time to
   catch the plane which takes him the 240 km to Antigua at an average
   speed of 320 km/h. Calculate his average speed for the whole journey
   from home to Antigua.

7 A liner steaming at 24 knots takes 18 days to travel between two
   ports. By how much must it increase its speed to reduce the length
   of the voyage by 2 days? (A knot is a speed of 1 nautical mile
   per hour.)

Getting information from travel graphs

The graph below shows the journey of a coach passing three service stations A, B and C on a motorway. B is 60 km north of A and C is 20 km north of B. Use the graph to answer the following questions:

a At what time does the coach leave A?

b At what time does the coach arrive at C?

c At what time does the coach pass B?

d How long does the coach take to travel from A to C?

e What is the average speed of the coach for the whole journey?

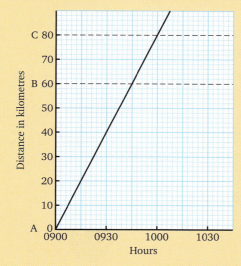

a The coach leaves A at 0900.

b It arrives at C at 1000. (Go from C on the distance axis across to the graph then down to the time axis).

c It passes through B at 0945.

d Time taken to travel from A to C is 1000 − 0900, i.e. 1 hour.

e Distance from A to C = 80 km (reading from the vertical axis)
Time taken to travel from A to C = 1 hour.

$$\text{average speed} = \frac{\text{distance travelled}}{\text{time taken}} = \frac{80\,\text{km}}{1\,\text{hour}} = 80\,\text{km/h}$$

1 The graph on page 282 shows the journey of a car through three towns, Axeter, Bexley and Canton, which lie on a straight road. Axeter is 100 km south of Bexley and Canton is 60 km north of it.

Make sure that you understand what the subdivisions on the scales represent.

Use the graph to answer the following questions:

a   At what time does the car
  i   leave Axeter
  ii   pass through Bexley
  iii   arrive at Canton?
b   How long does the car take to travel from Axeter to Canton?
c   How long does the car take to travel
  i   the first 80 km of the journey
  ii   the last 80 km of the journey?
d   What is the average speed of the car for the whole journey?

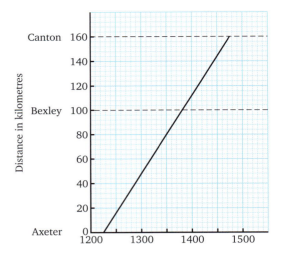

2   A car leaves Kingston at noon on its journey to Port Antonio via Morant Bay. The graph shows its journey.

a   How far it is from
  i   Kingston to Morant Bay
  ii   Morant Bay to Port Antonio?
b   How long does the car take to travel from Kingston to Port Antonio?
c   What is the car's average speed for the whole journey?

You need the difference between these two values on the distance axis.

d   How far does the car travel between 1.30 p.m. and 2.30 p.m.?
e   How far is the car from
  i   Kingston
  ii   Morant Bay, after travelling for $1\frac{1}{2}$ hours?

Go up from 1.30 p.m. on the time axis to the graph then across to the distance axis. Do the same for 2.30 p.m. Then find the difference between these readings on the distance axis.

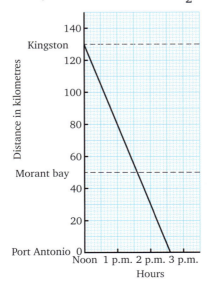

3 Father used the family car to transport the
children from their home to a summer
camp and then returned home.
The graph shows the journey.

The 'down hill' section of the graph
represents the return journey.

  a How far is it from home to the camp?
  b How long did it take the family to get to
      the camp?
  c What was the average speed of the car
      on the journey to the camp?
  d How long did the car take for the return
      journey?
  e What was the average speed for the return
      journey?
  f What was the car's average speed for the
      round trip?

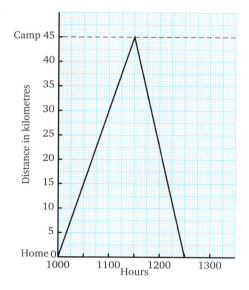

4 The graph shows the journey of a car through
three towns A, B and C

  a Where was the car at
      i 0900 h   **ii** 0930 h?
  b What was the average speed of the car between
      i A and B   **ii** B and C?
  c For how long does the car stop at B?
  d How long did the journey take?
  e What was the average speed of the car
      for the whole journey? Give your answer
      correct to 1 s.f.

The car arrives at B at the
point where the graph
stops going uphill and
leaves B at the point
where the graph starts
going uphill again.

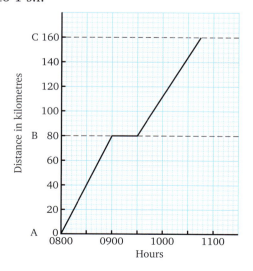

Exercise 17g

The graph shows Mrs Webb's journey on a bicycle to go shopping in the nearest town. Use it to answer the following questions:

a How far is town from home?

b How long did she take to get to town?

c How long did she spend in town?

d At what time did she leave for home?

e What was her average speed on the outward journey?

a Town is where Mrs Webb stops moving away from home, i.e. where the graph stops going uphill. The graph shows that it is 6 km from home to town.

b Mrs Webb left home at 1320 h and arrived in town at 1350 h. The journey therefore took 30 minutes.

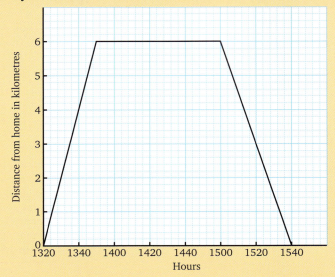

c Mrs Webb left town at the point where the graph starts going downhill. She arrived in town at 1350 h and left at 1500 h. She therefore spent 1 hour 10 min there.

d Mrs Webb left for home at 1500.

e On the outward journey:

$$\text{Average speed} = \frac{\text{distance travelled}}{\text{time taken}}$$

$$= \frac{6\,\text{km}}{30\,\text{min}}$$

$$= \frac{6\,\text{km}}{\frac{1}{2}\,\text{hour}} \quad \text{(time must be in hours)}$$

$$= 6 \times \frac{2}{1}\,\text{km/h}$$

$$= 12\,\text{km/h}$$

1   The graph shows the journey of a plane from St Vincent to Martinique
    and back again. Use the graph to answer the following questions:
    a   How far is St Vincent from Martinique?
    b   How long did the outward journey take?
    c   What was the average speed for the outward journey?
    d   How long did the plane remain in Martinique?
    e   At what time did the plane leave Martinique, and how long did the
           return journey take?
    f   What was the average speed on the return journey?

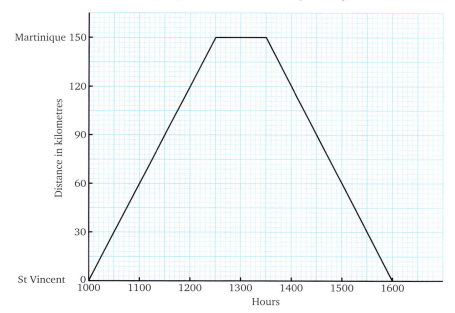

2   The graph represents the journey
    of a motorist from Kingston
    to Mandeville and back again.
    Use this graph to find
    a   the distance between the two cities
    b   the time the motorist spent in
           Mandeville
    c   his average speed on the outward
           journey
    d   the average speed on the homeward
           journey (including the stop).

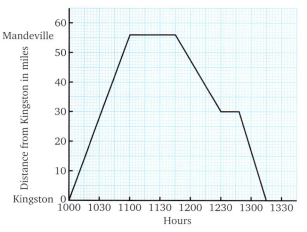

3   Overleaf is the travel graph for two motorists travelling between
    Kingston and Montego Bay which are 110 miles apart. The first leaves
    Montego Bay at 0900 h for Kingston, having a short break en route.
    The second leaves Kingston at 1015 h and travels non-stop to Montego
    Bay.

Use your graph to find

a the average speed of each motorist for the complete journey

 **b** when and where they pass

c their distance apart at 1200 h.

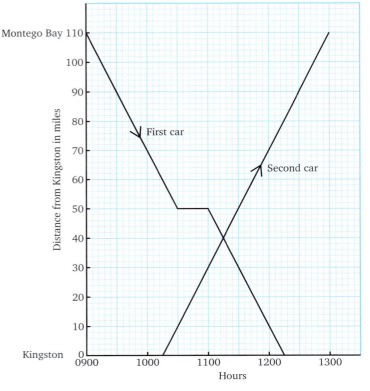

This is where the graphs intersect.

4 The graph below shows Judith's journeys between home and school.

a At what time did she leave home

    i in the morning   **ii** in the afternoon?

b How long was she in school during the day?

c How long was she away from school for her mid-day break?

d What was the average speed for each of these journeys?

e Find the total time for which she was away from home.

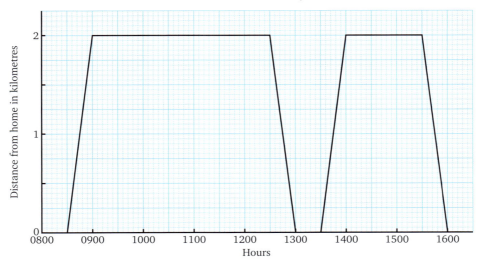

5 The graph shows the journeys of two cars between two towns, A and B, which are 180 km apart. Use the graph to find

   a the average speed of the first motorist and his time of arrival at B

   b the average speed of the second motorist and the time at which she leaves B

   c when and where the two motorists pass

   d their distance apart at 1427 h.

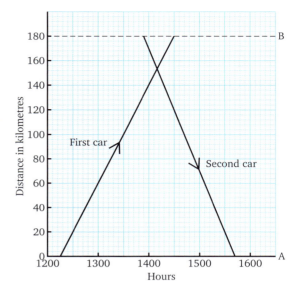

6 The graph represents the bicycle journeys of three school friends, Audrey, Betty and Chris, from the village in which they live to Spanish Town, the nearest main town, which is 30 km away.

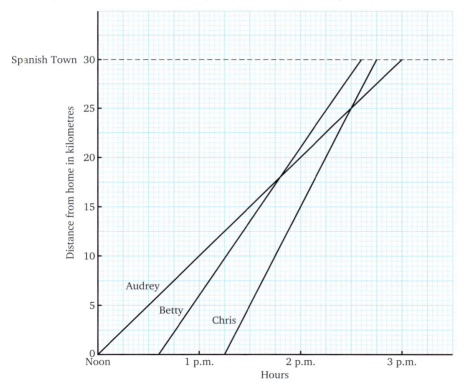

Use the graph to find:

   a their order of arrival at Spanish Town

   b Audrey's average speed for the journey

   c Betty's average speed for the journey

d Chris's average speed for the journey

e where and when Chris passes Audrey

f how far each is from town at 2 p.m.

7 Jane leaves home at 1 p.m. to walk at a steady 4 m.p.h. towards
Grand Bay, which is 16 miles away, to meet her boyfriend Tim.
Tim leaves Grand Bay at 2.18 p.m. and jogs at a steady 6 m.p.h.
to meet her. Draw a graph for each of these journeys taking
4 cm ≡ 1 hour on the time axis and 1 cm ≡ 1 mile on the distance
axis. From your graph find:

a when and where they meet

b their distance apart at 3 p.m.

8 A and B are motorway service areas 110 miles apart. A car leaves A at
2.16 p.m. and travels at a steady 63 m.p.h. towards B while a motorcycle
leaves B at 2.08 p.m. and travels towards A at a steady 45 m.p.h.
Draw a graph for the journeys taking 6 cm ≡ 1 hour and 1 cm ≡ 5 miles.
From your graph find:

a when and where they pass

b where the motorcycle is when the car starts

c where the motorcycle is when the car arrives at B.

Mixed exercises

Exercise 17h

1 The graph shows John's walk from home to his grandparents' home.

a How far away do they live?

b How long did the journey take him?

c What was his average walking speed?

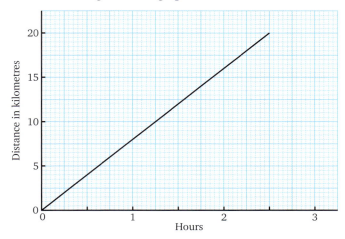

2 Jenny runs at 16 km/h. Draw a graph to show her running for $2\frac{1}{2}$ hours. Use your graph to find:

   a how far she has travelled in $1\frac{3}{4}$ hours

   b how long she takes to run the first 25 km.

3 A ship travels at 18 nautical miles per hour. How long will it take to travel:

   a 252 nautical miles

   b 1026 nautical miles?

4 Find the average speed in km/h of a journey of 48 km in 36 min.

5 I left Antigua airport at 1147 h to travel the 315 miles to Barbados. If I arrived at 1232 h, what was the plane's average speed?

6 I walk $\frac{2}{3}$ mile in 10 min and then run $\frac{1}{3}$ mile in 2 min. What is my average speed for the whole journey?

7 The graph shows Paul's journey in a sponsored walk from A to B. On the way his sister, who is travelling by car in the opposite direction from B to A, passes him.

   a How far does Paul walk?

   b How long does he take?

   c How much of this time does he spend resting?

   d What is his average speed for the whole journey?

   e What is his sister's average speed?

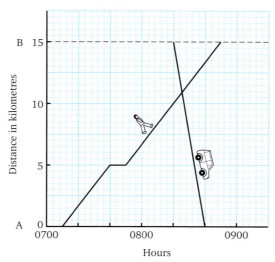

Exercise 17i

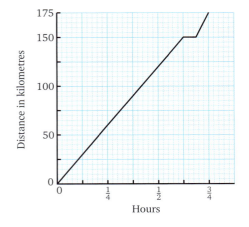

1   The graph shows the journey of a
    flight from my island to Castries.
    a   How far is my home from Castries?
    b   How long did the journey take?
    c   What happened during the journey
        that was not intended?
    d   What was the average speed of
        the flight for the first part of the
        journey?

2   Draw a travel graph to show a journey of 440 km in
    4 hours.

3   A horse runs at 15 m/sec. How far will it run in
    a   1 min   b $1\frac{3}{4}$ min?   Express its running speed in km/h.

4   How long will a coach travelling at 72 km/h take to travel
    a   216 km   b   126 km?

5   Which speed is the faster, and by how much: 50 m/sec or 200 km/h?

6   Find the average speed (km/h) for an 1800 m journey in 9 min.

7   A motorist wants to make a 300 mile journey in $5\frac{1}{2}$ h. He travels the
    first 60 miles at an average speed of 45 m.p.h., and the next 200 miles
    at an average speed of 60 m.p.h. What must be his average speed for
    the remaining part of the journey if he is to arrive on time?

? Puzzle

A train, 400 m long and travelling at 120 km/h, enters a tunnel that is
5.6 km long. For what time is any part of the train in the tunnel?

! Investigation

Did you know that

$3025 = (30 + 25)^2$ and that $2025 = (20 + 25)^2$?

Does this work for 4025 and 1025?

In this chapter you have seen that...

✔ a journey at constant speed can be represented by a straight line on a graph

✔ when you read values from a graph you need to make sure that you understand the meaning of the subdivisions on the scales on the axes

✔ the formula 'distance = speed × time' can be used to find one quantity when the other two are known

✔ when you are working out speeds, you must make sure that the units are consistent, e.g. to find a speed in kilometres per hour, the distance must be in kilometres and the time must be in hours

✔ the average speed for a journey is equal to the total distance travelled divided by the total time taken.

18 Practical applications of graphs

At the end of this chapter you should be able to...

1 Plot points from a given table of values.
2 Use a graph to find values corresponding to given values of a second variable.
3 Use graphs to convert from one unit to another.
4 Use graphs to make reliable estimates.
5 Use graphs to make deductions from data.

Did you know?

When you change Jamaican dollars into another currency, you will be given an exchange rate called 'sell' and when you change that currency back into Jamaican dollars you will be given a different exchange rate called 'buy'? This means that if you change Jamaican $ into US $ and then change them straight back to Jamaican $, you will get fewer dollars than you started with. The banks make money this way – and they generally also charge commission!

You need to know:

✔ how to draw and read graphs

Key words

cube, diameter, rate of exchange, rectangle, speed, the symbol ≡

Graphs involving straight lines

If you were to go to England for a holiday, you would probably have a little difficulty in knowing the cost of things in dollars and cents. If we know the rate of exchange, we can use a simple straight line graph to convert a given number of pounds into dollars or a given number of dollars into pounds.

Given that $100 converts to £0.75, we can draw a graph to convert values from, say, $0–$1500 into pounds.

Take $1\,cm \equiv \$100$ and $1\,cm \equiv £1$.

($\equiv$ means 'is equivalent to'.)

Because $\$600 \equiv £4.50$ and $\$1000 \equiv £7.50$ we can now plot these points and join them with a straight line.

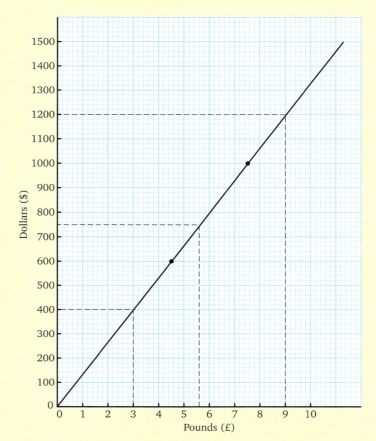

From the graph: $\$750 \equiv £5.60$ (going from $750 across the graph then
down to the £ axis)

$\$1200 \equiv £9$

$£3 \equiv \$400$ (going up from £3 then across to the $ axis)

$£9 \equiv \$1200$

Exercise 18a

1   The table gives temperatures in degrees Fahrenheit (°F) and the equivalent values in degrees Celsius (°C).

Temperature in °F	57	126	158	194
Temperature in °C	14	52	70	90

Plot these points on a graph for Celsius values from 0 to 100 and Fahrenheit values from 0 to 220. Let 2 cm represent 20 units on each axis.

Use your graph to convert:

a   97°F into °C   **b**   172°F into °C

c   25°C into °F   **d**   80°C into °F.

2   The table shows the conversion from US dollars to £s for various amounts of money.

US dollars	50	100	200
£s	35	70	140

Plot these points on a graph and draw a straight line to pass through them. Let 4 cm represent 50 units on both axes.

Use your graph to convert:

a   160 dollars into £s   **b**   96 dollars into £s

c   £122 into dollars   **d**   £76 into dollars.

> When you choose scales for axes, make them easy to read. On graph paper, you have 5 subdivisions between each centimetre, so choose a multiple of 5, e.g. in question **3** you could choose 2 cm = 5 km and in question **4** you could choose 2 cm ≡ 500 km.

3   Marks in an examination range from 0 to 65. Draw a graph that enables you to express the marks in percentages from 0 to 100. Note that a mark of 0 is 0% while a mark of 65 is 100%.

Use your graph

a   to express marks of 35 and 50 as percentages

b   to find the original mark for percentages of 50% and 80%.

4   Deductions from the wages of a group of employees amount to $35 for every $100 earned. Draw a graph to show the deductions made from gross pay in the range $0–$40 000 per week.

How much is deducted from an employee whose gross weekly pay is

a $12 500   **b** $24 000   **c** $33 500?

How much is earned each week by an employee whose weekly deductions amount to   **d** $4000   **e** $8800?

5   The table shows the fuel consumption figures for a car in both miles per gallon (X) and in kilometres per litre (Y).

m.p.g. (X)	30	45	60
km/litre (Y)	10.5	15.75	21

Plot these points on a graph taking 2 cm ≡ 10 units on the X-axis and 4 cm ≡ 5 units on the Y-axis. Your scale should cover 0–70 for X and 0–25 for Y.

Use your graph to find:

a   12 km/litre in m.p.g.   **b**   64 m.p.g. in km/litre

c   22.5 km/litre in m.p.g.   **d**   23 m.p.g. in km/litre.

6   The table gives various speeds in kilometres per hour with the equivalent values in metres per second.

Speed in km/h (S)	0	80	120	200
Speed in m/s (V)	0	22.2	33.3	55.5

Plot these values on a graph taking $4\,cm \equiv 50$ units on the S-axis and $4\,cm \equiv 10$ units on the V-axis.

Use your graph to convert:

a   140 km/h into m/s   **b**   46 m/s into km/h

c   18 m/s into km/h   **d**   175 km/h into m/s.

7   A number of rectangles, measuring l cm by b cm, all have a perimeter of 24 cm. Copy and complete the following table:

l	1	2	3	4	6	8
b			9			4

Draw a graph of these results using your own scale. Use your graph to find l if b is

a   2.5 cm   **b**   6.2 cm

and to find b if l is

c   5.5 cm   **d**   2.8 cm.

 Puzzle

Trains leave London for Edinburgh every hour on the hour. Trains leave Edinburgh for London every hour on the half-hour. The journey takes five hours each way.

Carrie takes a train from London to Edinburgh. How many trains from Edinburgh bound for London pass her train? Do not count any trains that may be in the stations at either end of the journey.

Graphs involving curves

When two quantities that are related are plotted one against the other, we often find that the points do not lie on a straight line. They may, however, lie on a smooth curve.

Consider the table below, which gives John's height on his birthday over a period of 8 years.

Age in years	11	12	13	14	15	16	17	18	19
Height in cm	138	140	144	150	158	165	170	172	173

These points can be plotted on a graph and joined to give a smooth curve through the points, as shown.

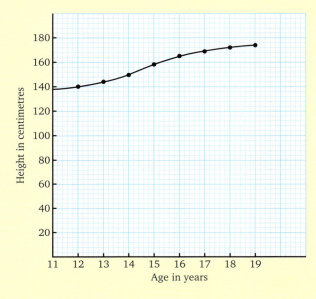

The graph enables us to estimate that:

a   he was 162 cm tall when he was $15\frac{1}{2}$ years old

b   he was 146 cm tall when he was 13 years 5 months

c   when he was 17 years 6 months, he was 171 cm tall.

We can also deduce that:

i   the fastest increase in height was between his fourteenth and fifteenth birthdays – the curve is steepest between these two birthdays

ii he grew very little between his eighteenth and nineteenth birthdays – the curve is quite flat in this region.

We could obtain more accurate results if we took 100 cm or 120 cm as the lowest height on the vertical axis and used a larger scale.

Exercise 18b

1 The masses of lead spheres of various diameters are shown in the table.

Diameter in mm (*D*)	4	5.2	6.4	7.2	7.9	8.8
Mass in grams (*M*)	380	840	1560	2230	2940	4070

Plot this information on a graph and draw a smooth curve through the points. Use 2 cm ≡ 1 unit on the *D*-axis and 2 cm ≡ 500 units on the *M*-axis.

Use your graph to estimate
a the mass of a lead sphere of diameter 6 mm
b the diameter of a lead sphere of mass 2 kg.

2 Recorded speeds of a motor car at various times after starting from rest are shown in the table.

Time in seconds	0	5	10	15	20	25	30	35	40
Speed in km/h	0	62	112	148	172	187	196	199	200

Taking 2 cm ≡ 5 sec and 1 cm ≡ 10 km/h, plot these results and draw a smooth curve to pass through these points.

Use your graph to estimate
a the time which passes before the car reaches
   i 100 km/h   ii 150 km/h
b its speed after
   i 13 seconds   ii 27 seconds.

3 The mass of a puppy at different ages is given in the table.

Age in days (*A*)	10	20	40	60	80	100	120	140
Mass in grams (*W*)	50	100	225	425	750	875	950	988

Draw a graph to represent this data, taking 1 cm ≡ 10 days on the *A*-axis and 1 cm ≡ 50 g on the *W*-axis.

Hence estimate
a the mass of the puppy after i 50 days ii 114 days
b the age of the puppy when its mass is i 500 g ii 1000 g
c the weight it puts on between day 25 and day 55
d its birth mass.

4 The speed of a particle (v metres per second) at various times
(t seconds) after starting is given in the table.

t	0	1	2	3	4	5	6	7
v	0	35	60	76.5	83	83	76	57

Plot this information on a graph using $2\,\text{cm} \equiv 1$ unit on the t-axis and
$2\,\text{cm} \equiv 10$ units on the v-axis.

Use your graph to find:
a the greatest speed of the particle and the time at which it occurs
b its speed after **i** 3.5 sec **ii** 6.8 sec
c when its speed is 65 m/sec.

5 In the United Kingdom the cost of fuel ($£C$) per nautical mile for a
ship travelling at various speeds (v knots) is given in the table.

V	12	14	16	18	20	22	24	26	28
C	18.15	17.16	16.67	16.5	16.5	16.67	16.94	17.36	17.82

Draw a graph to show how cost changes with speed.
Use $1\,\text{cm} \equiv 1$ knot and $10\,\text{cm} \equiv £1$. (Take £16 as the lowest value for C.)

Use your graph to estimate:
a the most economical speed for the ship and the corresponding cost
per nautical mile
b the speeds when the cost per nautical mile is £17
c the cost when the speed is **i** 13 knots **ii** 24.4 knots.

6 Cubes made from a certain metal with edges of the given lengths have
masses as given in the table.

Length of edge in cm (L)	1	2	3	4	5	6
Mass of cube in grams (M)	9	72	243	576	1125	1944

Plot this information on a graph, joining the points with a smooth
curve. Take $2\,\text{cm} \equiv 1$ unit on the L-axis and $1\,\text{cm} \equiv 100\,\text{g}$ on the M-axis.

From your graph find:
a the mass in grams of a cube with edge
i 3.5 cm **ii** 5.3 cm
b the length of the edge of a cube with mass
i 500 g **ii** 1500 g.

7 The temperatures, taken at 2-hourly intervals, at my home on a certain day were as given in the table.

Draw a graph to show this data taking $1\,\text{cm} \equiv 1$ hour and $1\,\text{cm} \equiv 1\,°\text{C}$.

Use your graph to estimate:

a the temperature at 11 a.m. and at 11 p.m.

b the times at which the temperature was 29 °C.

Time	Temperature in °C
midnight	26.6
2 a.m.	26.0
4 a.m.	25.8
6 a.m.	26.0
8 a.m.	27.4
10 a.m.	28.2
noon	29.4
2 p.m.	30.2
4 p.m.	30.0
6 p.m.	29.6
8 p.m.	28.8
10 p.m.	28.0
midnight	27.6

8 The time of sunset at Kingston on different dates, each two weeks apart, is given in the table.

	Sept		Oct		Nov		Dec	
Date (D)	5	29	12	26	10	24	7	21
Time (T)	18.18	17.57	17.47	17.38	17.32	17.30	17.31	17.37

Using $1\,\text{cm} \equiv 1$ week on the D-axis and $2\,\text{cm} \equiv 10$ minutes on the T-axis, plot these points on a graph and join them with a smooth curve. Take 1720 as the lowest value for T.

From your graph estimate:

a the time of sunset on 17 Nov   b the months in which the sun sets at 1734.

9 A rectangle measuring l cm by b cm has an area of $24\,\text{cm}^2$. The table gives different values of l with the corresponding values of b.

l	1	2	3	4	6	8	12	16
b	24		8		4		2	1.5

Complete the table and draw a graph to show this information, joining the points with a smooth curve. Take $1\,\text{cm} \equiv 1$ unit on the l-axis and $1\,\text{cm} \equiv 2$ units on the b-axis.

Use your graph to estimate the value of

a l when b is   i   14 cm   ii   2.4 cm

b b when l is   i   18 cm   ii   2.8 cm

In this chapter you have seen that...

✔ you can draw a graph and use it to convert a quantity from one unit to another.

19 Coordinates and the straight line

At the end of this chapter you should be able to...

1 Plot points with given coordinates.

2 Write the equation of a line through the origin.

3 Find the y- or x-coordinate of a point on the line $y = mx$, given the x- or y-coordinate of the point.

4 Determine, by calculation, whether a given point lies on the line $y = mx$.

5 Draw the graph of a line $y = mx + c$.

6 Calculate the gradient of a line, given two points on the line.

7 Determine if a line, whose equation is given, makes an acute or obtuse angle with the positive direction of the x-axis.

8 State the gradient and intercept on the y-axis of the line $y = mx + c$.

9 Determine whether two lines, whose equations are given, are parallel.

10 Identify lines parallel to the x-axis or y-axis given their equations.

11 Solve linear simultaneous equations graphically.

You need to know...

✔ how to draw graphs
✔ how to read the coordinates of points
✔ how to solve linear equations
✔ the meaning of acute and obtuse angles
✔ how to work with directed numbers

Key words

acute angle, component, coordinate, equation, gradient, intercept, intersect, obtuse angle, ordered number pair, origin, parallel, simultaneous

The equation of a straight line

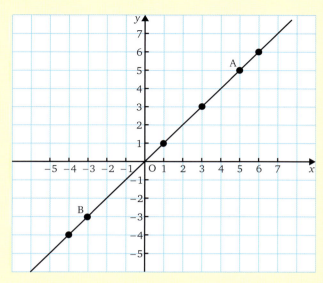

If we plot the points with coordinates (−4, −4), (1, 1), (3, 3) and (6, 6), we can see that a straight line can be drawn through these points that also passes through the origin.

For each point the y-coordinate is the same as the x-coordinate.

This is also true for any other point on this line,

e.g. the coordinates of A are (5, 5) and of B are (−3, −3).

Hence $\qquad$ y-coordinate $= x$-coordinate

or simply $\qquad\qquad$ $y = x$

This is called the equation of the line.

We can also think of a line as a set of points, i.e. this line is the set of points, or *ordered number pairs*, such that $\{(x, y)\}$ satisfies the relation $y = x$.

It follows that if another point on the line has an x-coordinate of −5, then its y-coordinate is −5 and if a further point has a y-coordinate of 4, its x-coordinate is 4.

In a similar way we can plot the points with coordinates (−2, −4), (1, 2), (2, 4) and (3, 6).

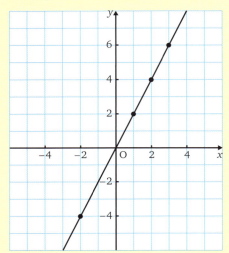

These points also lie on a straight line passing through the origin.

In each case the y-coordinate is twice the x-coordinate.

The equation of this line is therefore $y = 2x$ and we often refer, briefly, to 'the line $y = 2x$'.

If another point on this line has an x-coordinate of 4,

its y-coordinate is 2×4, i.e. 8,

and if a further point has a y-coordinate of −5,

its x-coordinate must be $-2\frac{1}{2}$.

Exercise 19a

1 Find the y-coordinates of points on the line $y = x$ that have x-coordinates of

   a 2   **b** 3   **c** 7   **d** 12.

2 Find the y-coordinates of points on the line $y = x$ that have x-coordinates of

   a −1   **b** −6   **c** −8   **d** −20.

3 Find the y-coordinates of points on the line $y = -x$ that have x-coordinates of

   a $3\frac{1}{2}$ **b** $-4\frac{1}{2}$ **c** 6.1   **d** −8.3

4   Find the x-coordinates of points on the line $y = -x$ that have y-coordinates of

   a 7   **b** −2   **c** $5\frac{1}{2}$ **d** −4.2

5   Find the y-coordinates of points on the line $y = 2x$ that have x-coordinates of

   a 5   **b** −4   **c** $3\frac{1}{2}$ **d** −2.6

6   Find the x-coordinates of points on the line $y = -3x$ that have y-coordinates of

   a 3   **b** −9   **c** 6   **d** −4

7   Find the x-coordinates of points on the line $y = \frac{1}{2}x$ that have y-coordinates of

   a 6   **b** −12   **c** $\frac{1}{2}$ **d** −8.2

8   Find the x-coordinates of points on the line $y = -4x$ that have y-coordinates of

   a 8   **b** −16   **c** 6   **d** −3

9   The points $(-1, a)$ $(b, 15)$ and $(c, -20)$ lie on the straight line with equation $y = 5x$. Find the values of a, b and c.

> $A(-1, a)$ lies on $y = 5x$. Replace y by a and x by -1. Then solve the equation to find a.

10 The points $(3, a)$, $(-12, b)$ and $(c, -12)$ lie on the straight line with equation $y = -\frac{2}{3}x$. Find the values of a, b and c.

11 Using 1 cm to 1 unit on each axis, plot the points $(-2, -6)$, $(1, 3)$, $(3, 9)$ and $(4, 12)$. What is the equation of the straight line that passes through these points?

12 Using 1 cm to 1 unit on each axis, plot the points $(-3, 6)$ $(-2, 4)$, $(1, -2)$ and $(3, -6)$. What is the equation of the straight line that passes through these points?

13 Using the same scale on each axis, plot the points $(-6, 2)$, $(0, 0)$, $(3, -1)$ and $(9, -3)$. What is the equation of the straight line that passes through these points?

14 Using the same scale on each axis, plot the points $(-6, -4)$, $(-3, -2)$, $(6, 4)$ and $(12, 8)$. What is the equation of the straight line that passes through these points?

15 Which of the points $(-2, -4)$, $(2.5, 4)$, $(6, 12)$ and $(7.5, 10)$ lie on the line $y = 2x$?

16 Which of the points $(-5, -15)$, $(-2, 6)$, $(1, -3)$ and $(8, -24)$ lie on the line $y = -3x$?

<u>**17**</u>   Which of the following points lie

    a above the line $y = \frac{1}{2}x$ **b** below the line $y = \frac{1}{2}x$

    $(2, 2)$, $(-2, 1)$, $(3, 0)$, $(-4.2, -2)$, $(-6.4, -3.2)$?

Plotting the graph of a given straight line

If we want to draw the graph of $y = 3x$ for values of x from -3 to $+3$, then we need to find the coordinates of some points on the line.

As we know that it is a straight line, two points are enough. However, it is sensible to find three points. The third point acts as a check on our working. It does not matter which three points we find, so we will choose easy values for x, one at each extreme and one near the middle.

If $x = -3$, $y = 3 \times (-3) = -9$

If $x = 0$, $y = 3 \times 0 = 0$

If $x = 3$, $y = 3 \times 3 = 9$

These look neater if we write them in table form:

x	-3	0	3
y	-9	0	9

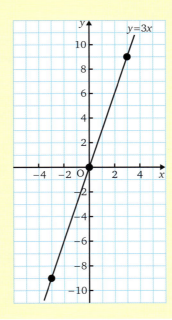

Exercise 19b

In questions **1** to **6**, draw the graphs of the given equations on the same set of axes. Use the same scale on both axes, taking values of x between -4 and 4, and values of y between -6 and 6. You should take at least three x values and record the corresponding y values in a table. Write the equation of each line somewhere on it.

1 $y = x$ **3** $y = \frac{1}{2}x$ **5** $y = \frac{1}{3}x$

2 $y = 2x$ **4** $y = \frac{1}{4}x$ **6** $y = \frac{3}{2}x$

In questions **7** to **12**, draw the graphs of the given equations on the same set of axes.

7 $y = -x$ **9** $y = -\frac{1}{2}x$ **11** $y = -\frac{1}{3}x$

8 $y = -2x$ **10** $y = -\frac{1}{4}x$ **12** $y = -\frac{3}{2}x$

13 Use GeoGebra to check your graphs.

14 Use GeoGebra to investigate how changing the value of m in $y = mx$ changes the line.

We can conclude from these exercises that the graph of an equation of the form $y = mx$ is a straight line that:

- passes through the origin
- gets steeper as m increases
- makes an acute angle with the positive x-axis if m is positive
- makes an obtuse angle with the positive x-axis if m is negative.

Gradient of a straight line

The gradient or slope of a line is defined as the amount the line rises vertically divided by the distance moved horizontally.

i.e.   gradient or slope of AB $= \dfrac{BC}{AC}$

The gradient of any line is defined in a similar way.

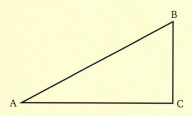

Considering any two points on a line, the gradient of the line is given by

$$\frac{\text{the increase in } y \text{ value}}{\text{the increase in } x \text{ value}}$$

If we plot the points O(0, 0), B(4, 4) and C(5, 5), all of which lie on the line with equation $y = x$, then:

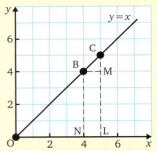

$$\text{gradient of OC} = \frac{CL}{OL} = \frac{5}{5} = 1$$

$$\text{gradient of OB} = \frac{BN}{ON} = \frac{4}{4} = 1$$

$$\text{gradient of BC} = \frac{CM}{BM} = \frac{5-4}{5-4} = \frac{1}{1} = 1$$

These show that, whichever two points are taken, the gradient of the line is 1.

Similarly, if we plot the points P(−3, 6), Q(−1, 2) and R(4, −8), all of which lie on the line with equation $y = -2x$, then:

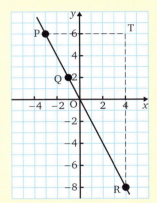

$$\text{gradient of PR} = \frac{\text{increase in } y \text{ value from P to R}}{\text{increase in } x \text{ value from P to R}}$$

$$= \frac{y\text{-coordinate of R} - y\text{-coordinate of P}}{x\text{-coordinate of R} - x\text{-coordinate of P}}$$

$$= \frac{(-8) - (6)}{(4) - (-3)}$$

$$= \frac{-8 - 6}{4 + 3} = \frac{-14}{7} = -2$$

Exercise 19c

Draw axes for x and y, for values between −6 and +6, taking 1 cm as 1 unit on each axis.

Plot the points A(−4, 4), B(2, −2) and C(5, −5), all of which lie on the line $y = -x$. Find the gradient of

a AB   **b** BC   **c** AC.

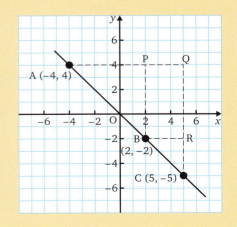

a Gradient of AB

$$= \frac{(-2) - (4)}{(2) - (-4)} = \frac{-6}{6} = -1$$

b Gradient of BC

$$= \frac{(-5) - (-2)}{(5) - (2)} = \frac{-3}{3} = -1$$

c Gradient of AC

$$= \frac{(-5) - (4)}{(5) - (-4)} = \frac{-9}{9} = -1$$

1. Using 2 cm to 1 unit on each axis, draw axes that range from 0 to 6 for x and from 0 to 10 for y. Plot the points A(2, 4), B(3, 6) and C(5, 10), all of which lie on the line $y = 2x$. Find the gradient of
   a   AB   b   BC   c   AC.

2. Draw the x-axis from −4 to 4 taking 2 cm as 1 unit, and the y-axis from −16 to 12 taking 0.5 cm as 1 unit. Plot the points X(−3, 12), Y(−1, 4) and Z(4, −16), all of which lie on the line $y = −4x$. Find the gradient of
   a   XY   b   YZ   c   XZ.

3. Choosing your own scale and range of values for both x and y, plot the points D(−2, −6), E(0, 0) and F(4, 12), all of which lie on the line $y = 3x$. Find the gradient of
   a   DE   b   EF   c   DF.

4. Taking 2 cm as 1 unit for x and 1 cm as 1 unit for y, draw the x-axis from −1.5 to 2.5 and the y-axis from −10 to 6. Plot the points A(−1.5, 6), B(0.5, −2) and C(2.5, −10), all of which lie on the line $y = −4x$. Find the gradient of
   a   AB   b   BC   c   AC.

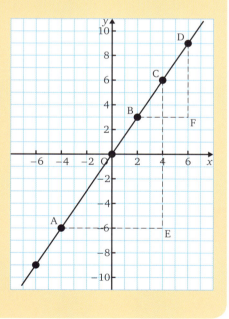

Copy and complete the following table and use it to draw the graph of $y = 1.5x$

x	−6	−4	0	2	4	6
y						

Choosing your own points, find the gradient of this line using two different sets of points.

x	−6	−4	0	2	4	6
y	−9	−6	0	3	6	9

Four points, A, B, C and D, have been chosen.

Gradient of line $= \dfrac{CE}{AE} = \dfrac{6 - (-6)}{4 - (-4)} = \dfrac{12}{8} = 1.5$

Gradient of line $= \dfrac{DF}{BF} = \dfrac{9-3}{6-2} = \dfrac{6}{4} = 1.5$

(Finding the gradient using any other two points also gives a value of 1.5)

5. Copy and complete the following table and use it to draw the graph of $y = 2.5x$

x	−3	−1	0	2	4
y					

Choose your own pairs of points to find the gradient of this line at least twice.

6   Copy and complete the following table and use it to draw the graph of $y = -0.5x$.

x	−6	−2	3	4
y				

Choose your own pairs of points to find the gradient of this line at least twice.

7   Determine whether the straight lines with the following equations have positive or negative gradients:

   a $y = 5x$ **b** $y = -7x$ **c** $y = 12x$

   d $3y = -x$ **e** $y = -\frac{1}{4}x$ **f** $5y = -12x$

These exercises, together with the worked examples, confirm our conclusions on page 305, namely that

- the larger the value of m the steeper is the slope
- lines with positive values for m make an acute angle with the positive x-axis
- lines with negative values for m make an obtuse angle with the positive x-axis.

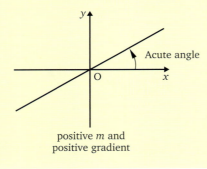

positive m and
positive gradient

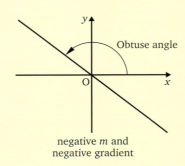

negative m and
negative gradient

Exercise 19d

For each of the following pairs of lines, state which line is the steeper. Show both lines on the same sketch.

1   $y = 5x,$ $y = \frac{1}{5}x$

2   $y = 2x,$ $y = 5x$

3   $y = \frac{1}{2}x,$ $y = \frac{1}{3}x$

4   $y = -2x,$ $y = -3x$

5   $y = 10x,$ $y = 7x$

6   $y = -\frac{1}{2}x,$ $y = -\frac{1}{4}x$

7   $y = -6x,$ $y = -3x$

8   $y = 0.5x,$ $y = 0.75x$

Determine whether each of the following straight lines makes an acute angle or an obtuse angle with the positive *x*-axis.

9 $y = 4x$ **12** $y = 3.6x$ **15** $y = 10x$ **18** $y = -\frac{2}{3}x$

10 $y = -3x$ **13** $y = \frac{1}{3}x$ **16** $y = 0.5x$ **19** $y = -\frac{3}{4}x$

11 $y = -\frac{1}{2}x$ **14** $y = 0.7x$ **17** $y = -6x$ **20** $y = -0.4x$

21   Estimate the gradient of each of the lines shown in the sketch.

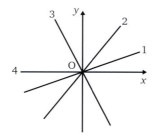

? Puzzle

Here is a very ingenious method of guessing the values of three dice rolled by a friend, without seeing them.

Tell him to think of the number showing on the first die.

Multiply by 2. Add 5. Multiply by 5.

Add the value of the number showing on the second die.

Multiply by 10. Add the value of the number showing on the third die.

Now ask the total. From this total subtract 250.

The three digits of your answer will be the values of his three dice.

As an example, if the answer was 706, then 706 − 250 = 456. The three dice were therefore 4, 5 and 6. Try it and see. Why does it work?

Lines that do not pass through the origin

If we plot the points (−3, −1), (1, 3), (3, 5), (4, 6) and (6, 8), and draw the straight line that passes through these points, we can use it to find
a   the equation of the line
b   its gradient
c   the distance from the origin to the point where the line crosses the *y*-axis.

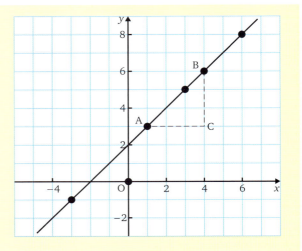

a In each case, the y-coordinate is 2 more than the x-coordinate, i.e. all the points lie on the line with equation $y = x + 2$

b Using the points A and B, the gradient of the line is given by

$\dfrac{BC}{AC}$, i.e. $\dfrac{3}{3} = 1$.

c The line crosses the y-axis at the point (0, 2) which is 2 units above the origin. This quantity is called the y-intercept.

Exercise 19e

Draw the graph of $y = -4x + 3$ for values of x between -4 and $+4$. Hence find

a the gradient of the line

b its y-intercept.

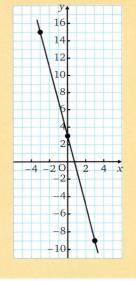

x	-3	0	3
y	15	3	-9

a Moving from the point (0, 3) to (−3, 15) the gradient is

$= \dfrac{3-15}{0-(-3)} = \dfrac{-12}{3} = \dfrac{-4}{1} = -4$

b The y-intercept is 3.

In the following questions, draw the graph of the given equation using the given x values. Hence find the gradient of the line and its intercept on the y-axis. Use 1 cm as 1 unit on each axis with x values ranging from -8 to $+8$ and y values ranging from -10 to $+10$. (You can use GeoGebra to draw these graphs.)

Compare the values you get for the gradient and the y-intercept with the numbers in the right-hand side of each equation.

1 $y = 3x + 1$; x values -3, 1, 3

Use your graph to find the value of y when x is   **a** -2 **b** 2.

2 $y = -3x + 4$; x values -2, 2, 4

Use your graph to find the value of y when x is   **a** -1 **b** 3.

3 $y = \frac{1}{2}x + 4$; x values -8, 0, 6

Use your graph to find

a the value of y when x is -2 **b** the value of x when y is 6.

4 $y = x - 3$; x values -4, 2, 8

Use your graph to find the value of x when y is **a** 4 **b** -5

5 $y = \frac{3}{4}x + 3$; x values -4, 0, 8

Use your graph to find the value of x when y is **a** 6 **b** 4.5

Draw the graph of $y = -2x + 3$ for values of x between -4 and $+4$.

Hence find

a the gradient of the line **b** its y-intercept.

Compare the values for the gradient and the y-intercept with the number of xs and the number term on the right-hand side of the equation.

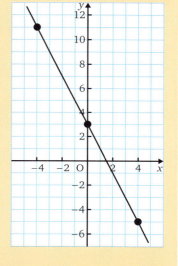

x	-4	0	4
y	11	3	-5

a Gradient of line $= \dfrac{3 - 11}{0 - (-4)} = -\dfrac{8}{4}$

$\qquad\qquad\qquad\quad = -2$

b The y-intercept is 3.

The number of xs on the right-hand side of $y = -2x + 3$ is -2, which is the same as the gradient of the line.

The number term on the right-hand side of $y = -2x + 3$ is 3, which is the same as the y-intercept.

In the following questions, draw a graph for each of the given equations. In each case find the gradient and the y-intercept for the resulting straight line. Take 1 cm as 1 unit on each axis, together with suitable values of x within the range -4 to $+4$. Choose your own range for y when you have completed the table.

Compare the values you get for the gradient and the y-intercept with

a the number of xs

b the number term on the right-hand side of the equation.

6 $y = 2x - 2$ **8** $y = 3x - 4$ **10** $y = -\frac{3}{2}x + 3$ **12** $y = -2x - 7$

7 $y = -2x + 4$ **9** $y = \frac{1}{2}x + 3$ **11** $y = 2x + 5$ **13** $y = -3x + 2$

The equation $y = mx + c$

The results of Exercise **19e** show that we can 'read' the gradient and the y-intercept of a straight line from its equation.

For example, the line with equation $y = 3x - 4$ has a gradient of 3 and its y-intercept is -4.

In general we can conclude that the equation $y = mx + c$ gives a straight line where m is the gradient of the line and c is the y-intercept.

Exercise 19f

Write down the gradient, m, and the y-intercept, c, for the straight line with equation $y = 5x - 2$

Comparing the line $y = 5x - 2$ with $y = mx + c$ gives

$$m = 5 \quad \text{and} \quad c = -2$$

Write down the gradient, m, and y-intercept, c, for the straight line with the given equation.

1 $\quad y = 4x + 7$ **3** $\quad y = 3x - 2$ **5** $\quad y = 7x + 6$ **7** $\quad y = \frac{3}{4}x + 7$

2 $\quad y = \frac{1}{2}x - 4$ **4** $\quad y = -4x + 5$ **6** $\quad y = \frac{2}{5}x - 3$ **8** $\quad y = 4 - 3x$

Sketch the straight line with equation $y = 5x - 7$

Comparing $y = 5x - 7$ with $y = mx + c$ shows that we want a line with gradient 5 and y-intercept -7.

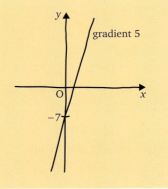

gradient 5

Sketch the straight lines with the given equations.

9 $\quad y = 2x + 5$ **12** $\quad y = -2x - 3$ **15** $\quad y = -5x - 3$

10 $\quad y = 7x - 2$ **13** $\quad y = -\frac{2}{3}x + 8$ **16** $\quad y = 3x + 7$

11 $\quad y = \frac{1}{2}x + 6$ **14** $\quad y = 4x + 2$ **17** $\quad y = \frac{3}{4}x - 2$

Sketch the straight line with equation $y = 2 - 3x$

First rearrange the equation in the form

$y = mx + c$

i.e. $\qquad y = -3x + 2$

Comparing $y = -3x + 2$ with $y = mx + c$ shows that we want a line with gradient -3 and y-intercept 2.

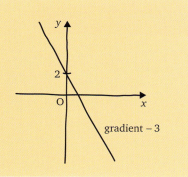

gradient -3

18	$y = 4 - x$	**21**	$y = -3 - x$	**24**	$y = -5(x - 1)$
19	$y = 3 - 2x$	**22**	$y = 2(x + 1)$	**25**	$y = 3(4 - x)$
20	$y = 8 - 4x$	**23**	$y = 3(x - 2)$	**26**	$y = -2(2x + 3)$

Parallel lines

Lines with the same gradient are said to be parallel.

The diagram shows the lines $y = x + 2$ and $y = x - 3$.

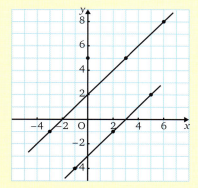

These lines have the same gradient, i.e. they are parallel.

Now consider a third line, parallel to the first two lines and passing through the point A(0, 5).

Its gradient is the same as that of the first lines, i.e. $m = 1$.

It crosses the y-axis at (0, 5) so its y intercept is 5, i.e. $c = 5$.

Therefore the equation of the third line is $y = x + 5$.

Similarly the equation of another parallel line passing through the point (0, −5) is $y = x - 5$.

Exercise 19g

1   Draw the graphs of $y = 3x+1$ and $y = 3x-4$ taking x values of -2, 2 and 3.
   (Let x range from -5 to $+5$ and y from -10 to $+10$.
   Take 1 cm to represent 1 unit on each axis.)

   What do you notice about these lines?
   What do you notice about their m values?

2   Draw the graphs of $y = -2x+3$ and $y = -2x-3$ taking x values of -3,
   0 and 3. (Take 1 cm to represent 1 unit on each axis. Let x range from
   -6 to $+6$ and y from -10 to $+10$.)

   What do you notice about these lines?
   What do you notice about their m values?

By finding the gradient of each line, determine whether or not the given
pairs of equations represent parallel lines.

3   $y = 4x+2$, $y = 4x-7$ 7 $y = -x+4$, $y = -x-3$

4   $y = \frac{1}{2}x+6$, $y = \frac{1}{2}x+10$ 8 $y = -5x+2$, $y = -5x-13$

5   $y = x+4$, $y = 2x+4$ 9 $y = \frac{2}{3}x+3$, $y = \frac{1}{3}x-4$

6   $y = 3x+5$, $y = x+7$ 10 $y = \frac{1}{2}x-4$, $y = 0.5x+2$

Find the gradient of each of the lines $x+y = 4$ and $y = -x+2$.
Hence determine whether or not the two lines are parallel.

$x+y = 4$ (1)

$y = -x+2$ (2)

Rearrange (1) in the form $y = mx+c$

Equation (1) gives $y = -x+4$
the gradient of this line is -1

The gradient of the line $y = -x+2$ is -1
i.e. the lines have the same gradient and are therefore parallel.

Find the gradient of each of the lines in each question. Hence determine
whether or not the two lines are parallel.

11   $y = 2x+3$, $2y = 4x-7$ 14 $3y = 5x+7$, $6y = 10x-3$

12   $3y = 9x-2$, $y = 3x+13$ 15 $5y = x+2$, $3y = x+2$

13   $x+y = 5$, $y = -2x+3$ 16 $x+y = 4$, $y = -x+6$

Lines parallel to the axes

We began by considering the equation $y = mx$,

i.e. the equation $y = mx + c$ when $c = 0$.

This equation gave a straight line passing through the origin.

Now we will see what happens when $m = 0$.

Think, for example, of the equation $y = 3$.

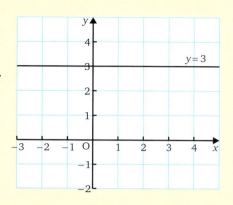

For every value of x the y-coordinate is 3.
This means that the graph of $y = 3$ is a straight line
parallel to the x-axis at a distance 3 units above it.

$y = c$ is therefore the equation of a straight line parallel to the
x-axis at a distance c away from it. If c is positive, the line is
above the x-axis, and if c is negative, the line is below the x-axis.

Similarly $x = b$ is the equation of a straight line parallel to
the y-axis at a distance b units from it.

Exercise 19h

Draw, on the same diagram, the straight line graphs of $x = -3$, $x = 5$,
$y = -2$ and $y = 4$.

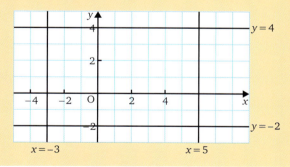

In the following questions, take both x and y in the range -8 to $+10$.
Let 1 cm be 1 unit on each axis.

1   Draw the straight line graphs of the following equations in a single
    diagram:
    $x = 2$, $x = -5$, $y = \frac{1}{2}$, $y = -3\frac{1}{2}$

2   Draw the straight line graphs of the following equations in a single
    diagram:
    $y = -5$, $x = -3$, $x = 6$, $y = 5.5$

3   On one diagram, draw graphs to show the following equations:

$x = 5$, $y = -5$, $y = 2x$

Write down the coordinates of the three points where these lines intersect.

What kind of triangle do they form?

4   On one diagram, draw the graphs of the straight lines with equations

$x = 4$, $y = -\frac{1}{2}x$, $y = 3$

Write down the coordinates of the three points where these lines intersect.

What kind of triangle is it?

5   On one diagram, draw the graphs of the straight lines with equations

$y = 2x + 4$, $y = -5$, $y = 4 - 2x$

Write down the coordinates of the three points where these lines intersect.

What kind of triangle is it?

Simultaneous equations

When we are given an equation we can draw a graph.
Any of the equations that occur in this chapter give
us a straight line. Two equations give us two straight
lines that usually cross one another.

Any equation that gives a straight line is called a linear equation.

Consider the two equations $x + y = 4$ $y = 1 + x$

Suppose we know that the x-coordinate of the point of intersection
is in the range $0 \leqslant x \leqslant 5$:

$x + y = 4$

x	0	4	5
y	4	0	-1

$y = 1 + x$

x	0	2	5
y	1	3	6

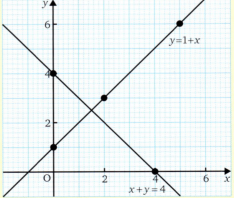

At the point where the two lines cross, the values of x and y
are the same for both equations, so they are the solutions of
the pair of equations.

From the graph we see that the solution is $x = 1\frac{1}{2}$, $y = 2\frac{1}{2}$.

$x = 1\frac{1}{2}$ and $y = 2\frac{1}{2}$ satisfy both equations.
We call a pair of equations like this, simultaneous equations.

Exercise 19i

Solve the following pairs of simultaneous equations graphically. In each case draw axes for x and y and use values in the ranges indicated, taking 2 cm to 1 unit:

1   $x + y = 6$ $0 \leqslant x \leqslant 6$, $0 \leqslant y \leqslant 6$,
    $y = 3 + x$

2   $x + y = 5$ $0 \leqslant x \leqslant 6$, $0 \leqslant y \leqslant 6$
    $y = 2x + 1$

3   $y = 4 + x$ $0 \leqslant x \leqslant 6$ $0 \leqslant y \leqslant 6$
    $y = 1 + 3x$

4   $x + y = 1$ $-3 \leqslant x \leqslant 2$, $-2 \leqslant y \leqslant 4$
    $y = x + 2$

5   $2x + y = 3$ $0 \leqslant x \leqslant 3$, $-3 \leqslant y \leqslant 3$
    $x + y = 2\frac{1}{2}$

6   $y = 5 - x$ $0 \leqslant x \leqslant 5$, $0 \leqslant y \leqslant 7$
    $y = 2 + x$

7   $3x + 2y = 9$ $0 \leqslant x \leqslant 4$, $-2 \leqslant y \leqslant 5$
    $2x - 2y = 3$

8   $2x + 3y = 4$ $-2 \leqslant x \leqslant 2$, $0 \leqslant y \leqslant 4$
    $y = x + 2$

9   $x + 3y = 6$ $0 \leqslant x \leqslant 5$, $0 \leqslant y \leqslant 5$
    $3x - y = 6$

10   $x = 2y - 3$ $-2 \leqslant x \leqslant 3$, $0 \leqslant y \leqslant 4$
    $y = 2x + 1$

Exercise 19j

Try to solve the following equations graphically. Why do you think the method breaks down?

1   $x + y = 9$ $0 \leqslant x \leqslant 9$
    $x + y = 4$ $0 \leqslant y \leqslant 9$

2   $y = 2x + 3$ $0 \leqslant x \leqslant 4$
    $y = 2x - 1$ $-1 \leqslant y \leqslant 11$

3   $2x + y = 3$ $0 \leqslant x \leqslant 3$
    $4x + 2y = 7$ $-3 \leqslant y \leqslant 4$

4   $y = 2x - 4$ $0 \leqslant x \leqslant 4$
    $2x = y + 4$ $-4 \leqslant y \leqslant 4$

There are other ways of solving two simultaneous equations. This work is done in Grade 9.

Forming equations

When we know enough information about two unknown quantities, we can form a pair of equations. We can then draw graphs to solve the equations and find the values of the unknowns.

Exercise 19k

One number added to twice another number is 7. Twice the first number added to the second number is 8.

a Use this information to form a pair of equations.

b Draw graphs representing your equations and hence find the two numbers.

a First allocate letters to the unknown numbers.

Let x be the first number and y be the second number.

Then the first sentence can be written as $x + 2y = 7$

and the second sentence can written as $2x + y = 8$

b Now make tables of values: $x + 2y = 7$: $2x + y = 8$:

x	–3	0	3
y	5	3.5	2

x	–2	0	2
y	12	8	4

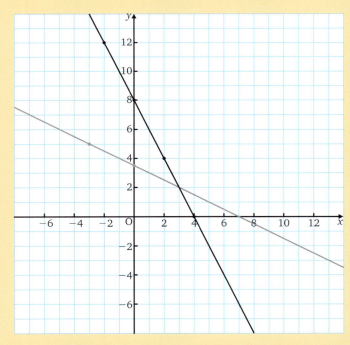

The graphs intersect where $x = 3$ and $y = 2$ so the numbers are 2 and 3.

In questions **1** to **6**, form two equations to show the information.

1 The sum of two numbers is 8. Twice the smaller number minus the larger number is 1.

2 Two numbers are such that if I double the first and subtract the second, I get 5 but if I add double the second to the first I get 10.

3 The perimeter of this rectangle is 12. The length is twice the width.

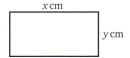

4 The sum of two numbers is 18 and one number is twice the other.

5 The equation of a straight line is $y = mx + c$. When $x = 2$, $y = 8$ and when $x = -1$, $y = 2$.

6 In the diagram, one angle is twice the size of the other.

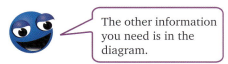

The other information you need is in the diagram.

7 Use a graph drawing package such as GeoGebra to draw graphs to solve your equations.

Mixed exercises

Exercise 19I

1 Find the x-coordinates of the points on the line $y = 3x$ that have y-coordinates of

   a 6   **b** −12   **c** 2.

2 If the points $(6, a)$, $(-\frac{1}{2}, b)$ and $(c, 1)$ lie on the straight line with equation $3y = -2x$, find the values of a, b and c.

3 Determine whether the straight lines with the given equations have positive or negative gradients:

   a $y = 4x$ **b** $y = -2x + 2$ **c** $y = \frac{2}{3}x - 7$

4 Copy and complete the following table and use it to draw the graph of $y = 2x - 3$:

x	−3	0	4
y			

Find the gradient of this line.

5 Determine in each case whether the straight line with the given equation makes an acute angle or an obtuse angle with the positive x-axis.

a $y = -\frac{2}{3}x$ **b** $y = 5x + 2$

c $2x + y = 3$ **d** $3y = -4x + 7$

6 Draw on the same axes, using 1 cm as 1 unit in each case, the graphs of $y = 2x - 4$ and $2x + y + 8 = 0$. Write down the coordinates of the point where these lines intersect.

Exercise 19m

1 Find the y-coordinates of the points on the line $y = 5x$ that have x-coordinates of

a 2   **b** 3   **c** $\frac{1}{2}$.

2 If the points $(-1, a)$, $(b, 15)$ and $(c, -20)$ lie on the straight line with equation $y = 5x$, find the values of a, b and c.

3 Determine whether the straight lines with the given equations have positive or negative gradients:

a $y = 6x$ **b** $y = -3x + 2$ **c** $x + y = 4$

4 Write down the gradients and y-intercepts for the straight lines with the given equations:

a $y = 4x - 7$ **b** $2y = 5x + 2$ **c** $y - 3x = 2$ **d** $3y = -x - 12$

5 Determine whether or not the given pairs of equations represent parallel lines:

a $y = -x + 2$, $x + y = 3$ **b** $2y = 4x + 3$, $y + 2x = 5$

6 Draw, on the same axes, the graphs of $x = -3$, $y = \frac{1}{2}x$ and $y = 4$, for values of x between -4 and $+8$. Write down the coordinates of the three points where these lines intersect.

Exercise 19n

1 Find the y-coordinates of the points on the line $y = 7x + 4$ that have x-coordinates of

a 1   **b** -2 **c** -5

2 If the points $(3, a)$, $(-2, b)$ and $(c, -10)$ lie on the straight line with equation $y = 5 - 3x$, find the values of a, b and c.

3 Sketch on the same axes the graphs of the straight lines with equations

a $y = -3x$ **b** $y = 2x + 4$.

4   Draw the graph of $y = 5x - 2$ for values of x between -4 and 4. Use 2 cm as 1 unit on the x-axis and 1 cm as 1 unit on the y-axis. From your graph, or otherwise, find
    a   the gradient of the line   **b**   its y-intercept.

5   Write down the equations of the straight lines that have the given gradients and y-intercepts:
    a   gradient 2, y-intercept -4 **b**   gradient $\frac{1}{2}$, y-intercept 5
    c   gradient -4, y-intercept -3

6   Draw, on the same axes, the graphs of $x = 1$, $y = -2x - 2$, $y = 4$ for values of x between -4 and $+4$. Write down the coordinates of the three points where these lines intersect.

7   Draw on the same axes, using 1 cm as 1 unit in each case, the graphs of $2x - y = 4$ and $2x + y = -8$. Add together the two equations. Draw the graph of the new equation on the same set of axes. Do you notice anything special about these three lines?

8   Draw on the same axes the graphs of $x + 2y = 8$, $x + y = 4$ and $2x + 3y = 12$. What do you notice about these three lines?

In this chapter you have seen that...

✔ you can find the missing coordinate of a point on a line, given the equation of the line and one coordinate, by substituting the given coordinate into the equation and solving it

✔ you can draw a straight line graph, given its equation by finding the coordinates of three points on the line

✔ the gradient of a straight line whose equation is $y = mx + c$ is m

✔ a positive gradient means that the line makes an acute angle with the positive x-axis and a negative gradient means that the line makes an obtuse angle with the positive x-axis

✔ a straight line whose equation is $y = mx + c$ crosses the y-axis where $y = c$

✔ if you know the gradient and y-intercept of a straight line you can substitute these values for m and c in $y = mx + c$ and hence give the equation of the line

✔ two straight lines are parallel if their gradients are equal

✔ the equation of a straight line parallel to the x-axis is $y = a$ and the equation of a straight line parallel to the y-axis is $x = b$

✔ two linear simultaneous equations can be solved by drawing the graphs of the equations and finding where they intersect.

20 Inequalities

Did you know?

There are an infinite number of counting numbers: 1 is the 1st, 2 is the 2nd, 3 is the 3rd, 4 is the 4th, and so on.

There are exactly the same number of positive even numbers: 2 is the 1st, 4 is the 2nd, 6 is the 3rd, 8 is the 4th, and so on.

you need to know...

✔ the equations of lines parallel to the axes

✔ the meaning of the symbols $<$, $\leqslant$, $>$ and $\geqslant$

✔ how to draw a number line

✔ how to work with positive and negative numbers

Key words

boundary line, inequality, range, two-dimensional space, xy-plane, symbols $<$, $\leqslant$, $>$ and $\geqslant$

Consider the statement

$$x > 5$$

This is an *inequality* (as opposed to $x = 5$ which is an equality or equation).

This inequality is true when x stands for any number that is greater than 5. So there is a range of numbers that x can stand for and we can illustrate this range on a number line.

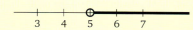

The circle at the left hand end of the range is 'open', because 5 is not included in the range.

Exercise 20a

Use a number line to illustrate the range of values of x for which $x < -1$

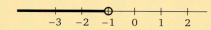

(The open circle means that -1 is not included. All values smaller than -1 are to the left of it on the number line.)

Use a number line to illustrate the range of values of x for which each of the following inequalities is true:

1	$x > 7$	**4**	$x > 0$	**7**	$x < 5$
2	$x > 4$	**5**	$x < -2$	**8**	$x < 0$
3	$x > -2$	**6**	$x > \frac{1}{2}$	**9**	$x < 1.5$

10 State which of the inequalities given in questions **1** to **9** are satisfied by a value of x equal to

    a 2   **b** -3 **c** 0   **d** 1.5   **e** 0.0005

11 For each of the questions **1** to **9** give a number that satisfies the inequality and is

    a a whole number   **b** not a whole number

12 Consider the true inequality $3 > 1$

    a Add 2 to each side.   **b** Add -2 to each side.

    c Take 5 from each side.   **d** Take -4 from each side.

    In each case state whether or not the inequality remains true.

13 Repeat question **12** with the inequality $-2 > -3$

14 Repeat question **12** with the inequality $-1 < 4$

15 Try adding and subtracting different numbers on both sides of a true inequality of your own choice.

Solving inequalities

From the last exercise we can see that

an inequality remains true when the *same* number is added to, or subtracted from, *both* sides.

Now consider the inequality $x - 2 < 3$

Solving this inequality means finding the range of values of x for which it is true.

Adding 2 to each side gives $x < 5$

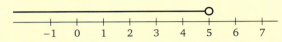

We have now solved the inequality.

Exercise 20b

Solve the following inequalities and illustrate your solutions on a number line:

1	$x - 4 < 8$	**5**	$x + 4 < 2$	
2	$x + 2 < 4$	**6**	$x - 5 < -2$	
3	$x - 2 > 3$	**7**	$x - 3 < -6$	
4	$x - 3 > -1$	**8**	$x + 7 < 0$	**9** $x + 2 < -3$

Add 4 to each side first.

Solve the inequality $4 - x < 3$

$$4 - x < 3$$

(Aim to get the x term on one side of the inequality and the number term in the other.)

Add x to each side $4 < 3 + x$

Take 3 from each side $1 < x$ or $x > 1$

An inequality remains true if the sides are reversed but you must remember to reverse the inequality sign.

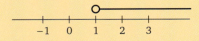

Solve the following inequalities and illustrate your solutions on a number line:

10	$4-x>6$	**16**	$3-x>2$	**22**	$3-x<3$
11	$2<3+x$	**17**	$6<x+8$	**23**	$5<x-2$
12	$7-x>4$	**18**	$2+x<-3$	**24**	$7>2-x$
13	$5<x+5$	**19**	$2>x-3$	**25**	$3>-x$
14	$5-x<8$	**20**	$4<5-x$	**26**	$4-x>-9$
15	$2>5+x$	**21**	$1<-x$	**27**	$5-x<-7$

28 Consider the true inequality $12<36$

    a Multiply each side by 2.   **b** Divide each side by 4.

    c Multiply each side by 0.5.   **d** Divide each side by 6.

    e Multiply each side by –2.   **f** Divide each side by –3.

In each case state whether or not the inequality remains true.

29 Repeat question **28** with the true inequality $36>-12$

30 Repeat question **28** with the true inequality $-18<-6$

31 Repeat question **28** with a true inequality of your own choice.

32 Can you multiply both sides of an inequality by any one number and be confident that the inequality remains true?

An inequality remains true when both sides are multiplied or divided by the same *positive* number.

Multiplication or division of an inequality by a negative number should be avoided, because it destroys the truth of the inequality.

Exercise 20c

Solve the inequality $2x-4>5$ and illustrate the solution on a number line.

$$2x-4>5$$

Add 4 to both sides $2x>9$

Divide both sides by 2 $x>4\frac{1}{2}$

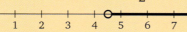

Solve the inequalities and illustrate the solutions on a number line:

1	$3x-2<7$	**3**	$4x-1>7$	**5**	$5+2x<6$	**7**	$4x-5<4$
2	$1+2x>3$	**4**	$3+5x<8$	**6**	$3x+1>5$	**8**	$6x+2>11$

Solve the inequality $3-2x \leqslant 5$ and illustrate the solution on a number line.
($\leqslant$ means 'less than or equal to')

(As with equations, we collect the letter term on the side with the greater number to start with. In this case we collect on the right.)

$$3-2x \leqslant 5$$

Add $2x$ to each side $\qquad\qquad 3 \leqslant 5+2x$

Take 5 from each side $\qquad\qquad -2 \leqslant 2x$

Divide each side by 2 $\qquad\qquad -1 \leqslant x$ i.e. which in reverse is $x \geqslant -1$

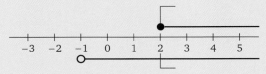

(A solid circle is used for the end of the range because -1 *is* included.)

Solve the inequalities and illustrate each solution on a number line:

9 $3 \leqslant 5-2x$	**12** $4 \geqslant 9-5x$	**15** $x-1 > 2-2x$	**18** $2x+1 \leqslant 7-4x$
10 $5 \geqslant 2x-3$	**13** $10 < 3-7x$	**16** $2x+1 \geqslant 5-x$	**19** $1-x > 2x-2$
11 $4-3x \leqslant 10$	**14** $8-3x \geqslant 2$	**17** $3x+2 \leqslant 5x+2$	**20** $2x-5 > 3x-2$

Find, where possible, the range of values of x which satisfy both of the inequalities
  a $x \geqslant 2$ and $x > -1$ **b** $x \leqslant 2$ and $x > -1$ **c** $x \geqslant 2$ and $x < -1$
  a

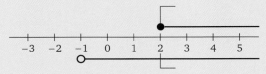

(Illustrating the ranges on a number line, we can see that both inequalities are satisfied for values on the number line where the ranges overlap.)

$\therefore x \geqslant 2$ and $x > -1$ are both satisfied for $x \geqslant 2$

  b

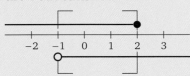

$x \leqslant 2$ and $x > -1$ are both satisfied for $-1 < x \leqslant 2$.

  c

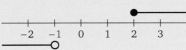

There are no values of x for which $x \geqslant 2$ and $x < -1$ are both satisfied.
(The lines do not overlap.)

Find, where possible, the range of values of x for which the two inequalities are both true:

21 a $x > 2$ and $x > 3$

   b $x \geqslant 2$ and $x \leqslant 3$

   c $x < 2$ and $x > 3$

22 a $x \geqslant 0$ and $x \leqslant 1$

   b $x \leqslant 0$ and $x \leqslant 1$

   c $x < 0$ and $x > 1$

23 a $x \leqslant 4$ and $x > -2$

   b $x \geqslant 4$ and $x < -2$

   c $x \leqslant 4$ and $x < -2$

24 a $x < -1$ and $x > -3$

   b $x < -1$ and $x < -3$

   c $x > -1$ and $x < -3$

Solve each of the following pairs of inequalities and then find the range of values of x which satisfy both of them:

25 $x - 4 < 8$ and $x + 3 > 2$

26 $3 + x \leqslant 2$ and $4 - x \leqslant 1$

27 $x - 3 \leqslant 4$ and $x + 5 \geqslant 3$

28 $2x + 1 > 3$ and $3x - 4 < 2$

29 $5x - 6 > 4$ and $3x - 2 < 7$

30 $3 - x > 1$ and $2 + x > 1$

31 $1 - 2x \leqslant 3$ and $3 + 4x < 11$

32 $0 > 1 - 2x$ and $2x - 5 \leqslant 1$

Find the values of x for which $x - 2 < 2x + 1 < 3$

($x - 2 < 2x + 1 < 3$ represents two inequalities,

i.e. $x - 2 < 2x + 1$ and $2x + 1 < 3$, so solve each one separately.)

$$x - 2 < 2x + 1 \qquad\qquad 2x + 1 < 3$$

$$-2 < x + 1 \qquad\qquad 2x < 2$$

$$-3 < x \text{ i.e. } x > -3 \qquad\qquad x < 1$$

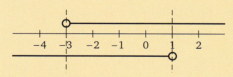

$$-3 < x < 1$$

Find the range of values of x for which the following inequalities are true:

33 $x + 4 > 2x - 1 > 3$

34 $x - 3 \leqslant 2x \leqslant 4$

35 $3x + 1 < x + 4 < 2$

36 $2 - x < 3x + 2 < 8$

37 $2 - 3x \leqslant 4 - x \leqslant 3$

38 $x - 3 < 2x + 1 < 5$

39 $2x < x - 3 < 4$

40 $4x - 1 < x - 4 < 2$

41 $4 - 3x < 2x - 5 < 1$

42 $x < 3x - 1 < x + 1$

? Puzzle

Find two numbers, one of which is twice the other, such that the sum of their squares is equal to the cube of one of the numbers.

Using two-dimensional space

So far we have discussed inequalities in a purely algebraic way. Now we look at them in a more visual way, using graphs.

If we have the inequality $x \geqslant 2$, x can take any value greater than or equal to 2. This can be represented by the following diagram.

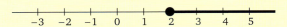

On this number line, x can take any value on the heavy part of the line including 2 itself, as indicated by the solid circle at 2.

If $x > 2$ then the diagram is as shown below.

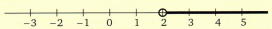

In this case, x cannot take the value 2 and this is shown by the open circle at 2.

It is sometimes more useful to use two-dimensional space with x and y axes, rather than a one-dimensional line. We represent $x \geqslant 2$ by the set of points whose x coordinates are greater than or equal to 2. (y is not mentioned in the inequality so y can take any value.)

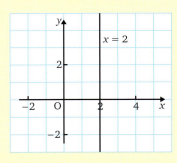

The boundary line represents all the points for which $x = 2$ and the region to the right contains all points with x coordinates greater than 2.

To indicate this, and to make future work easier, we use a continuous line for the boundary when it is included and we shade the region we do *not* want.

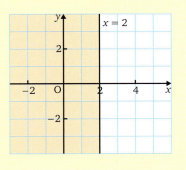

The inequality $x > 2$ tells us that x may not take the value 2. In this case we use a broken line for the boundary.

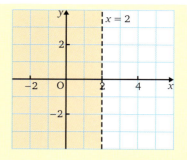

We can draw a similar diagram for $y > -1$

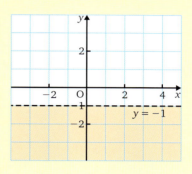

Exercise 20d

Draw diagrams to represent the inequalities

a $x \leqslant 1$ **b** $2 < y$

a $x \leqslant 1$

The boundary line is $x = 1$ (included).
The unshaded region represents $x \leqslant 1$

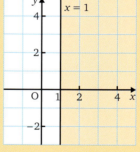

b $2 < y$

The boundary line is $y = 2$ (not included).
The unshaded region represents $2 < y$

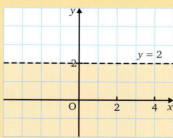

Draw diagrams to represent the following inequalities:

1 $x \geqslant 2$ **3** $x > -1$ **5** $x \geqslant 0$ **7** $x \leqslant -4$

2 $y \leqslant 3$ **4** $y < 4$ **6** $0 > y$ **8** $2 < x$

Draw a diagram to represent $-3 < x < 2$ and state whether or not the points (1, 1) and (−4, 2) lie in the given region.

$-3 < x < 2$ gives two inequalities, $-3 < x$ and $x < 2$ so the boundary lines are $x = -3$ and $x = 2$ (neither included). Shade the regions not wanted.

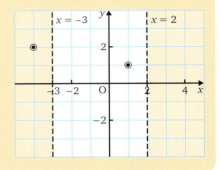

The unshaded region represents $-3 < x < 2$

Plot the points, then you see that (−4, 2) does not lie in the given region.

(1, 1) lies in the given region.

Draw diagrams to represent the following pairs of inequalities:

9 $2 \leqslant x \leqslant 4$ **12** $4 < y < 5$ **15** $-\frac{1}{2} \leqslant x \leqslant 1\frac{1}{2}$

10 $-3 < x < 1$ **13** $0 \leqslant x < 4$ **16** $-2 \leqslant y < -1$

11 $-1 \leqslant y \leqslant 2$ **14** $-2 < y \leqslant 3$ **17** $3 \leqslant x < 5$

18   In each of the questions **9** to **11**, state whether or not the point (1, 4) lies in the unshaded region.

Give the inequality that defines the unshaded region.

Boundary line $x = 2$ (included)

Inequality is $x \leqslant 2$

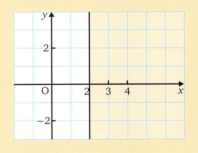

Give the inequalities that define the unshaded region

Boundary lines $y = 4$ (not included)

and $y = -1$ (included)

The inequalities are $y < 4$ and $y \geqslant -1$ or $-1 \leqslant y < 4$

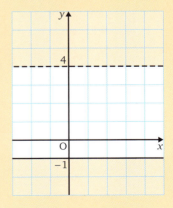

Give the inequalities that define the unshaded regions:

19

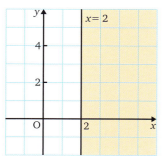

22

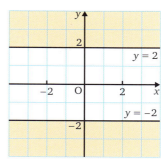

20

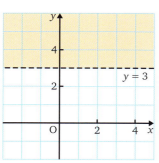

23

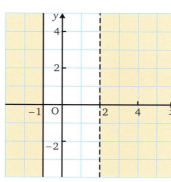

21

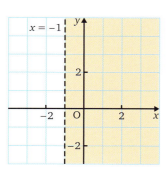

24

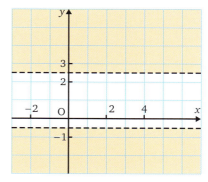

25 In each of the questions **19** to **24** state whether or not the point (2, −1) is in the unshaded region.

Give the inequalities that define the *shaded* regions:

26

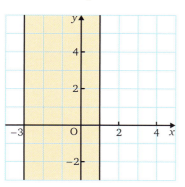

27

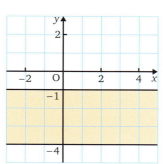

28

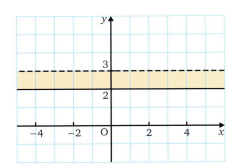

29

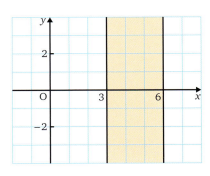

30 In each of the questions **26** to **29** state whether or not the point (0, 2) is in the shaded region.

Exercise 20e

Draw a diagram to represent the region defined by the set of inequalities
$-1 \leqslant x \leqslant 2$ and $-5 \leqslant y \leqslant 0$

There are four inequalities here: $-1 \leqslant x$, $x \leqslant 2$, $-5 \leqslant y$ and $y \leqslant 0$.

The boundary lines are

$x = -1$: for $-1 \leqslant x$, shade the region on the left of the line

$x = 2$: for $x \leqslant 2$, shade the region on the right of the line

$y = -5$: for $-5 \leqslant y$, shade the region below this line

$y = 0$: for $y \leqslant 0$, shade the region above the line

The unshaded region represents the inequalities.

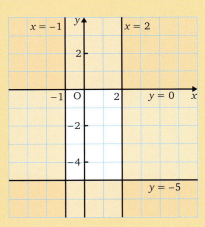

Draw diagrams to represent the regions described by the following sets of inequalities. In each case, draw axes for values of x and y from -5 to 5.

1 $2 \leqslant x \leqslant 4, -1 \leqslant y \leqslant 3$

2 $-2 < x < 2, -2 < y < 2$

3 $-3 < x \leqslant 2, -1 \leqslant y$

4 $0 \leqslant x \leqslant 4, 0 \leqslant y \leqslant 3$

5 $-4 < x < 0, -2 < y < 2$

6 $-1 < x < 1, -3 < y < 1$

7 $x \geqslant 0, y \geqslant 0$

8 $x \geqslant 1, -1 \leqslant y \leqslant 2$

Give the sets of inequalities that describe the unshaded regions:

9

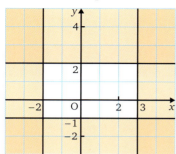

10

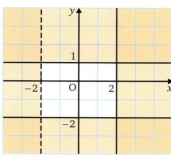

11 Is the point $(2\frac{1}{2}, 0)$ in either of the unshaded regions in questions **9** and **10**?

Give the sets of inequalities that describe the unshaded regions:

12

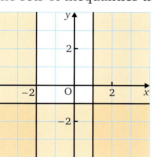

13

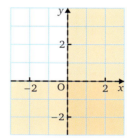

In this chapter you have seen that...

✔ an inequality remains true when the same number is added to, or subtracted from, both sides

✔ an inequality remains true when both sides are multiplied or divided by the same **positive** number. Do not multiply or divide an inequality by a negative number. It destroys the inequality.

21 Statistics

At the end of this chapter you should be able to...

1 Understand line graphs and be able to extract information from them.

2 Calculate the arithmetic mean of a set of data.

3 Find the missing number from a set, given the mean and the remaining numbers.

4 Find the mode of a set of numbers.

5 Find the median of a set of numbers.

6 Find the mode, mean and median from data given in a frequency table.

You need to know...

✔ how to add, subtract, multiply and divide whole numbers

✔ how to add, subtract, multiply and divide decimals

✔ how to put a set of numbers in ascending or descending order

Key words

arithmetic average, bar chart, frequency table, line graph, mean, median, mode

Line graphs

In Chapter 20 of Grade 7 we saw that we could represent data using bar charts, pie charts and pictographs. Another way is to draw line graphs. This kind of graph shows upward and downward trends clearly in data that varies over time.

Exercise 21a

The line graph below shows how the population of a town at the end of each year varied over a decade.

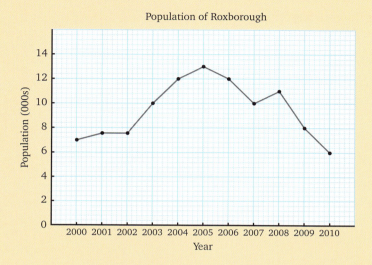

Population of Roxborough

a In which year was the population   **i** greatest   **ii** least?
b In which year did the population remain stationary?
c What was the increase in population between 2001 and 2005?
d Which year saw the most rapid decline in the population?
   How much was this?

a **i** The population was greatest in 2005 when it stood
        at 13 000.
   ii It was least in 2010 when it stood at 6000.
b The population remained stationary in 2002.
c Between 2001 and 2005 the population increased from 7500 to
   13 000 i.e. by 5500.
d The most rapid decline in the population is shown by the steepest
   line sloping downwards. This is for the year 2009.

1 The line graph shows the numbers of male
absentees (solid line) and female absentees (broken
line) in the workforce at a factory.

  a How many men were absent on

  i Tuesday   **ii** Saturday?

  b How many women were absent on

  i Wednesday   **ii** Friday?

  c On how many days were more women than men
absent?

  d How many more men than women were
absent on

  i Monday   **ii** Saturday?

  e How many person days production were lost
through absenteeism
during the week?

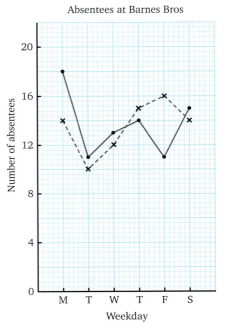

Absentees at Barnes Bros

2 Mr Worrell keeps a plant nursery. He is growing a new
plant and measures the height of the plant at the end of each week.
These heights are shown on the graph.

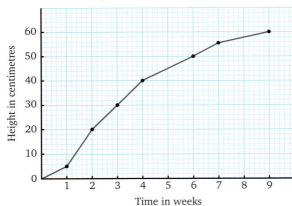

Height of a new plant over 9 weeks

  a How high was the plant after

  i 3 weeks

  ii 7 weeks?

  b How much did the plant grow

  i in the 4th week

  ii from the end of the first week to the end of the
5th week?

  c During which week did the plant grow fastest?

3   The line graph shows the quarterly sales figures for a manufacturing company

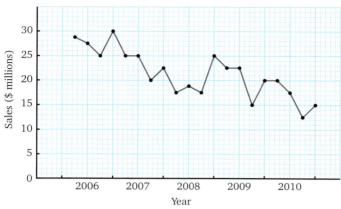

Sales figures at Aberdale & Co.

a   What were the 4th quarter sales in
   i   2007   **ii**   2009?
b   What were the 2nd quarter sales in
   i   2006   **ii**   2010?
c   Which quarter
   i   regularly sees an increase in sales
   ii   was the best quarter during the 5-year period?
d   Is there a pattern for the annual sales? If so, describe it.
e   If you were the owner of the business would you be concerned or pleased with the company's sales figures? Give a reason for your answer.

4   The line graph shows the average monthly precipitation, in centimetres, for the cities of Calcutta (solid line) and Miami (broken line).

Monthly precipitation for Calcutta and Miami

    a Which city has the greater range in precipitation?

    b For Calcutta find

       i the greatest monthly precipitation

      ii the lowest monthly precipitation.

    c Repeat part **b** for Miami.

    d Find the difference in precipitation for   **i** April   **ii** August.

    e You would like to have a rain-free visit to one of these cities.
        Which one would you choose and when?

5   This line graph shows the minimum monthly temperatures for Moscow
      (solid line) and Buenos Aires (broken line) over a period of a year.

Minimum monthly temperature in °C for Moscow and Buenos Aires

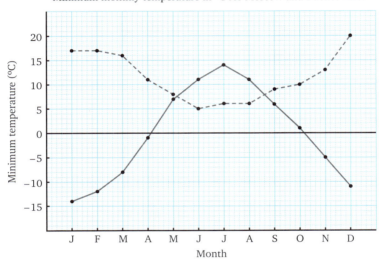

  a Write down the minimum temperature

     i in Buenos Aires in September

    ii in Moscow in June.

  b What is the difference in the minimum temperatures of the two
      cities in July?

  c In which month is the difference in the minimum temperatures

     i greatest   **ii** least?

  d For how many months is the minimum temperature in Buenos
      Aires lower than the minimum temperature in Moscow?

6   The table shows the gross profits of a company in millions of dollars.

Year	2001	2002	2003	2004	2005	2006	2007	2008	2009	2010
Profit $m	14.6	16.3	15.2	16.8	16.2	17.3	16.9	17.5	17.0	17.9

  a Draw a line graph to show this data.
      Use 1 cm to represent 2 units on the vertical axis and for 1 year on
      the horizontal axis.

b Are these the profits for a successful company? Give a reason for your answer.

7 The numbers of unemployed persons, expressed as a percentage of people of working age, in the city of Verzon for the first decade of the present century are given in the following table.

	2000	2001	2002	2003	2004	2005	2006	2007	2008	2009	2010
Males	4.8	4.3	4.7	5.6	6.7	5.7	5.9	4.2	4.1	4.2	3.9
Females	3.7	3.8	3.5	3.9	3.7	5.4	5.7	6.0	6.2	6.0	5.9

a Draw line graphs to illustrate this data.
   Use 1 cm to represent 1 unit on the vertical axis and for 1 year on the horizontal axis.
   Use a solid line for males and a broken line for females.
b Describe the trend that the graph shows.
c Can you suggest a possible reason for the shape of each of these line graphs.

8 The table shows the average monthly sea temperature in degrees Celsius for Nice and New York.

Month	J	F	M	A	M	J	J	A	S	O	N	D
Nice	14	13	13	14	17	21	23	24	23	20	18	15
New York	3	2	4	8	13	18	22	23	21	17	11	6

a Draw line graphs to illustrate this data.
   Use 1 cm to represent 5 units on the vertical axis and for 1 month on the horizontal axis.
   Use a solid line for Nice and a broken line for New York.
b Which sea has the more variable temperature? Give a reason for your answer.
c You would like to go swimming on holiday in Nice and don't want the water to be too cold. In which months would you consider going? Give a reason for your answer.

Averages

We are frequently looking for ways of representing a set of figures in a simple form. Can we choose a single number that will adequately represent a set of numbers?

We try to do this by using averages.

Three different types of averages are used, each with its own individual advantages and disadvantages.

They are the *arithmetic average* or *mean*, the *mode* and the *median*.

The arithmetic average or mean

Consider a group of five children. When they are asked to produce the money they are carrying the amounts collected are $56, $142, $96, $24 and $77 respectively. If the total of this money ($395) is shared equally amongst the five children, each will receive $79. This is called the arithmetic average or mean of the five amounts.

The arithmetic average or mean of a set of figures is the sum of the figures divided by the number of figures in the set.

For example, the average or mean of 12, 15, 25, 42 and 16 is

$$\frac{12+15+25+42+16}{5} = \frac{110}{5} = 22$$

One commonplace use of the arithmetic average is to compare the marks of pupils in a group or form. The pupils are given positions according to their average mark over the full range of subjects they study. An advantage is that we can compare the results of pupils who study 7 subjects with those who study 11 subjects. A disadvantage is that one very poor mark may pull the mean down significantly.

The mean may also be rather artificial, for example, giving $5\frac{1}{3}$ to each of a group of people, or having a mean shoe size of 5.1, or a mean family size of 2.24 children.

Exercise 21b

Find the arithmetic average or mean of the following sets of numbers:

1. 3, 6, 9, 14
2. 2, 4, 9, 13
3. 12, 13, 14, 15, 16, 17, 18
4. 23, 25, 27, 29, 31, 33, 35
5. 19, 6, 13, 10, 32
6. 34, 14, 39, 20, 16, 45
7. 1.2, 2.4, 3.6, 4.8
8. 18.2, 20.7, 32.5, 50, 78.6
9. 6.3, 4.5, 6.8, 5.2, 7.3, 7.1
10. 3.1, 0.4, 7.2, 0.7, 6.1
11. 38.2, 17.6, 63.5, 80.7
12. 0.76, 0.09, 0.35, 0.54, 1.36

John's examination percentages in 8 subjects were
83, 47, 62, 49, 55, 72, 58 and 62. What was his mean mark?

Mean mark for 8 subjects

$$= \frac{\text{Sum of the marks in the 8 subjects}}{\text{number of subjects}}$$

$$= \frac{83 + 47 + 62 + 49 + 55 + 72 + 58 + 62}{8}$$

$$= \frac{488}{8}$$

$$= 61$$

13   In the Christmas terminal examinations Lisa scored a total of 504 in 8 subjects. Find her mean mark.

14   A darts player scored 2304 in 24 visits to the board. What was his average number of points per visit?

15   A bowler took 110 wickets for 1815 runs. Calculate his average number of runs per wicket.

16   Peter's examination percentages in 7 subjects were 64, 43, 86, 74, 55, 53 and 66. What was his mean mark?

17   In six consecutive English examinations, Jane's percentage marks were 83, 76, 85, 73, 64 and 63. Find her mean mark.

18   A football team scored 54 goals in 40 league games. Find the average number of goals per game.

19   The first Hockey XI scored 14 goals in their first 16 matches. What was the average number of goals per match?

20   In an ice-dancing competition the recorded scores for the winners were 5.8, 5.9, 6.0, 5.8, 5.8, 5.8, 5.6 and 5.7. Find their mean score.

21   The recorded rainfall each day at a holiday resort during the first week of my holiday was 3 mm, 0, 4.5 mm, 0, 0, 5 mm and 1.5 mm. Find the mean daily rainfall for the week.

22   The masses of the members of a rowing eight were 82 kg, 85 kg, 86 kg, 86 kg, 84 kg, 88 kg, 92 kg and 85 kg. Find the average mass of the eight. If the cox had a mass of 41 kg, what was the average mass of the crew?

On average my car travels 28.5 miles on each gallon of petrol. How far will it travel on 30 gallons?

If the car travels 28.5 miles on 1 gallon of petrol it will travel 30×28.5 miles, i.e. 855 miles, on 30 gallons.

23 My father's car travels on average 33.4 miles on each gallon of petrol. How far will it travel on 55 gallons?

24 Olga's car travels on average 12.6 km on each litre of petrol. How far will it travel on 205 litres?

25 The average daily rainfall in Puddletown during April was 2.4 mm. How much rain fell during the month?

26 The daily average number of hours of sunshine during my 14 day holiday in Greece was 9.4. For how many hours did the sun shine while I was on holiday?

Elaine's average mark after 7 subjects is 56 and after 8 subjects it has risen to 58. How many does she score in her eighth subject?

We can find the total scored in 7 subjects and the total scored in 8 subjects. Then the score in the 8th subject is the difference between these two totals.

$$\text{Total scored in 7 subjects is } 56 \times 7 = 392$$
$$\text{Total scored in 8 subjects is } 58 \times 8 = 464$$

Score in her eighth subject

$$= \text{total for 8 subjects} - \text{total for 7 subjects}$$
$$= 464 - 392$$
$$= 72$$

Therefore Elaine scores 72 in her eighth subject.

27 Vivian Richard's batting average after 11 completed innings was 62. After 12 completed innings it had increased to 68. How many runs did he score in his twelfth innings?

28 Richard was collecting money for charity. The average amount collected from the first 15 houses at which he called was $30, while the average amount collected after 16 houses was $35. How much did he collect from the sixteenth house?

29 After six examination results Tom's average mark was 57. His next result increased his average to 62. What was his seventh mark?

30 Anne's average mark after 8 results was 54. This dropped to 49 when she received her ninth result which was for French. What was her French mark?

31 In seven consecutive innings a batsman scored 53, 4, 73, 104, 66, 44 and 83. What was his average? What does he score in his next innings if his average falls to 56?

His score in his 8th innings is the difference between his total score in all 8 innings and his total score in the first 7 innings.

32 During a certain week the number of lunches served in a school canteen were: Monday 213, Tuesday 243, Wednesday 237 and Thursday 239. Find the average number of meals served daily over the four days. If the daily average for the week (Monday–Friday) was 225, how many meals were served on Friday?

33 A paperboy's sales during a certain week were: Monday 84, Tuesday 112, Wednesday 108, Thursday 95 and Friday 131. Find his average daily sales. When he included his sales on Saturday his daily average increased to 128. How many papers did he sell on Saturday?

34 The number of hours of sunshine in Barbados for successive days during a certain week were 11.1, 11.9, 11.2, 12.0, 11.7, 12.9 and 11.8. Find the daily average. The following week the daily average was 11 hours. How many more hours of sunshine were there the first week than the second?

35 Jean's marks in the end of term examinations were 46, 80, 59, 83, 54, 67, 79, 82 and 62. Find her average mark. It was found that there had been an error in her mathematics mark. It should have been 74, not 83. What difference did this make to her average?

36 The heights of the 11 girls in a basketball team are 162 cm, 152 cm, 166 cm, 149 cm, 153 cm, 165 cm, 169 cm, 145 cm, 155 cm, 159 cm and 163 cm. Find the average height of the team. If the girl who was 145 cm tall was replaced by a girl 156 cm tall, what difference would this make to the average height of the team?

37 During the last five years the distances I travelled in my car, in miles, were 10 426, 12 634, 11 926, 14 651 and 13 973. How many miles did I travel in the whole period? What was my yearly average? How many miles should I travel this year to reduce the average annual mileage over the six years to 11 984?

38 The average mass of the 18 boys in a class is 63.2 kg. When two new boys join the class the average mass increases to 63.7 kg. What is the combined mass of the two new boys?

39 The average height of the 12 boys in a class is 163 cm and the average height of the 18 girls is 159 cm. Find the average height of the class.

You can find the total height of the boys and the total height of the girls. From this you can find the total height of all 30 pupils. Then you can work out the average height.

40 The average mass of the 15 girls in a class is 54.4 kg while the average mass of the 10 boys is 57.4 kg. Find the average mass of the class.

41 In a school the average size of the 14 lower school forms is 30, the average size of the 16 middle school forms is 25 and the average size of the 20 upper school forms is 24. Find the average size of form for the whole school.

42 Northshire has an area of 400 000 hectares and last year the annual rainfall was 274 cm, while Southshire has an area of 150 000 hectares and last year the annual rainfall was 314 cm. What was the annual rainfall last year for the combined area of the two counties?

43 After 10 three-day matches and 8 one-day matches, the average *daily* attendances for a cricket season were 2160 for three-day matches and 4497 for one-day matches. Calculate the average *daily* attendance for the 18 matches.

? Puzzle

How is it possible for a batsman, whose average is 53, to increase that average by scoring 35 in the next innings?

Mode

The mode of a set of numbers is the number that occurs most frequently, e.g. the mode of the numbers 6, 4, 6, 8, 10, 6, 3, 8 and 4 is 6, since 6 is the only number occurring more than twice.

It would obviously be of use for a firm with a chain of shoe shops to know that the mode or modal size for men's shoes in one part of the country is 8, whereas in another part of the country it is 7. Such information would influence the number of pairs of shoes of each size kept in stock.

If all the figures in a set of figures are different, there cannot be a mode, for no figure occurs more frequently than all the others. On the other hand, if two figures are equally the most popular, there will be two modes.

In Chapter 20 of Grade 7, we used bar charts to show such things as the spread of heights in a group of children, and the favourite colour of a group of people. These may be used to determine the mode of the group.

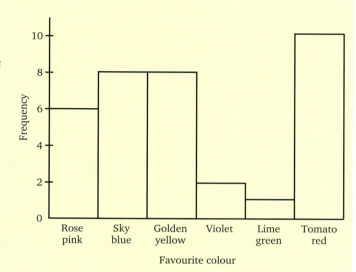

The bar chart is reproduced from Chapter 20 of Grade 7. It shows the colour selected by 35 people when asked to choose their favourite colours from a card showing six colours.

It shows that the most popular colour, or the modal colour, is tomato red.

Exercise 21c

What is the mode of each of the following sets of numbers:

1   10, 8, 12, 14, 12, 10, 12, 8, 10, 12, 4

2   3, 9, 7, 9, 5, 4, 8, 2, 4, 3, 5, 9

3   1.2, 1.8, 1.9, 1.2, 1.8, 1.7, 1.4, 1.3, 1.8

4   58, 56, 59, 62, 56, 63, 54, 53

5   5.9, 5.6, 5.8, 5.7, 5.9, 5.9, 5.8, 5.7

6   26.2, 26.8, 26.4, 26.7, 26.5, 26.4, 26.6, 26.5, 26.4

7   The table shows the number of goals scored by a football club last season.

Number of goals	0	1	2	3	4	5	6
Frequency	12	16	7	4	2	0	1

Draw a bar chart to show these results and find the modal score.

8   Given below are the marks out of 10 obtained by 30 girls in a history test.

8, 6, 5, 7, 8, 9, 10, 10, 3, 7, 3, 5, 4, 8, 7,

8, 10, 9, 8, 7, 10, 9, 9, 7, 5, 4, 8, 1, 9, 8

Draw a bar chart to show this information and find the mode.

9   The heights of 10 girls, correct to the nearest centimetre, are:

155, 148, 153, 154, 155, 149, 162, 154, 156, 155

What is their modal height?

10   The number of letters in the words of a sentence were:

2, 4, 3, 5, 2, 3, 8, 2, 5, 7, 9, 3, 6, 3, 7, 3,
4, 9, 2, 3, 8, 3, 5, 2, 10, 3, 4, 6, 2, 3, 4

How many words were there in the sentence? What is the mode?

11   The shoe sizes of pupils in a class are:

4, 4, 7, 6, 5, 5, 6, 6, 6, 4, 5, 8, 6, 7, 4, 7, 9, 6,
5, 7, 6, 7, 8, 6, 4, 4, 4, 5, 5, 7, 7, 7, 5, 8, 6, 5

How many pupils are there in the class?
What is the modal shoe size?

Median

The median value of a set of numbers is the value of the middle number when they have been placed in ascending (or descending) order of magnitude.

Imagine nine children arranged in order of their height.

The height of the fifth or middle child is 154 cm,
i.e. the median height is 154 cm.

Similarly 24 is the median of 12, 18, 24, 37 and 46. Two numbers are smaller than 24 and two are larger.

To find the median of 16, 49, 53, 8, 32, 19 and 62, rearrange the numbers in ascending order:

8, 16, 19, 32, 49, 53, 62

then we can see that the middle number of these is 32,
i.e. the median is 32.

If there is an even number of numbers, the median is found by finding the average or mean of the two middle values after they have been placed in ascending or descending order.

To find the median of 24, 32, 36, 29, 31, 34, 35, 39, rearrange in ascending order:

24, 29, 31, <u>32</u>, <u>34</u>, 35, 36, 39

Then the median is $\dfrac{32+34}{2} = \dfrac{66}{2}$

i.e. the median is 33.

Exercise 21d

Find the median of each of the following sets of numbers.

1   1, 2, 3, 5, 7, 11, 13

2   26, 33, 39, 42, 64, 87, 90

3   13, 24, 19, 13, 6, 36, 17

4   4, 18, 32, 16, 9, 7, 29

5   1.2, 3.4, 3.2, 6.5, 9.8, 0.4, 1.8

6   5, 7, 11, 13, 17, 19

7   34, 46, 88, 92, 104, 116, 118, 144

8   34, 42, 16, 83, 97, 24, 18, 38

9   1.92, 1.84, 1.89, 1.86, 1.96, 1.98, 1.73, 1.88

10   15.2, 6.3, 14.8, 9.5, 16.3, 24.9

 Investigation

Investigate whether or not it is possible to write down a set of seven different whole numbers such that their mean, mode and median are all the same.

Uses for the mean, mode and median

Each of the three averages have uses in different situations.

The mean is useful because it includes all the values and can be used to compare similar sets of data. Different sets of examination marks are usually compared using the mean mark for each set.

The disadvantage of the mean is that it can be affected by a few very large (or very small) values. For example, the mean of these values, 4, 6, 7, 9, 200 is 45.2 and this does not represent most of the values. However, the median is 7 and this is a better representative.

The median is useful when there are a few extreme values that will distort the mean.

The mode is useful when we need to know the most frequent value. When a store manager wants to order stock, it helps to know what items sell the most.

Mixed examples

Exercise 21e

Find **a** the mean **b** the mode and **c** the median, of each of the following sets of numbers:

1   21, 16, 25, 21, 19, 32, 27

2   67, 71, 69, 82, 70, 66, 81, 66, 67

3   43, 46, 47, 45, 45, 42, 47, 49, 43, 43

4   84, 93, 13, 16, 28, 13, 32, 63, 45

5   30, 27, 32, 27, 28, 27, 26, 27

6   In seven rounds of golf, a golfer returns scores of: 72, 87, 73, 72, 86, 72 and 77. Find the mean, mode and median of these scores.

7   The heights (correct to the nearest centimetre) of a group of girls are: 159, 155, 153, 154, 157, 162, 152, 160, 161, 157.
Find **a** their mean height   **b** their modal height   **c** their median height.

8   The marks, out of 100, in a geography test for the members of a class were: 64, 50, 35, 85, 52, 47, 72, 31, 74, 49, 36, 44, 54, 48, 32, 52, 53, 48, 71, 52, 56, 49, 81, 45, 52, 80, 46.
Find **a** the mean mark   **b** the modal mark   **c** the median mark.

9   Find the mean, mode and median of the following golf scores: 85, 76, 91, 83, 88, 84, 84, 82, 77, 79, 80, 83, 86, 84.

10   The table shows how many pupils in a form were absent for various numbers of sessions during a certain school week.

Number of sessions absent	0	1	2	3	4	5	6	7	8	9	10
Frequency	20	2	4	0	2	0	1	2	0	0	1

Find **a** the mode   **b** the median   **c** the mean.
Which of these averages would you choose to represent the number of absences? Give a reason for your answer.

11 The table shows the number of children per family in the families of the pupils in a class.

Number of children	1	2	3	4	5	6	7
Frequency	1	3	9	5	5	2	1

Find **a** the mode **b** the median **c** the mean.
Which of these averages would you choose to represent the number of children in a family? Give a reason for your answer.

? Puzzle

Find the missing number in this set.

6	10	5	3
7	6	7	8
8	9	8	

Finding the mode from a frequency table

The frequency table shows the number of houses in a village that are occupied by different numbers of people:

Number of people living in one house	0	1	2	3	4	5	6
Frequency	2	10	8	15	25	12	4

The highest frequency is 25 so there are more houses with four people living in them than any other number, i.e. the mode is 4.

Finding the mean from a frequency table

The pupils in a class were asked to state the number of children in their own family and the following frequency table was made:

Number of children per family	1	2	3	4	5
Frequency	7	15	5	2	1

Total no. of families = 30

All the numbers are here so we can total this set. We have seven families with one child giving seven children, fifteen with two children giving thirty children and so on, giving the total number of children as
$$(7 \times 1) + (15 \times 2) + (5 \times 3) + (2 \times 4) + (1 \times 5) = 65$$

There are 30 numbers in the set, so the mean is

$\frac{65}{30} = 2.2$ (to 1 d.p.)

i.e. there are, on average, 2.2 children per family.

To avoid unnecessary errors, this kind of calculation needs to be done systematically and it helps if the frequency table is written vertically.

We can then add a column for the number of children in each group and sum the numbers in this column for the total number of children.

Number of children per family x	Frequency f	fx
1	7	7
2	15	30
3	5	15
4	2	8
5	1	5
	No. of families = 30	No. of children = 65

mean $= \frac{65}{30} = 2.2$ (to 1 d.p.)

Exercise 21f

1

Number of tickets bought per person for a football match	1	2	3	4	5	6	7
Frequency	250	200	100	50	10	3	1

Find the mean number of tickets bought per person.

2

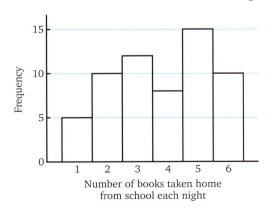

Frequency / Number of books taken home from school each night

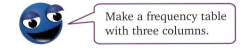

Draw a table, like the one above, with a third column.

Make a frequency table with three columns.

Find the mean for the number of books taken home each night.

3   This table shows the results of counting the number of prickles per leaf on 44 leaves.

Number of prickles	1	2	3	4	5	6
Frequency	4	2	2	7	20	9

Find

a   the mean number of prickles per leaf

b   the mode.

4   A six-sided die was thrown 50 times. The table gives the number of times each score was obtained.

Score	1	2	3	4	5	6
Frequency	7	8	10	8	5	12

Find

a   the mean score per throw

b   the mode.

5   Three coins were tossed together 30 times and the number of heads per throw was recorded.

Number of heads	0	1	2	3
Frequency	3	12	10	5

Find

a   the mean number of heads per throw

b   the mode.

Finding the median from a frequency table

Exercise 21g

1   A group of students gathered this information about themselves.

Number of children in each family	Frequency
1	8
2	12
3	4
4	2

Find the median number of children per family.

You want to find the middle value.

First find the number of families (add up the frequencies).

You can find where in the order the middle value is by adding 1 to the total, then dividing this by 2.

This tells you which family or families you want.

Is it in the first 8 families? This would give a median of 1 child.

Is it in the next 12 families, i.e. 9 to 20?

2 Once every five minutes, Debbie counted the number of people queuing at a checkout. Her results are shown in this table.

Number of people queuing at a supermarket checkout	Frequency
0	4
1	6
2	5
3	2
4	2

Write down the median number of people queuing.

3 This frequency table shows the distribution of scores when a die is rolled 20 times.

Score	1	2	3	4	5	6
Frequency	3	2	5	3	3	4

Find the median score.

4 In a shooting competition a competitor fired 50 shots at a target and got the following scores

Score	1	2	3	4	5
Frequency	3	4	18	16	9

Find
    a the median score   **b** the mode   **c** the mean.

5 The table shows the distribution of goals scored by the home teams one Saturday.

Score	0	1	2	3	4	5
Frequency	3	8	4	3	5	2

Find
    a the median score   **b** the mode   **c** the mean.

? Puzzle

A man left an estate of $604 500 to be divided among his widow, 4 sons and 5 daughters. He directed that every daughter should have three times as much as a son and that every son twice as much as their mother. What was the widow's share?

Investigation

a   Count the number of letters in the surname of each of the teachers in your school.

Enter this information in a frequency table like this.

Number of letters in surname	3	4	5	6	7	8		
Frequency								

You do not have to start your table with 3. Start with the number of letters in the shortest surname. Go up as far as you need to.

b   Now find, from these data,
   i the mode   **ii** the median   **iii** the mean.

c   Repeat this investigation for the students in your class. Compare the mode, median and mean for your class with that for the teachers. Are the values similar or quite different?

In this chapter you have seen that...

✔ a line graph shows upward and downward trends of data over time

✔ you find the mean by adding all the values then dividing this sum by the number of values

✔ the mode is the value or values that occur most often, i.e. with the greatest frequency

✔ the median is the middle value after a set of values has been arranged in order of size; when there are two middle values the median is half-way between them

✔ the mean, mode and median can all be found from an ungrouped frequency table.

22 Sets

At the end of this chapter you should be able to...

1 Use correctly the given symbols $\in$, $\notin$, $\subset$, $\subseteq$, $\cup$, $\cap$, $\varnothing$ and { }.

2 Classify sets as finite or infinite.

3 Determine when two sets are equal.

4 Identify empty sets and use the correct symbol for a set.

5 Write down the subsets of a given set.

6 Give a suitable universal set for a given set.

7 Find the union or intersection of sets.

8 Draw Venn diagrams to show the union or intersection of sets.

9 Solve simple problems using Venn diagrams.

Did you know?

Venn diagrams are named after John Venn (1834–1923), an Englishman born in Yorkshire who studied logic at Cambridge University.

Key words

disjoint sets, element, empty set, equal set, finite set, infinite set, intersection of sets, member, null set, proper subset, set, subset, union of sets, universal set, Venn diagram, the symbols $\in$, $\notin$, $\subset$, $\subseteq$, $\cup$, $\cap$, $\varnothing$ and { }

Set notation

A *set* is a collection of things having something in common.

Things that belong to a set are called *members* or *elements*. These elements are usually separated by commas and written down between curly brackets or braces.

Instead of writing 'the set of Jamaican reggae stars', we write {Jamaican reggae stars}.

The symbol ∈ means 'is a member of' so that 'History is a member of the set of school subjects' may be written History ∈ {school subjects}

Similarly the symbol ∉ means 'is not a member of'.

'Elm is not a breed of dog' may be written Elm ∉ {breeds of dogs}

Exercise 22a

1 Use the correct set notation to write down the following sets.
   a the set of teachers in my school
   b the set of books I have read.

2 Write down two members from each of the sets given in question **1**.

Describe in words the set {2, 4, 6, 8, 10, 12}.

{2, 4, 6, 8, 10, 12} = {even numbers from 2 to 12 inclusive}

3 Write down in words the given sets:
   a {1, 3, 5, 7, 9}
   b {Monday, Tuesday, Wednesday, Thursday, Friday}
   Note that these descriptions must be very precise, e.g. it is correct to say
   {1, 2, 3, 4, 5} = {first five natural numbers}
   but it is incorrect to say
   {alsatian, boxer} = {breeds of dogs}
   because there are many more breeds than the two that are given.

4 Describe a set that includes the given members of the following sets and state another member of each.
   a Hungary, Poland, Slovakia, Bulgaria   **b** 10, 20, 30, 40, 50

Write each of the following statements in set notation.

5 John is a member of the set of boys' names.

6 English is a member of the set of school subjects.

7 June is not a day of the week.

8 Monday is not a member of the set of domestic furniture.

State whether the following statements are true or false.

9 32 ∈ {odd numbers}

10 Washington ∈ {American states}

11   Washington ∈ {capital cities}

12   1 ∉ {prime numbers}

Finite, infinite, equal and empty sets

When we can write down all the members of a set, the set is called a *finite set*, e.g. A = {days of the week} is a finite set because there are seven days in a week. If we denote the number of members in the set A by $n(A)$, then $n(A) = 7$.

Similarly if B = {5, 10, 15, 20, 25, 30}, $n(B) = 6$
and if C = {letters in the alphabet}, $n(C) = 26$.

If there is no limit to the number of members in a set, the set is called an *infinite set* e.g. {even numbers} is an infinite set because we can go on adding 2 time and time again.

Two sets are *equal* if they contain exactly the same elements, not necessarily in the same order,

e.g. if A = {prime numbers greater than 2 but less than 9}

and B = {odd numbers between 2 and 8}

then $A = B$, i.e. they are equal sets.

A set that has no members is called an *empty* or *null* set. It is denoted by ø or { }.

Exercise 22b

Are the following sets finite or infinite sets?
1   {odd numbers}

2   {the number of leaves on a particular tree}

3   {trees more than 60 m tall}

4   {the decimal numbers between 0 and 1}

Find the number of elements in each of the following sets.
5 A = {vowels}

6 C = {prime numbers less than 20}

If $n(A)$ is the number of elements in set A. find $n(A)$ for each of the following sets.
7 A = {5, 10, 15, 20, 25, 30}

8 A = {the consonants}

9 A = {players in a soccer team}

State whether or not the following sets are equal.

10 A = {8, 4, 2, 12}, B = {2, 4, 6, 8}

11 C = {letters of the alphabet except consonants}, D = {i, o, u, a, e}

12 X = {integers between 2 and 14 that are exactly divisible by 3 or 4},
        Y = {3, 4, 6, 8, 9, 12}

Determine whether or not the following sets are null sets.

13   {animals that have travelled in space}

14   {multiples of 11 between 12 and 20}

15   {prime numbers less than 2}

16   {consonants}

Universal sets

Think of the set {pupils in my class}.

With this group of pupils in mind we might well think of several other sets,

i.e. A = {pupils wearing spectacles}

      B = {pupils wearing brown shoes}

      C = {pupils with long hair}

      D = {pupils more than 150 cm tall}

We call the set {pupils in my class} *a universal set* for the sets A, B, C and D.

All the members of A, B, C and D must be found in a universal set, but a universal set may contain other members as well.

We denote a universal set by U or $\mathscr{E}$.

{pupils in my year at school} or {pupils in my school} would also be suitable universal sets for the sets A, B, C and D given above.

Exercise 22c

Suggest a universal set for {5, 10, 15, 20} and {6, 18, 24}

U = {integers}

In questions **1** to **3** suggest a universal set for:

1 {knife, dessert spoon}, {fork, spoon}

2 {10, 20, 30, 40}, {15, 25, 35}

3 {8, 12, 16, 20, 24}, {9, 12, 15, 18, 21, 24}

4 $U = $ {integers from 1 to 20 inclusive}

$A = $ {prime numbers} $B = $ {multiples of 3}

Find $n(A)$ and $n(B)$.

5 $U = $ {positive integers less than 16}

$A = $ {factors of 12} $B = $ {prime numbers}

$C = $ {integers that are exactly divisible by 2 and by 3}

List the sets A, B and C.

6 $U = \{x$, a whole number, such that $4 \leqslant x \leqslant 20\}$

$A = $ {multiples of 5} $B = $ {multiples of 7} $C = $ {multiples of 4}

Find $n(A)$, $n(B)$ and $n(C)$.

Subsets

If all the members of a set B are also members of a set A. then the set B is called *a subset* of the set A. This is written $B \subseteq A$. We use the symbol $\subseteq$ rather than $\subset$ if we don't know whether B could be equal to A.

Subsets that do not contain all the members of A are called *proper subsets*. If B is such a subset we write $B \subset A$.

Exercise 22d

If $A = $ {David, Edward, Fritz, Harold}, write down all the subsets of A with exactly three members.

The subsets of A with exactly three members are

{David, Edward, Fritz}

{David, Edward, Harold}

{David, Fritz, Harold}

{Edward, Fritz, Harold}

1   If A = {John, Jill, Peter, Audrey, Janet}, write down all the subsets of A with exactly two female members.

2   If N = {positive integers from 1 to 15 inclusive}, list the following subsets of N:

   A = {even numbers from 1 to 15 inclusive}

   B = {prime numbers less than 15}

   C = {multiples of 3 that are less than or equal to 15}

   Do sets A and B have any element in common?

3   If A = {even numbers from 2 to 20 inclusive}, list the following subsets of A:

   B = {multiples of 3}

   C = {prime numbers}

   D = {numbers greater than 12}

 Puzzle

During the day, because of the heat, the pendulum of a clock lengthens, causing it to gain half a minute during daylight hours. During the night the pendulum cools, causing it to lose one-third of a minute. The clock shows the correct time at dawn on the first of August. When will it be five minutes fast?

Venn diagrams

In the Venn diagram the universal set (U) is usually represented by a rectangle and the subsets of the universal set by circles within the rectangle.

If U = {families}, A = {families with one car} and B = {families with more than one car} the Venn diagram would be as shown.

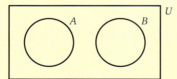

No family can have just one car and, at the same time, more than one car, i.e. A and B have no members in common.

Two such sets are called *disjoint sets*.

1

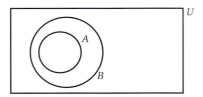

You are given the following information. U = {pupils in my year}

A = {pupils in my class who are my friends} B = {pupils in my class}

a Shade the region that shows the pupils in my class that are not my friends.

b Are all my friends in my class?

For each of questions **2** to **5** draw the diagram given below.

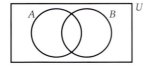

U = {pupils who attend my school}

A = {pupils who like coming to my school},

B = {pupils who are my friends}

In each case describe, in words, the shaded area.

2

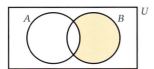

4

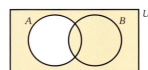

3

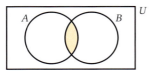

5

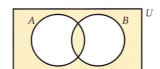

Union and intersection of two sets

If we write down the set of all the members that are in either set A or set B we have what we call the *union* of the sets A and B.

The union of A and B is written $A \cup B$.

The set of all the members that are members both of set A and of set B is called the *intersection* of A and B, and is written $A \cap B$.

Exercise 22f

$U = \{1, 2, 3, 4, 5, 6, 7, 8\}$

If $A = \{2, 4, 6, 8\}$ and $B = \{1, 2, 3, 4, 5\}$ find $A \cup B$ illustrating these sets on a Venn diagram.

$A \cup B = \{1, 2, 3, 4, 5, 6, 8\}$

We could show this on a Venn diagram as follows.

The shaded area represents the set $A \cup B$

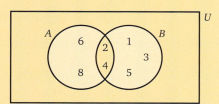

In questions **1** to **3** find the union of the two given sets, illustrating your answer on a Venn diagram.

1   $U = \{$girls' names beginning with the letter J$\}$

   $A = \{$Janet, Jill, Jamila$\}$ $B = \{$Judith, Janet, Jacky$\}$

2   $U = \{$positive integers from 1 to 16 inclusive$\}$

   $X = \{4, 8, 12, 16\}$ $Y = \{2, 6, 10, 14, 16\}$

3   $U = \{$letters of the alphabet$\}$

   $P = \{$letters in the word GEOMETRY$\}$

   $Q = \{$letters in the word TRIGONOMETRY$\}$

4   Draw suitable Venn diagrams to show the unions of the following sets, and describe these unions in words as simply as possible.

   a $U = \{$quadrilaterals$\}$ $A = \{$parallelograms$\}$ $B = \{$trapeziums$\}$

   b $U = \{$angles$\}$ $P = \{$obtuse angles$\}$ $Q = \{$reflex angles$\}$

$U = \{$integers from 1 to 12 inclusive$\}$

If $A = \{1, 2, 3, 4, 5, 6, 7, 8\}$ and $B = \{1, 2, 3, 5, 7, 11\}$ find $A \cap B$ and show it on a Venn diagram.

$A \cap B = \{1, 2, 3, 5, 7\}$

The shaded area represents $A \cap B$

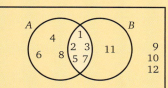

Draw suitable Venn diagrams to show the intersections of the following sets. In each case write down the intersection in set notation.

5 U = {integers from 4 to 12 inclusive}

  X = {4, 5, 6, 7, 10} Y = {5, 7, 11}

6 U = {colours of the rainbow}

  A = {red, orange, yellow} B = {blue, red, violet}

7 U = {positive whole numbers}

  C = {positive whole numbers that divide exactly into 24}

  D = {positive whole numbers that divide exactly into 28}

8 U = {integers less than 25}

  A = {multiples of 3 between 7 and 23}

  B = {multiples of 4 between 7 and 23}

Simple problems involving Venn diagrams

Exercise 22g

If U = {girls in my class}

A = {girls who play netball} = {Helen, Bina, Moira, Sara, Lana} and
B = {girls who play tennis} = {Kath, Sara, Helen, Maria}

Illustrate A and B on a Venn diagram. Use this diagram to write down the following sets:

a   {girls who play both netball and tennis}

b   {girls who play netball but not tennis}

c   If $n(U)$ = 30 find the number of girls who play neither netball nor tennis.

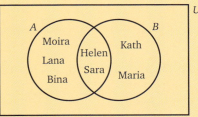

From the Venn diagram

a   {girls who play both netball and tennis} = {Helen, Sara}

b   {girls who play netball but not tennis} = {Moira, Lana, Bina}

c n(girls who play neither netball nor tennis) = $30 - 7 = 23$

1   U = {the pupils in a class}

   X = {pupils who like history}

   Y = {pupils who like geography}

   List the set of pupils who
   a   like history but not geography
   b   like geography but not history
   c   like both subjects.

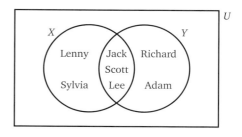

2   U = {boys in my class}

   A = {boys who play soccer}

   B = {boys who play rugby}

   Write down the sets of boys who
   a   play soccer
   b   play both games
   c   play rugby but not soccer.

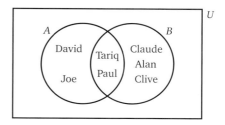

3   U = {my friends}

   P = {friends who wear glasses}

   Q = {friends who wear brown shoes}

   List all my friends who
   a   wear glasses
   b   wear glasses but not brown shoes
   c   wear both glasses and brown shoes.

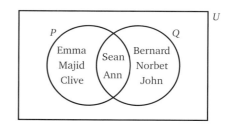

4   U = {whole numbers from 1 to 14 inclusive}

   A = {even numbers between 3 and 13}

   B = {multiples of 3 between 1 and 14}

   Illustrate this information on a Venn diagram and hence write down
   a   the even numbers between 3 and 13 that are multiples of 3
   b $n(A)$ and $n(B)$.

5   U = {letters of the alphabet}

   P = {different letters in the word SCHOOL}

   Q = {different letters in the word SQUASH}

   Show these on a Venn diagram and hence write down
   a $n(P)$
   b $n(P \cup Q)$
   c $n(P \cap Q)$.

U = {months of the year}

A = {months of the year beginning with the letter J}

B = {months of the year ending with the letter Y}

a   Find n(U), $n(A)$ and $n(B)$

Hence find

b $n(A \cap B)$

c $n(A \cup B)$

a $n(U) = 12$ (there are 12 months in a year)

A = {January, June, July} so $n(A) = 3$

B = {January, February, May, July} so $n(B) = 4$

b   We can illustrate these sets with a Venn diagram using the numbers
in each region, rather than the members.

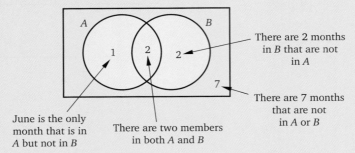

There are 2 months
in B that are not
in A

There are 7 months
that are not
in A or B

June is the only
month that is in
A but not in B

There are two members
in both A and B

This shows that $n(A \cap B) = 2$

c $n(A \cup B) = 5$

Alternatively, we know that $A \cup B$ is the set of months in both A and B.
However two months, January and July are in both A and B. This means
that we cannot find $n(A \cup B)$ just by adding $n(A)$ and $n(B)$, because that
includes the two months in $(A \cap B)$ twice.

Hence $n(A \cup B) = n(A) + n(B) - n(A \cap B)$

$= 3 \ + \ 4 \ - \ 2 = 5$

For any two sets, A and B, $n(A \cup B) = n(A) + n(B) - n(A \cap B)$

6 U = {letters of the alphabet}

P = {letters used in the word LIBERAL}

Q = {letters used in the word LABOUR}

a   Find $n(U)$, $n(P)$ and $n(Q)$

b   Show these on a Venn diagram.

Hence find   **i** $n(P \cap Q)$ **ii** $n(P \cup Q)$ describing each of these sets.

7   $U = \{$counting numbers less than 12$\}$

     $C = \{$prime numbers$\}$ $D = \{$odd numbers$\}$

     a   Find $n(U)$, $n(C)$ and $n(D)$.

     b   Show these on a Venn diagram.

     Hence find   **i** $n(C \cap D)$ **ii** $n(C \cup D)$.

8   $U = \{$whole numbers from 1 to 35 inclusive$\}$

     $R = \{$multiples of 4$\}$ $S = \{$multiples of 6$\}$

     a   Find $n(U)$, $n(R)$ and $n(S)$.

     b   Find   **i** $n(R \cap S)$ **ii** $n(R \cup S)$.

9   A and B are two sets such that $n(A) = 8$, $n(B) = 5$ and $n(A \cap B) = 3$.
    Find $n(A \cup B)$.

In this chapter you have seen that...

✔ an infinite set has no limit on the number of members in it

✔ in a finite set, all the members can be counted or listed

✔ the intersection of two sets contains the elements that are in both sets

✔ a proper subset of a set A contains some, but not all, of the members of A

✔ the union of two sets contains all the members of the first set together
with the members of the second set that have not already been included

✔ when two sets have exactly the same members, they are said to be equal

✔ a set that has no members is called an empty or null set and is written $\{ \, \}$ or $\varnothing$.

23 Number bases

 Investigation

Use a search engine to find out what you can about hexadecimal numbers and why they are important.

You need to know...

✔ your multiplication tables – and this means instant recall

✔ the meaning of a positive index

Key words

base, binary, denary system, number base

Denary system (base ten)

We have ten fingers. This is probably why we started to count in tens and developed a system based on ten for recording large numbers. For example

$$3125 = 3 \text{ thousands} + 1 \text{ hundred} + 2 \text{ tens} + 5 \text{ units}$$
$$= 3 \times 10^3 \qquad + 1 \times 10^2 \qquad + 2 \times 10^1 + 5 \times 1$$

Each column is ten times the value of its right-hand neighbour. The base of this number system is ten and it is called the *denary system*.

Base five

If man had started to count using just one hand, we would probably have a system based on five.

Suppose we had eleven stones. Using one hand to count with, we could arrange them like this:

i.e.   two handfuls + one ⎱

or   2   fives   + 1 ⎰

The next logical step is to use a single marker to represent each group of five. We also need to place these markers so that they are not confused with the marker representing the one unit. We do this by having separate columns for the fives and the units, with the column for the fives to the left of the units.

Fives   Units

We can write this number as 21_5 and we call it 'two one to the base five'. We do *not* call it 'twenty-one to the base five', because the word 'twenty' means 'two tens'.

To cope with larger numbers, we can extend this system by adding further columns to the left such that each column is five times the value of its right-hand neighbour. Thus, in the column to the left of the fives column, each marker is worth twenty-five, or 5^2.

twenty-fives   fives   units

For example.

The markers here represent 132_5 and it means

$$(1 \times 5^2) + (3 \times 5) + 2$$

Exercise 23a

Write in figures the numbers represented by the markers:

The number is 1302_5

(The '5' column is empty, so we write zero in this column.)

5^3	5^2	5	units
	•		
	•		•
•	•		•

Write in figures the numbers represented by the markers in questions **1** to **4**:

	5^4	5^3	5^2	5	units
1			• •	•	• • • •
2		• •		•	• • • •
3	• • • •	•	•	• • • •	
4	• • •		• •		•

Write 120_5 in headed columns.

5^2	5	Units
1	2	0

Write the following numbers in headed columns.

5	31_5	**7**	410_5	**9**	34_5	**11**	204_5
6	42_5	**8**	231_5	**10**	10_5	**12**	400_5

Write 203_5 as a number to the base 10.

$$203_5 = (2 \times 5^2) + (0 \times 5) + 3$$
$$= 50_{10} \qquad + 3_{10}$$
$$= 53_{10}$$

(Although we do not normally write fifty-three as 53_{10}, it is sensible to do so when dealing with other bases as well.)

Write the following numbers as denary numbers, i.e. to base 10:

13	31_5	**16**	121_5	**19**	32_5	**22**	400_5
14	24_5	**17**	204_5	**20**	20_5	**23**	240_5
15	40_5	**18**	43_5	**21**	4_5	**24**	300_5

Write 38_{10} as a number to the base 5.

(To write a number to the base 5 we have to find how many ... 125s, 25s, 5s and units the number contains.)

Method 1   (Starting with the highest value column.)

38 contains no 125s.

$38 \div 25 = 1$ remainder 13   i.e. $38 = 1 \times 5^2 + 13$

$13 \div 5 = 2$ remainder 3   i.e. $13 = 2 \times 5 + 3$

$\therefore$ $38 = 1 \times 5^2 + 2 \times 5 + 3$

$= 123_5$

Method 2   (Starting with the units.)

5) 38

5) 7   remainder 3 (units)

5) 1   remainder 2 (fives)

    0   remainder 1 (twenty-fives)

$\therefore$ $38 = 123_5$

Write the following numbers in base 5:

25	8_{10}	**28**	39_{10}	**31**	7_{10}	**34**	128_{10}
26	13_{10}	**29**	43_{10}	**32**	21_{10}	**35**	82_{10}
27	10_{10}	**30**	150_{10}	**33**	30_{10}	**36**	100_{10}

Addition, subtraction and multiplication

Numbers with a base of 5 do not need to be converted to base 10; provided that they have the same base they can be added, subtracted and multiplied in the usual way, as long as we remember which base we are working with.

For example, to find $132_5 + 44_5$ we work in fives, not tens. To aid memory, the numbers can be written in headed columns:

Twenty-fives	Fives	Units	
1	3	2	
①	4 ①	4	+
2	3	1	
	⑧	⑥	

Adding the units gives 6 units:

$6 \text{ (units)} = 1 \text{ (five)} + 1 \text{ (unit)}$

We put **1** in the units column and carry the single five to the fives column.

Adding the fives gives 8 fives

$$8 \text{ (fives)} = 5 \text{ (fives)} \qquad +3 \text{ (fives)}$$
$$= 1 \text{ (twenty-five)} \qquad +3 \text{ (fives)}$$

We put 3 in the fives column and carry the 1 to the next column.

Adding the numbers in the last column gives 2,

i.e. $\qquad 132_5 + 44_5 = 231_5$

If the numbers are not to the same base we cannot add them in this way.

For example, 432_7 and 621_8 cannot be added directly.

Exercise 23b

Find $243_5 + 434_5$

5^3	5^2	5	units
	2	4	3
	4①	3①	4
1	2	3	2
	⑦	⑧	⑦

∴ $\qquad 243_5 + 434_5 = 1232_5$

Find

1	$12_5 + 31_5$	**4**	$13_5 + 44_5$	**7**	$134_5 + 424_5$	**10**	$413_5 + 40_5$
2	$21_5 + 33_5$	**5**	$221_5 + 33_5$	**8**	$432_5 + 42_5$	**11**	$342_5 + 222_5$
3	$42_5 + 12_5$	**6**	$32_5 + 312_5$	**9**	$232_5 + 123_5$	**12**	$243_5 + 403_5$

Find $132_5 - 13_5$

(There are two methods of doing subtraction and we show both of them here.)

First method

5^2	5	Units
1	$3^{②}$	$2^{⑦}$
	1	3
1	1	4

(We cannot take 3 from 2 so we take one 5 from the fives column, change it to 5 units and add it to the 2 units.)

Second method

If you use the 'pay back' method of subtraction, the calculation looks like this

5^2	5	Units
1	3	2 ⑦
	1 ①	3
1	1	4

In either case $132_5 - 13_5 = 114_5$

Find:

13	$43_5 - 12_5$	**16**	$23_5 - 14_5$	**19**	$242_5 - 23_5$	**22**	$322_5 - 233_5$
14	$32_5 - 13_5$	**17**	$124_5 - 23_5$	**20**	$333_5 - 243_5$	**23**	$432_5 - 234_5$
15	$34_5 - 24_5$	**18**	$142_5 - 34_5$	**21**	$221_5 - 114_5$	**24**	$342_5 - 243_5$

Find $342_5 \times 4_5$

5^3	5^2	5	Units	
	3	4	2	
			4	×
3 ③	0 ③	2 ①	3	

15 = 3 × 5   17 = 3 × 5 + 2   8 = 5 × 3

25	$4_5 \times 3_5$	**27**	$13_5 \times 3_5$	**29**	$33_5 \times 2_5$	**31**	$44_5 \times 3_5$
26	$4_5 \times 4_5$	**28**	$24_5 \times 3_5$	**30**	$42_5 \times 3_5$	**32**	$32_5 \times 4_5$

Find:

33	$31_5 - 13_5$	**36**	$33_5 \times 3_5$	**39**	$43_5 \times 4_5$	**42**	$21_5 \times 4_5$
34	$43_5 \times 3_5$	**37**	$413_5 + 404_5$	**40**	$32_5 - 14_5$	**43**	$321_5 - 44_5$
35	$323_5 + 140_5$	**38**	$423_5 + 40_5$	**41**	$133_5 - 14_5$	**44**	$323_5 + 134_5$

45 Find the base in which the following calculations have been made:

- **a** $13 \times 2 = 31$
- **b** $12 \times 3 = 36$
- **c** $13 - 4 = 4$
- **d** $16 - 9 = 7$

Binary numbers

Numbers with a base of two are called *binary numbers*. We have singled out binary numbers for special attention because of the wide application that they have, especially in the world of computers. Binary numbers use just two symbols, 0 and 1.

Binary numbers are written in columns of powers of 2. Each column is twice the value of the column on its right.

$32\,(2^5)$	$16\,(2^4)$	$8\,(2^3)$	$4\,(2^2)$	2	units
	•	•		•	
•		•	•	•	•

The number represented by the first row of markers is 11010_2

As a denary number, 11010_2 is $1 \times 16 + 1 \times 8 + 1 \times 2 = 26$.

The number represented by the second row of markers is 101111_2

As a denary number, 101111_2 is $1 \times 32 + 1 \times 8 + 1 \times 4 + 1 \times 2 + 1$

$\therefore\ 101111_2$ as a denary number is $32 + 8 + 4 + 2 + 1 = 47$

The relationship between denary and binary numbers is:

Binary number	1	10	11	100	101	110	111	1000	1001	1010	1011	1100 ...
Denary numbers	1	2	3	4	5	6	7	8	9	10	11	12 ...

Exercise 23c

Write in figures the binary number represented by the markers:

$32\,(2^5)$	$16\,(2^4)$	$8\,(2^3)$	$4\,(2^2)$	2	units
	•	•		•	

The binary number represented by the markers is 11010_2

Write in figures the numbers represented by the markers in questions **1** to **8**:

	$32\,(2^5)$	$16\,(2^4)$	$8\,(2^3)$	$4\,(2^2)$	2	units
1			•		•	
2	•		•		•	
3	•	•				•
4		•	•	•		
5			•		•	•
6		•		•		
7		•		•	•	•
8	•	•	•			•

9 Write these binary numbers in headed columns.

    a 1101_2 **b** 10001_2 **c** 110101_2 **d** 101011_2 **e** 1010101_2

Change these binary numbers into denary numbers.

a 10110_2 **b** 11001111_2

a Binary numbers are numbers expressed to the base 2

so $10110_2 = 2^4 + 2^2 + 2^1 + 1 = 16 + 4 + 2 + 1 = 2310$

b $11001111_2 = 2^7 + 2^6 + 2^3 + 2^2 + 2^1 + 1 = 128 + 64 + 8 + 4 + 2 + 1 = 207$

---

10 Express each binary number as a denary number to the base 10:

    a 10111_2 **b** 101_2 **c** 1111_2 **d** 111001_2 **e** 1110111_2

11 Express these binary numbers as denary numbers:

    a 1010   **b** 11011   **c** 1110   **d** 111100

    e 110001   **f** 1010111   **g** 101100   **h** 1011110

12 State, for each pair of binary numbers, which is the larger:

    a 101, 110   **b** 1011, 1101   **c** 101011, 111000   **d** 110011, 110101

We can write denary numbers as binary numbers by reversing this process, i.e. by writing the number as the sum of different powers of 2.

For example $29 = 16 + 13$

$$= 16 + 8 + 5$$
$$= 16 + 8 + 4 + 1$$
$$= 2^4 + 2^3 + 2^2 + 1$$

i.e. $29_{10} = 11101_2$

and $71 = 64 + 4 + 2 + 1$

$$= 2^6 + 2^2 + 2 + 1$$

so $71_{10} = 1000111_2$ (There is no 2^5, 2^4 or 2^3 so we put zeros to show this.)

---

13 Express each denary number as the sum of powers of 2 and then write it as a binary number.

    a 12   **b** 19   **c** 25   **d** 38   **e** 60

14 Convert the following denary numbers to binary numbers:

    a 23   **b** 27   **c** 34   **d** 70   **e** 91

Addition, subtraction and multiplication of binary numbers

Exercise 23d

Find

a $1011_2 + 111_2 + 100_2$ **b** $11011_2 + 1101_2 + 101011_2$

a

2^4	2^3	2^2	2	unit
	1	0	1	1
		1	1	1
		1	0	0
1	0	1	1	0
	①		①	①

These are the carried figures.

Remember: 0+0=0, 0+1=1, 1+0=1 and 1+1=10

b

2^6	2^5	2^4	2^3	2^2	2	unit
			1	1	0	1	1
				1	1	0	1
	1	0	1	0	1	1
1	0	1	0	0	1	1
		①	②	①	①	①

Find in the binary scale:

1 **a** 111
          101

   b 101
          111
          101

   c 1101
          111
          1001

2 **a** 1011
          100
          101

   b 10111
          1011
          1101

   c 11011
          11110
          1011

3 **a** $101 + 110$ **b** $1100 + 111$ **c** $1011 + 1010$
4 **a** $10101 + 1110$ **b** $11100 + 1011$ **c** $10111 + 10111$
5 **a** $1110111 + 1011 + 110011$ **b** $101100 + 10110 + 110110$

Find $10110_2 - 1011_2$

2^4	2^3	2^2	2	unit
1	0	1	1	0
	1	0	1	1
	1	0	1	1

Remember: when you 'borrow one' you are borrowing 2 not 10.

Find in the binary scale:

6　a　$1010 - 101$　　　　b　$10100 - 1011$　　　　c　$1110 - 111$

7　a　$10101 - 1101$　　　b　$11010 - 1111$　　　　c　$11011 - 1101$

8　a　$110011 - 100010$　b　$100101 - 10111$　　　c　$110110 - 100111$

Find　　a　$1101_2 \times 11_2$　　　　　　b　$10111_2 \times 1011_2$

a	2^5	2^4	2^3	2^2	2	unit
		1	1	0	1	
×				1	1	
		1	1	0	1	
+		1	1	0	1	
	1	0	0	1	1	1

b	2^7	2^6	2^5	2^4	2^3	2^2	2	unit
				1	0	1	1	1
×					1	0	1	1
				1	0	1	1	1
			0	1	0	1	1	1	0
	1	0	1	1	1	0	0	0
	1	1	1	1	1	1	0	1

9　Find in the binary scale:

　　a　101×101　　　b　1101×11　　　c　1011×111

10　Find in the binary scale:

　　a　111×101　　　b　11011×111　　c　11011×1011

11　For each set of binary numbers state which is the smallest:

　　a　110, 101, 100　b　10101, 11011, 11100　　c　1100011, 111111, 1110000

12　Write the denary numbers 3, 5, 7, 9, 11, as binary numbers.
　　What is the last digit every time?

13　Write the denary numbers 2, 4, 6, 8, 10, 12 as binary numbers.
　　What is the last digit every time?

14　Which of these binary numbers are odd?

　　a　10110　b　11001　　c　10111　　d　1100110　e　111000111

15　What can you say about a binary number that ends in **a** two zeros **b** three zeros?

16　What is　a　the largest　b　the smallest, binary number that can be
　　represented by 4 digits?

17　Which of these binary numbers are multiples of 4?
　　1000, 1100, 1111, 10101, 11100

18　Which of these binary numbers are multiples of 8?
　　1000, 10000, 1100100, 101010, 111011

19　Express 20_{10} as a binary number. Hence write down the binary numbers
　　for 40_{10} and 80_{10}.

20　Express 25_{10} as a binary number. Hence write down the binary numbers
　　for 50_{10} and 100_{10}.

21 Express these binary numbers as base 5 numbers:
a 10010_2 b 101101_2 c 1100_2 d 11101_2

22 Express these base 5 numbers as binary numbers:
a 32_5 b 14_5 c 40_5 d 23_5 e 303_5 f 234_5

23 Find $124_5 + 10111_2$, giving your answer as a binary number.

Mixed exercises

Exercise 23e

1 Convert these denary numbers to binary numbers:
a 15   b 28   c 34   d 53

2 Complete this statement which expresses 300 as the sum of different powers of 2.
$300 = 2^8 + 2^5 + ...$ $= 256 + ...$
Hence write 300_{10} as a binary number.

3 How many different symbols are needed to represent all the numbers in base 5?

4 Convert these denary numbers to base 5 numbers:
a 9   b 16   c 44   e 133

5 Find a $111001_2 + 100111_2$ b $110001_2 - 10111_2$ c $11001_2 \times 1101_2$

6 Find a $321_5 + 333_5$ b $341_5 - 333_5$ c $432_5 \times 3_5$

7 Look at these binary numbers:
101011, 110011, 111001, 11111, 1000000.
Which number is a the largest b the smallest?

8 Look at these base 5 numbers:
$4420_5, 1422_5, 444_5, 1331_5, 2221_5$
Which number is a the largest b the smallest?

9 If a number to the base of ten ends in 0, what does the same number to the base 5 end in?

10 a How many digits are there in 5^3 written in base 5?
b Express 5^3 as a denary number.
c Convert your answer to part b into a binary number.

In this chapter you have seen that...

✔ numbers can be expressed in different bases.

✔ if a number is in the base 5, the first column from the right is units, the second is the number of 5s, the third the number of 5^2 s, the fourth the number of 5^3 s and so on.

For example 134_5 means $1 \times 5^2 + 3 \times 5 + 4$

✔ you can add, subtract and multiply in base 5 but it is sensible to write the digits in headed columns to keep track of your working

✔ if a number is in the base 2, the first column from the right is units, the second is the number of 2s, the third the number of 2^2, the fourth the number of 2^3 s and so on.

For example 11001_2 means $1 \times 2^4 + 1 \times 2^3 + 0 \times 2^2 + 0 \times 2 + 1$

✔ you can add, subtract and multiply binary numbers and numbers to the base 5 without changing them to denary numbers.

REVIEW TEST 3: CHAPTERS 15–23

In questions 1 to 9 choose the letter for the correct answer.

1 A minibus travels between two towns at an average speed of 40 km per hour. How far will it travel in $2\frac{1}{2}$ hours?

   A 100 km   **B** $80\frac{1}{2}$ km   **C** 80 km   **D** $42\frac{1}{2}$ km

2 A cyclist rides a distance of 9 km in 15 minutes. What is his average speed in kilometres per hour?

   A 36   **B** 80   **C** 144   **D** 180

3 0, 1, 3, 3, 5, 5, 5, 10.
The mean of these numbers is

   A 3.4   **B** 4   **C** 5.3   **D** 7

4 What is the volume of a cylinder of radius $3\frac{1}{2}$ cm and length 4 cm?
$\left(\text{Take } \pi = \frac{22}{7}\right)$

   A 14 cm³   **B** 44 cm³   **C** 88 cm³   **D** 154 cm³

5 The mode of the set of cricket scores 54, 30, 60, 54, 18, 62, 30, 54 is

   A 62   **B** 54   **C** 36   **D** 18

6 The scores in six rounds of golf are 73, 75, 80, 84, 73, 73.
The median score is

   A 84   **B** 82   **C** 74   **D** 73

7 The solution of the inequality $8 < 3 - 5x$ is

   A $x > -1$ **B** $x < 1$ **C** $x > 1$ **D** $x < -1$

8 What is the gradient of the line segment joining the points $(-3, 2)$ and $(5, 7)$?

   A $\frac{9}{2}$ **B** $\frac{5}{2}$ **C** $\frac{9}{8}$ **D** $\frac{5}{8}$

9 A line parallel to the x-axis at a distance k away from it has the equation

   A $x = k$ **B** $y = k$ **C** $x + y = k$ **D** $y = x - k$

10 **a** Change **i** 2 m³ to cm³   **ii** 425 mm³ to cm³   **iii** 0.0045 m³ to cm³.

    b Find the volume, in cm³, of a cuboid of length 1.2 m, width 35 cm and height 25 cm.

11 This shape consists of a semicircle on a rectangle. Calculate the area of the figure, correct to 2 decimal places. (Take $\pi = 3.142$)

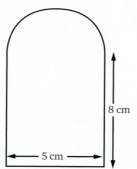

8 cm

5 cm

12 A prism of length 20 cm has its cross-section in the shape of a right-angled triangle with sides 6 cm, 8 cm and 10 cm. Calculate the volume of the prism.

13 **a** Find the exact value of   **i** 3.42^2 **ii** $\sqrt{0.16}$ **iii** $\sqrt{5.9049}$

    b The frequency table shows the marks scored by students in a test.

mark	1	2	3	4	5	6	7	8	9
frequency	2	1	4	3	6	10	12	10	5

      i How many students took the test?

      ii What is the median mark?

      iii Find the mean mark.

14 A vehicle is travelling at a constant speed of 20 cm/s. Copy and complete the following table showing the time t and distance s.

t	0	1	2	3	4	6	8	10
s	0	20	40			120		

Using a scale of 1 cm to represent 2 seconds on the t-axis and 1 cm to represent 20 cm on the s-axis, draw a graph showing the time and distance travelled. From the graph find

    a the distance travelled in 7 seconds   **b** the time taken to travel 90 cm.

15 A man travelling to a town 4 kilometres away travels the first 2 kilometres in 30 minutes. He then rests for 20 minutes and afterwards continues his journey, reaching the town 30 minutes later.

Show this information on a distance-time graph. Use the graph to find the average speed for the last 30 minutes.

16 **a** Describe in words

    i Set A

    ii Set B.

  b Find

    i $n(A \cup B)$

    ii $n(A \cap B)$.

17 Find

  a $21_5 + 43_5$

  b $234_5 + 132_5$

  c $43_5 - 14_5$

  d $23_5 \times 3_5$

  e $111\,00_2 + 111_2$

  f $111\,00_2 - 111_2$

  g $111\,00_2 \times 111_2$

18 **a** Express as a denary number

    i 324_5 **ii** $101\,10_2$

  b Express the denary number 169

    i as a number to base 5   **ii** as a binary number.

REVIEW TEST 4: CHAPTERS 1–23

Choose the letter for the correct answer.

1 The value of $-2-(-4)+(-3)$ is

   A −9   **B** −7   **C** −1   **D** 5

2 The expression $3(4-x)-(7-x)$ simplifies to

   A $5-2x$ **B** $5-4x$ **C** $-2x$ **D** $5+2x$

3 The value of x that satisfies the equation $5-2x=15-7x$ is

   A $\frac{20}{9}$ **B** $\frac{1}{2}$ **C** −2   **D** 2

4 An item bought for $12 000 is sold to make a profit of 20%. The selling price is

   A $9600   **B** $12 500   **C** $14 000   **D** $14 400

5 The angles marked p and q are

   A alternate angles

   B corresponding angles

   C vertically opposite angles

   D none of these

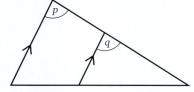

6

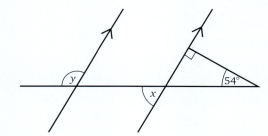

The values of x and y are respectively

   A 54° and 126°   **B** 36° and 144°   **C** 36° and 164°   **D** 54° and 146°

7 The largest integer x such that $17-x>3x-11$ is

   A 3   **B** 6   **C** 7   **D** 8

8 If 10 is the mean of x, 10, 12 and 14, then $x=$

   A 10   **B** 8   **C** 6   **D** 4

9 In a game, the score occurs with the frequency shown in the table.

x	1	2	4	6	12
Frequency	12	6	3	2	1

The mean score is

A 1   B 2.5   C 11   D 12

10

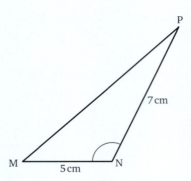

In the triangle MNP shown above, MN = 5 cm, NP = 7 cm and $\hat{N}$ is obtuse.
The following statements are made:

  i MP is greater than 7 cm

 ii Triangle MNP is isosceles

The true statement(s) is/are

A i only   B ii only   C i and ii   D Neither

11 The median of the set of numbers 6, 1, 2, 3, 7 and 5 is

A 2   B 2.5   C 3   D 4

12 The value of $2^3 + 3^2$ is

A 12   B 17   C 36   D 72

13

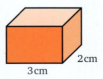

The volume of this cuboid is 18 cm³. The height is

A 9 cm   B 6 cm   C 3.6 cm   D 3 cm

14 The lines $y = -3$ and $x = 2$ intersect at the point

A (−3, 2)   B (3, −2)   C (−2, 3)   D (2, −3)

15 10% of 20 differs from 0.5 of 2 by

 A 1   **B** 2   **C** 4   **D** 5

16 The line ST is parallel to the line
$y = 2x - 1$.
The most likely equation for ST is

 A $y = 3x - 3$ **B** $y = 3x + 2$

 C $y = x + 2$ **D** $y = 2x + 3$

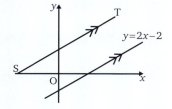

17 The area of this shape is

 A $46 \, \text{cm}^2$ **B** $94 \, \text{cm}^2$

 C $102 \, \text{cm}^2$ **D** $80 \, \text{cm}^2$

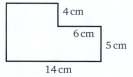

18 The equation of the straight line which passes through the
point (0, 3) with gradient $\frac{1}{2}$ is

 A $y = 2x + 3$ **B** $y = 2x + 6$ **C** $2y = x + 3$

 D $2y = x + 6$

19 The value of x that satisfies the equation $5x - 4 = 2(x + 7)$ is

 A 6   **B** 4   **C** 3   **D** 2

20 An item of jewellery appreciates by 5% each year. If its original
cost was $2000, its value at the end of two years will be

 A $2400   **B** $2205   **C** $2200   **D** $2010

21 The area, in square centimetres of a square of side 5 mm is

 A 25   **B** 0.25   **C** 20   **D** 625

22 Josh and Emma share some money in the ratio 2 : 3. Emma gets $36. Josh gets

 A $20   **B** $24   **C** $30   **D** $60

23 The domain of the relation {(1, 2), (1, 3), (2, 2)} is

 A {1, 3}   **B** {1, 2, 3}   **C** {2, 3}   **D** {1, 2}

24 In a certain town of 12 000 people, two-thirds are children
and one-half of the remainder are women. A person is chosen
at random. The probability that this person is a woman is

 A $\frac{2}{3}$ **B** $\frac{1}{2}$ **C** $\frac{1}{4}$ **D** $\frac{1}{6}$

25 $P = \{1, 2, 3, \ldots 10\}$. In the set P, the largest odd number exceeds the largest prime number by

    A   0   **B**   1   **C**   2   **D**   3

26 Which one of the following statements defines an infinite set?

    A   The fifth form students of a certain school.

    B   Plane figures bounded by three or more straight lines.

    C   Vehicles that carry passengers.

    D   Sixth form students who can sing well.

27 Before a sale a store owner raises the price of an item by 10%. In the sale he offers a 10% discount. The cost of the item in the sale is

    A   the same as before he raised the price

    B   less than before he raised the price

    C   more than before he raised the price

    D   more of these.

28 The order of rotational symmetry of a square is

    A   1   **B**   2   **C**   3   **D**   4

29 The square root of 0.000 049 is

    A   0.07   **B**   0.007   **C**   0.0022   **D**   0.0007

30 Quadrants of radius 1 cm are cut from each corner of the square ABCD to leave the figure shaded in the diagram.

The area of the shaded region, in cm², is

    A $4 - \pi$ **B** $1 - \pi$ **C** $8 - 2\pi$ **D** $4 - 2\pi$

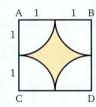

31 The range of values for which the inequalities $-3 - x < 2x + 3 \leqslant 9$ are true is

    A $2 < x \leqslant 3$ **B** $0 \leqslant x < 3$ **C** $-2 < x \leqslant 3$ **D** $-6 < x \leqslant 3$

32 The set of inequalities that describe
the unshaded region are

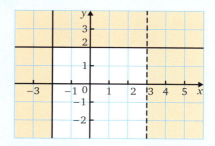

    A $x \geqslant -2, x \leqslant 3, y \leqslant 2$

    B $x > -2, x < 3, y < 2$

    C $x > 2, x \leqslant 3, y < 2$

    D $x \geqslant -2, x < 3, y \leqslant 2$

33 In the diagram AB = BC = BD.

The transformation that will not transform ABC to BCD is

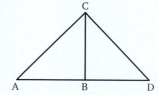

    A   a rotation through 90° clockwise about B

    B   a reflection in BC

    C   rotation of 270° anticlockwise about B

    D   rotation about C through 90° clockwise.

34

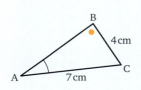
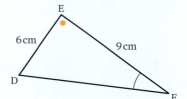

In Δs ABC and DEF, ∠A = ∠F, ∠E = ∠B, BC = 4 cm,
AC = 7 cm, DE = 6 cm and EF = 9 cm. The length of AB is

    A   9 cm   **B**   8 cm   **C**   7 cm   **D**   6 cm

35 The interior angle of a polygon is twice its exterior angle.
The number of sides of the polygon is

    A   4   **B**   6   **C**   8   **D**   10

Answers

CHAPTER 1

Exercise 1a   page 3

1 2, 3, 6, 1, −5, −3, 5, −3, −5, 5, 0
2 2, −2, 5, −4, 2, 5, −5, 0
3 5 below
4 3 above
5 1 below
6 10 above
7 on *x*-axis
8 4 below
9 3 right
10 5 left
11 2 right
12 7 left
13 on *y*-axis
14 9 left
15 A(−2, 3),   B(3, l),   C(2, −2),   D(−3, 1),
    E(6, l),   F(−2, −2),   G(−4, −4),   H(l, 2),
    I(4, −4),   J(−4, 3)
18 square
19 isosceles triangle
20 rectangle
21 right-angled

Exercise 1b   page 4

1 6   **3** 6   **5** 2   **7** 5   **9** 11
2 8   **4** 2   **6** 7   **8** 7   **10** 11
11 (−1, 1)   **18** (−1, 2)   **25** (−5, −2)
12 (1, −2)   **19** (−1, 3)   **26** $\left(4, \frac{3}{2}\right)$
13 (−1, 3)   **20** (1, 0)   **27** (−1, 3)
14 (−6, −1)   **21** (4, 2)   **28** (−1, 0)
15 (−5, 1)   **22** (2, −1)   **29** (0, 0)
16 (0, −1)   **23** $\left(-\frac{7}{2}, 3\right)$ **30** (−1, 0)
17 (3, 2)   **24** (−3, −1)

Exercise 1c   page 5

1 a (1, 2), (3, 6), (−3, −6), (−2, −4), (2, 4)
  b 10
  c 16, 20, −8, 6, 9, −5, 2*a*
2 a (2, 2), (4, 3), (6, 4), (10, 6), (−4, −1), (−8, −3), (0, 1)
  b *y*-coordinate = $\frac{1}{2}$(*x*-coordinate) + 1
  c 5
  d 7, 11, 16, −5, 16, $\frac{1}{2}a + 1$
3 a (3, −1), (5, −3), (6, −4), (8, −6), (−2, 4),
    (−4, 6), (1, 1)
  b −5, −8, −10, −18, 9, 11, −8, 10, −10

Exercise 1d   page 7

1 a parallelogram   **d** both
  c no   **e** no
2 a square   **d** both
  c yes   **e** yes
3 a trapezium   **d** neither
  c no   **e** no
4 a rhombus   **d** both
  c no   **e** yes
5 a rectangle   **d** both
  c yes   **e** no
6 rectangle, square
7 rhombus, square
8 parallelogram, rectangle, rhombus, square

Exercise 1e   page 9

1 +10°   **3** −3°   **5** −8°
2 −7°   **4** +5°   **6** 0°
7 2° below
8 3° above
9 4° above
10 10° below
11 8° above
12 freezing point
13 10°
14 12°
15 4°
16 −3°
17 2°
18 −2°
19 1°
20 3°

21 −7°
22 −2°
23 **A** 75 above,   **B** 50 above,   **C** 75 above,
    D 25 above,   **E** sea level,   **F** 50 below,
    G sea level,   **H** 25 below
24 −5 s   **30** −$50   **36** +21 °C
25 +5 s   **31** +5 paces   **37** +150 m
26 $50   **32** −5 paces   **38** −3 °C
27 $50   **33** +200 m   **39** +$25
28 −1 min   **34** −5 m   **40** 6 paces in front
29 +$500   **35** −3 °C

Exercise 1f   page 11

1 >   **9** >   **17** 0, −3
2 >   **10** <   **18** 5, 8
3 >   **11** <   **19** −7, −11
4 <   **12** >   **20** 16, 32
5 >   **13** 10, 12   **21** $\frac{1}{6}, \frac{1}{36}$
6 <   **14** −10, −12   **22** −4, −2
7 >   **15** −2, −4   **23** −8, −16
8 >   **16** 2, 4   **24** −2, −3

Exercise 1g   page 12

1 −3   **11** 5   **21** 3   **31** 1
2 3   **12** −2   **22** −3   **32** 2
3 −2   **13** −2   **23** −10   **33** 2
4 −2   **14** −1   **24** −5   **34** −2
5 2   **15** 4   **25** 4   **35** −1
6 7   **16** 6   **26** 6   **36** −2
7 1   **17** 2   **27** 3   **37** 1
8 2   **18** −3   **28** 0   **38** 2
9 −12   **19** −3   **29** −3   **39** 2
10 −1   **20** −1   **30** −5   **40** 16

Exercise 1h   page 14

1 2   **6** 3   **11** −14   **16** 7   **21** 13
2 −3   **7** −3   **12** 0   **17** −3   **22** 13
3 7   **8** 6   **13** 0   **18** 2   **23** −6
4 3   **9** −14   **14** 6   **19** −4   **24** 8
5 −9   **10** 10   **15** −6   **20** 5   **25** 1

Exercise 1i   page 14

1 1   **15** 1   **29** −3   **43** −2
2 −5   **16** 9   **30** −19   **44** 1
3 9   **17** −1   **31** 2   **45** 2
4 8   **18** 0   **32** 3   **46** −12
5 −12   **19** 2   **33** 0   **47** 3
6 7   **20** 16   **34** 0   **48** 18
7 4   **21** 5   **35** −1   **49** −2
8 10   **22** −4   **36** 0   **50** 1
9 15   **23** −8   **37** 9   **51** 2
10 2   **24** 19   **38** −7   **52** −15
11 5   **25** −4   **39** −4   **53** −9
12 −12   **26** −4   **40** 3   **54** −6
13 5   **27** 4   **41** −10   **55** −8
14 −9   **28** −3   **42** −3

Exercise 1j   page 16

1 −3   **7** −10   **13** −2
2 −2   **8** −3   **14** −2
3 −5   **9** −5   **15** −4
4 −4   **10** −4   **16** −9
5 −4   **11** −1   **17** −4
6 −2   **12** −2   **18** −2

Exercise 1k page 18

1 −15	**11** +27	**21** −6
2 −8	**12** −16	**22** +15
3 +14	**13** −35	**23** −18
4 +4	**14** +24	**24** +20
5 −42	**15** −15	**25** −24
6 +12	**16** −45	**26** −24
7 −18	**17** −24	**27** +45
8 +16	**18** +8	**28** −20
9 −5	**19** +3	**29** −28
10 +18	**20** −8	**30** +36

Exercise 1l page 19

1 $-6x+30$	**16** $-3x-2$
2 $-15c-15$	**17** $16-24x$
3 $-10e+6$	**18** $-6y+2x$
4 $-3x+4$	**19** $20x-5$
5 $-16+40x$	**20** $-5+20x$
6 $-7x-28$	**21** $24+30x$
7 $-6d+6$	**22** $-24-30x$
8 $-8-4x$	**23** $24-30x$
9 $-14+21x$	**24** $-24+30x$
10 $-4+5x$	**25** $-5a-5b$
11 $12x+36$	**26** $6x+4y+2$
12 $10+15x$	**27** $-25-10x$
13 $6x-18$	**28** $4x-4y$
14 $-14-7x$	**29** $-4c+5$
15 $-6x+2$	**30** $18x-9$

Exercise 1m page 19

1 $25x+12$	**16** $12x-14$
2 $27-6c$	**17** $4x-12$
3 $14m-20$	**18** $9x+19$
4 $3-6x$	**19** $x-21$
5 $6x-4$	**20** $31x-11$
6 $13-8g$	**21** $14x+11$
7 $x-2$	**22** $-6x-19$
8 $4f+12$	**23** $14x-19$
9 $4s-3$	**24** $-6x+11$
10 $19x-3$	**25** $15x-9$
11 $17x-1$	**26** $11x+7$
12 $9x-18$	**27** $-7-15x$
13 $9x+1$	**28** $2x+21$
14 $15-5x$	**29** $2x+15$
15 $12x+8$	**30** $5x-2$

Exercise 1n page 20

1 −2	**11** 7	**21** 3	**31** 1	**41** $\frac{2}{3}$
2 −5	**12** $-\frac{3}{4}$	**22** 2	**32** 2	**42** 3
3 −2	**13** 6	**23** 2	**33** $\frac{1}{2}$	**43** $-\frac{1}{2}$
4 −1	**14** 5	**24** 1	**34** 2	**44** $\frac{3}{10}$
5 −2	**15** 7	**25** 6	**35** 2	**45** −1
6 −4	**16** 2	**26** −4	**36** −2	**46** 3
7 4	**17** 1	**27** 3	**37** 1	**47** $2\frac{1}{2}$
8 1	**18** 3	**28** −3	**38** 0	**48** 1
9 3	**19** 1	**29** $1\frac{1}{3}$	**39** 2	**49** $\frac{1}{4}$
10 5	**20** 2	**30** 1	**40** −2	**50** 2

Exercise 1p page 22

1 −5°

2 a $<$	**b** $>$		
3 2	**6** 4	**9** −24	**12** −14
4 −5	**7** 0	**10** −12	**13** −24
5 −2	**8** 5	**11** 10	**14** 7

15 a $-6+15x$ **b** $22-7x$

Exercise 1q page 23

1 a $(2, 1)$ **b** $(-1, -1)$ **c** $(6, -3)$

2 a a parallelogram
  b i $(3, 1)$ **ii** $\left(1, -\frac{1}{2}\right)$

3 a $9-9b$ **b** $11x-13$

4 a $s=-5$ **b** $x=3$ **c** $x=-3$

CHAPTER 2

Exercise 2a page 25

1 4:5	**8** 1:6	**15** 5:6:8
2 5:4	**9** 16:17	**16** 1:8:7
3 2:3	**10** 1:1000	**17** 3:4:7
4 1:4	**11** 2:3:5	**18** 1:8:4
5 1:3	**12** 3:4:6	**19** 12:1:2
6 9:200	**13** 1:5:10	**20** 14:9:2
7 16:3	**14** 3:4:5	

Exercise 2b page 26

1 15:1	**8** 9:4	**15** 15:19
2 8:1	**9** 16:7	**16** 5:4
3 3:2	**10** 10:7	**17** 1:2:3
4 3:1	**11** 8:5:3	**18** 4:3
5 4:9	**12** 2:3	**19** 4:3:2
6 7:10	**13** 40:9	**20** 3:4:6
7 35:24	**14** 2:15	

Exercise 2c page 27

1 5:7	**5** $6:8 = 24:32 = \frac{3}{4}:1$
2 13:8	**6** $10:24 = \frac{5}{9}:\frac{4}{3}$
3 5:8	**7** $8:64 = \frac{1}{16}:\frac{1}{2}$
4 7:10	**8** $\frac{2}{3}:3 = 4:18$

Exercise 2d page 27

1 3:2, 2:5	**3** 2:5, 8:3	**5** 2:1
2 3:4, 9:16	**4** 3:2, 2:3	

6 a 3:2 **b** 9:5 **c** 18:13 **d** 1:1
7 8:12:9 **8** 3:7
9 a 12:3:5 **b** 2:3 **c** 5:3

Exercise 2e page 29

1 10	**3** 7	**5** 8	**7** 6	**9** 9
2 4	**4** 2	**6** 12	**8** 6	**10** 12

Exercise 2f page 29

1 2^3	**11** 32	**21** 10 000
2 3^4	**12** 27	**22** 10
3 5^4	**13** 25	**23** 2^2
4 7^5	**14** 8	**24** 3^2
5 2^5	**15** 9	**25** 2^3
6 3^6	**16** 49	**26** 3^3
7 13^3	**17** 81	**27** 7^2
8 19^2	**18** 16	**28** 5^2
9 2^7	**19** 100	**29** 2^5
10 6^4	**20** 1000	**30** 2^6

31 24 **32** 400 **33** 80 **34** 288 **35** 6400

Exercise 2g page 31

1 0.233... $0.2\dot{3}$
2 0.002727... $0.00\dot{2}\dot{7}$
3 0.571428571... $0.\dot{5}7142\dot{8}$
4 0.14333... $0.14\dot{3}$
5 0.004285714285... $0.00\dot{4}28571\dot{4}$
6 0.1222... $0.12\dot{2}$
7 0.444... $0.\dot{4}$
8 0.666... $0.\dot{6}$

Answers

9 0.1818... 0.1̇8̇
10 0.714 285 714... 0.7̇14 285̇
11 0.777... 0.7̇
12 1.142 857 142 8... 1.1̇42 857̇

Exercise 2h page 33
1 0.33	**11** 14	**21** 0.363	**31** 1.8
2 0.32	**12** 6	**22** 0.026	**32** 42.6
3 1.27	**13** 27	**23** 0.007	**33** 1.01
4 2.35	**14** 3	**24** 0.070	**34** 0.0094
5 0.04	**15** 4	**25** 0.001	**35** 0.735
6 0.69	**16** 7	**26** 0.084	**36** 1.64
7 0.84	**17** 110	**27** 0.084	**37** 1.6
8 3.93	**18** 6	**28** 0.325	**38** 2
9 0.01	**19** 74	**29** 0.033	**39** 3.50
10 4.00	**20** 4	**30** 4.000	**40** 3.5

Exercise 2i page 34
1 0.17	**19** 2.1
2 0.93	**20** 0.9
3 0.35	**21** 9.7
4 2.03	**22** 0.6
5 2.85	**23** 1.7
6 0.16	**24** 27.3
7 0.04	**25** 0.006
8 0.05	**26** 0.018
9 0.24	**27** 0.417
10 0.04	**28** 0.021
11 0.22	**29** 0.038
12 0.95	**30** 0.001
13 4.1	**31** 0.028
14 57.4	**32** 0.031
15 2.6	**33** 0.016
16 0.9	**34** 0.019
17 7.3	**35** 0.039
18 1.2	**36** 0.037

Exercise 2j page 35
1 0.625	**16** 0.857
2 0.075	**17** 1.143
3 0.1875	**18** 0.111
4 0.6	**19** 0.333
5 0.36	**20** 0.364
6 0.14	**21** 0.214
7 0.0625	**22** 0.235
8 1.375	**23** 0.462
9 0.52	**24** 0.190
10 0.0375	**25** 0.158
11 0.429	**26** 0.176
12 0.444	**27** 0.267
13 0.167	**28** 0.389
14 0.667	**29** 0.136
15 0.818	**30** 0.121

Exercise 2k page 37
1 0.2	**13** 800	**25** 0.8
2 0.02	**14** 360	**26** 900
3 8	**15** 0.012	**27** 0.31
4 20	**16** 0.01	**28** 0.16
5 4500	**17** 100	**29** 24.5
6 12	**18** 2.3	**30** 3.2
7 0.16	**19** 21	**31** 1.2
8 6	**20** 0.012	**32** 41
9 60	**21** 0.00171	**33** 7
10 5	**22** 52 000	**34** 1.2
11 13	**23** 0.004	**35** 9
12 120	**24** 60	**36** 0.08

Exercise 2l page 37
1 6.33	**11** 0.02	**21** 36
2 8.43	**12** 2.9	**22** 3.9
3 16.67	**13** 8.2	**23** 0.167
4 28.17	**14** 0.087	**24** 1.1
5 0.72	**15** 1.3333	**25** 2.3
6 41.67	**16** 32.9	**26** 4
7 0.03	**17** 20.3	**27** 0.72
8 0.93	**18** 0.032	**28** 0.2571
9 1.03	**19** 283.333	**29** 0.57
10 0.71	**20** 1.7	**30** 2.5

Exercise 2m page 38
1 0.144	**8** 0.14	**15** 4
2 1.6	**9** 6.72	**16** 4
3 0.0512	**10** 4.2	**17** 10
4 128	**11** 12.24	**18** 0.12
5 2.88	**12** 84	**19** 0.125
6 5.76	**13** 0.3	**20** 0.7
7 0.000 126	**14** 0.16	**21** 12

Exercise 2n page 39
1 $0.2, \frac{1}{4}$ **5** $\frac{7}{8}, \frac{8}{9}, 0.9$ **9** $\frac{3}{7}, \frac{5}{11}, \frac{6}{13}$

2 $\frac{2}{5}, \frac{4}{9}$ **6** $\frac{3}{4}, \frac{17}{20}$ **10** $\frac{8}{11}, 0.7̇$

3 $\frac{4}{9}, \frac{1}{2}$ **7** $0.35, \frac{9}{25}, \frac{3}{8}$ **11** $0.3̇, \frac{5}{12}$

4 $\frac{3}{11}, 0.3, \frac{1}{3}$ **8** $\frac{4}{7}, 0.59, \frac{3}{5}$ **12** $0.45, \frac{9}{19}, \frac{1}{2}$

Exercise 2p page 41
1 a 0.75 **b** 0.6 **c** 0.3
d 0.15 **e** 0.875 **f** 0.24

2 0.45, 0.47 is larger

3 a $\frac{3}{50}$ **b** $\frac{1}{250}$ **c** $15\frac{1}{2}$
d $2\frac{1}{100}$ **e** $3\frac{1}{4}$

4 $\frac{43}{50}$

5 $\frac{1}{20}$

6 a 30% **b** 20% **c** 70%
d 3.5% **e** 92.5%

7 a 132% **b** 150% **c** 240%
d 105% **e** 255.5%

8 a 0.45 **b** 0.6 **c** 0.95
d 0.055 **e** 0.125

9 a $\frac{2}{5}$ **b** $\frac{13}{20}$ **c** $\frac{27}{50}$ **d** $\frac{1}{4}$

10 a 40% **b** 15% **c** 42% **d** 30%

11 $\frac{3}{5}$ **13** $\frac{19}{20}$ **15** 85% **17** 60%

12 $\frac{7}{20}$ **14** $\frac{8}{25}$ **16** 34% **18** 12.5%

19

Fraction	Percentage	Decimal
$\frac{3}{5}$	60%	0.6
$\frac{4}{5}$	80%	0.8
$\frac{3}{4}$	75%	0.75
$\frac{7}{10}$	70%	0.7
$\frac{11}{20}$	55%	0.55
$\frac{11}{25}$	44%	0.44

20 a $\frac{1}{20}$ **b** 5%

21 a 42% **b** 0.42

22 a $\frac{7}{25}$ **b** 60% **c** 12%

23 a $\frac{3}{5}$ **b i** 35% **ii** 95% **c** 0.05 **d** 12 : 1

Exercise 2q   page 43

1 100	**11** 600	**21** 10
2 36	**12** 4.5	**22** 0.36
3 0.35	**13** 2	**23** 10
4 20	**14** 0.7	**24** 20
5 180 000	**15** 17	**25** 32
6 0.8	**16** 0.003	**26** 1.2
7 0.48	**17** 0.0056	**27** 15
8 3.6	**18** 80	**28** 0.25
9 1.3	**19** 90 000	**29** 0.12
10 3 500 000	**20** 1.5	**30** 140

Exercise 2r   page 45

1 7.08	**26** 36.8	**51** 49.0
2 7.55	**27** 1949.2	**52** 11243.3
3 7.02	**28** 38.0	**53** 83.6
4 8.54	**29** 1354.2	**54** 2.28
5 9.19	**30** 14395.0	**55** 0.672
6 7.71	**31** 2.70	**56** 9.83
7 7.49	**32** 0.0196	**57** 0.693
8 9.15	**33** 0.0549	**58** 0.742
9 1.61	**34** 526.4	**59** 0.128
10 1.56	**35** 4.65	**60** 10300
11 3.80	**36** 0.0481	**61** 6337.1
12 1.50	**37** 1.79	**62** 0.00608
13 2.94	**38** 0.00515	**63** 34.8
14 1.54	**39** 3.97	**64** 483736.6
15 1.44	**40** 0.548	**65** 0.361
16 1326.8	**41** 0.121	**66** 0.0203
17 8371.2	**42** 0.0825	**67** 0.000123
18 6581.5	**43** 0.393	**68** 631.1
19 15.5	**44** 0.103	**69** 0.000 000 0961
20 6.65	**45** 0.139	**70** 49.1
21 172.3	**46** 123.6	**71** 0.174
22 14.7	**47** 55.8	**72** 16.7
23 11.2	**48** 91.7	**73** 0.000146
24 1169.6	**49** 186	**74** 13.4
25 12567.6	**50** 956.7	

Exercise 2s   page 46

1 4:9	**5** $\frac{19}{200}$	**9** 2.88
2 5:8	**6** 60 000	**10** 144
3 0.048	**7** 0.0614	
4 0.667	**8** 3.714	

Exercise 2t   page 46

1 1:2	**4** 9.186	**7** 21 500
2 32:24	**5** 0.875	**8** 1000
3 $6\frac{2}{3}$	**6** 50 000	

9 a 8   **b** 7.8   **c** 7.782

10 a $\frac{7}{20}$ **b** 0.35

Exercise 2u   page 47

1 20:3	**7** 2
2 9:7	**8** 10,9.9
3 0.714285	**9** 4.70
4 a 0.75 **b** $\frac{3}{4}$	**10 a** 62.5% **b** 0.625
5 10 000	**11** 200
6 25	

CHAPTER 3

Exercise 3a   page 49

1 20, 40	**4** 48, 96
2 36, 64, 100, 169	**5** 21, 25
3 36	**6** −7, −11

7 26, 37

8 $\frac{1}{32}, \frac{1}{64}$

9 $\frac{5}{6}, \frac{6}{7}$

10 0.0001, 0.00001

11 −28, −37

12 2, 1

13 54, 79

14 −17, −23

15 18, 12

16 −35, −46

17 25, −36

18 32, −64

19 0.004, 0.0008

20 42, −56

21 5, 10, 20, 40, 80, 160

22 20, 17, 14, 11, 8, 5, 2

23 1, 2, 4, 8, 16

24 11, 8, 5, 2, −1, −4, −7

25 $\frac{1}{16}, \frac{1}{8}, \frac{1}{4}, \frac{1}{2}, 1, 2$

26

Number of dots ↕	2	4	6	8	10
Number of dots ↔	3	4	5	6	7
Number of dots	5	8	11	14	17

27

Number of dots	4	8	12	16	20

28 15, 21, 28, 36

29 a 56, 72   **b** The first number is even and an even number is always added.

30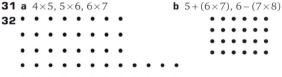

31 a $4\times5, 5\times6, 6\times7$ **b** $5+(6\times7), 6-(7\times8)$

32

33 $(3+4)\times5 = 35$
$(4+5)\times6 = 54$
$(5+6)\times7 = 77$
$(6+7)\times8 = 104$

34 $(3\times4)+5 = 17$
$(4\times5)+6 = 26$
$(5\times6)+7 = 37$
$(6\times7)+8 = 50$

35 $7-(5\times6)$
$8-(6\times7)$

36 $5\times6-8$
$6\times7-9$

37 1, 5, 13, 25, 41

38 Multiply the left and right numbers together, add the top number and subtract the bottom number. 18

Exercise 3b   page 53

1 a 10   **b** 12   **c** 18

2 a i 5 **ii** 9 **iii** 13   **b** 41

3 a 7, 11, 13, 17, 19, 23   **b** 8, 13, 21, 36, …
Add the two previous numbers

4 a and **b**

Pattern number	1	2	3	4	5	6
Number of black tiles	1	1	9	9	25	25
Number of white tiles	0	4	4	16	16	36
Total number of tiles	1	5	13	25	41	61

c yes   **d i** 7, 6   **ii** 10, 9

5 a

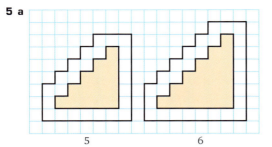

5   6

b

Diagram number	1	2	3	4	5	6
Number of black tiles	1	3	6	10	15	21
Number of white tiles	8	12	16	20	24	28

c 36

6 a 12 **b** 14   **c** 26

7 a $4^2 + 5^2 = 21^2 - 20^2$
$5^2 + 6^2 = 31^2 - 30^2$
$6^2 + 7^2 = 43^2 - 42^2$
$7^2 + 8^2 = 57^2 - 56^2$
$8^2 + 9^2 = 73^2 - 72^2$

b i $9^2 + 10^2 = 91^2 - 90^2$
ii $7^2 + 8^2 = 57^2 - 56^2$

8 2, 3, 7, 43, 1807, 3 263 443,   no, 1807 will divide by 13

9 30, 55

10 Gary is the better off by $260 the first year and every year.

Exercise 3c   page 56

1 5, 10, 15, 20, 25, 30
2 7, 11, 15, 19, 23
3 −1, 2, 5, 8, 11, 14
4 4, 11, 18, 25, 32
5 2.5, 3, 3.5, 4,
6 −5, −3, −1, 1, 3,
7 9, 6, 3, 0, −3
8 6, 9, 12, 15, 18, 21
9 6, 14, 22, 30, 38
10 4, 12, 24, 40, 60, 84
11 a 1, 5, 9, 13, 17, 21   **b** 57   **c** 21
12 a 1, 4, 7, 10, 13, 16,   **b** 34   **c** 68

CHAPTER 4

Exercise 4b   page 60

1 e and f
2 m and k, j and d
3 d and f, f and e, e and g, g and d
4 f and g
5 f and g, g and d, d and e, e and f
6 n and d, d and p, p and m, m and n
7 50°, 130°, 130°
8 60°, 120°, 120°
9 180°, 60°
10 105°, 180°
11 45°, 135°, 135°
12 180°, 155°
13 80°, 100°, 100°
14 165°, 180°

Exercise 4c   page 62

1 110°   **5** 180°   **9** 310°
2 60°   **6** 150°   **10** 60°
3 110°   **7** 100°
4 80°   **8** 120°

Exercise 4d   page 63

1 120°   **5** 150°, 60°
2 120°, 60°   **6** 50°
3 120°   **7** 40°
4 310°   **8** 120°, 60°, 120°, 60°

Exercise 4e   page 65

1 PR, PQ
2 a XY   **b** XZ   **c** $\hat{X}$

Exercise 4f   page 67

1 60°   **6** 30°   **11** 50°
2 85°   **7** 55°   **12** 90°
3 55°   **8** 60°   **13** 120°
4 110°   **9** 75°   **14** 55°
5 40°   **10** 25°   **15** 65°

Exercise 4g   page 69

1 60°, 50°   **4** 65°, 115°   **7** 60°
2 45°, 65°   **5** 85°, 30°   **8** 60°, 30°
3 70°   **6** 45°   **9** 90°, 45°

Exercise 4h   page 70

1 50°   **5** 60°   **9** 120°   **13** 60°, 120°
2 80°   **6** 40°   **10** 90°   **14** 80°, 70°
3 110°   **7** 90°   **11** 110°   **15** 80°, 115°
4 50°   **8** 60°   **12** 65°   **16** 50°, 130°

Exercise 4i   page 73

10 a scalene   **b** equilateral   **c** right-angled and isosceles
     d isosceles   **e** right-angled
13 70°   **19** 45°   **29** 55°, 70°
14 70°   **20** 70°   **30** 45°, 135°
15 65°   **21** 60°   **31** 80°, 80°
16 40°   **22** 20°   **32** 50°, 80°
17 90°   **23** 75°   **33** 40°, 140°
18 110°   **24** 86°   **34** 20°, 70°

Exercise 4j   page 77

1 65°   **3** 70°
2 80°

Exercise 4k   page 77

1 85°, 45°   **3** 55°, 125°
2 45°, 135°

Exercise 4l   page 78

1 60°, 30°   **3** 80°, 140°
2 65°, 65°, 60°

CHAPTER 5

Exercise 5b   page 82

1 g **3** d **5** f **7** d
2 e **4** e **6** f **8** g

Exercise 5d   page 85

1 60°   **4** 60°   **7** 110°   **10** 130°
2 110°   **5** 60°   **8** 120°   **11** 130°
3 75°   **6** 80°   **9** 30°

Exercise 5e   page 86

1 50°, 50°
2 130°, 130°, 50°
3 60°, 60°, 60°, 120°, 60°
4 50°, 80°, 50°
5 70°, 80°, 30°
6 115°, 115°
7 140°, 40°, 40°
8 70°, 110°, 70°, 70°
9 50°, 45°, 50°
10 55°, 125°, 55°
11 110°, 70°, 130°, 130°

12 40°, 100°
13 80°
14 90°, 90°, 50°
15 120°
16 40°
17 70°
18 60°
19 135°
20 55°
21 55°
22 120°
23 120°

24 45°

Exercise 5f   page 89

1 e
2 e
3 d
4 d
5 d
6 g
7 g
8 e
9 d

Exercise 5g   page 91

1 50°, 130°
2 130°, 50°
3 50°, 70°
4 260°, 40°, 60°
5 70°, 70°, 70°
6 45°, 90°
7 55°, 65°
8 60°
9 45°
10 30°
11 90°

Exercise 5h   page 93

1 e, g
2 e, d
3 e, g
4 e, d
5 h, f
6 d, g
7 70°, 110°, 180°
8 130°, 50°, 180°
9 140°, 40°, 180°
10 120°, 60°, 180°

Exercise 5i   page 94

1 120°
2 130°, 50°
3 85°
4 40°, 100°, 60°
5 55°, 125°
6 40°
7 80°, 80°
8 130°, 130°, 50°
9 80°, 100°, 80°, 100°
10 70°, 110°

Exercise 5j   page 95

1 65°
2 140°
3 55°
4 110°
5 70°
6 70°
7 45°

Exercise 5k   page 96

1 80°
2 60°
3 110°
4 40°
5 50°
6 40°
7 40°

CHAPTER 6

Exercise 6a   page 98

1 4
2 4
3 12
4 2
5 3
6 4
7 1
8 3
9 2
10 3
11 4
12 5
13 3
14 4
15 1
16 1
17 3
18 8
19 −2
20 1
21 2
22 4
23 2
24 1
25 1
26 1
27 4
28 1
29 $4\frac{1}{2}$
30 $1\frac{1}{2}$
31 2
32 1
33 $\frac{2}{3}$
34 2
35 3
36 2
37 $5\frac{1}{2}$
38 $\frac{4}{5}$
39 1
40 2

Exercise 6b   page 100

1 $6x+24$
2 $6x+3$
3 $4x-12$
4 $6x-10$
5 $12-8x$
6 $20x+10$
7 $6-9x$
8 $35-28x$
9 $10x-14$
10 $42+12x$
11 $8x+18$
12 $26x+13$
13 $34x-13$
14 $8x+4$
15 $21x+5$
16 $13x+19$
17 $28x+27$
18 $18x-44$
19 $4x+25$
20 $-30x+47$
21 $10x+5$
22 $-x+12$
23 $17x-33$
24 $-4x+14$
25 $-10x+14$
26 $36x+26$
27 $-7x+32$
28 $x+22$
29 $21x-19$
30 $-6x+2$
31 2
32 $1\frac{1}{2}$
33 1
34 2
35 $\frac{1}{2}$
36 $1\frac{1}{2}$
37 $1\frac{1}{4}$
38 1
39 −1
40 −4
41 $1\frac{1}{2}$
42 $-\frac{2}{3}$
43 3
44 1
45 3
46 $\frac{1}{2}$

Exercise 6c   page 101

1 15
2 8
3 48
4 12
5 $3\frac{5}{9}$
6 $22\frac{1}{2}$
7 14
8 $8\frac{1}{3}$
9 $\frac{1}{6}$
10 $\frac{3}{20}$
11 $\frac{3}{2}$
12 $\frac{5}{9}$
13 $1\frac{1}{3}$
14 $1\frac{1}{20}$
15 $\frac{5}{12}$
16 $\frac{7}{10}$
17 $2\frac{1}{4}$
18 $13\frac{3}{4}$
19 $3\frac{6}{13}$
20 $1\frac{11}{17}$
21 $6\frac{3}{4}$
22 $3\frac{1}{2}$
23 $12\frac{2}{3}$
24 20
25 $-1\frac{1}{4}$
26 $1\frac{5}{7}$
27 $\frac{3}{4}$
28 $1\frac{1}{3}$
29 11
30 $\frac{1}{14}$
31 $1\frac{6}{7}$
32 $\frac{23}{27}$
33 $\frac{18}{23}$
34 $\frac{1}{2}$
35 $1\frac{1}{7}$
36 $1\frac{2}{3}$
37 $1\frac{1}{6}$
38 $5\frac{5}{6}$
39 $1\frac{1}{22}$
40 $1\frac{18}{23}$
41 $\frac{1}{30}$
42 7
43 $2\frac{17}{26}$
44 2
45 $-\frac{1}{4}$
46 $\frac{28}{33}$

Exercise 6d   page 103

1 $150
2 40
3 30 cm
4 12
5 24 cm
6 5 cm
7 9
8 3
9 12
10 $1000

Exercise 6e   page 104

1 $2\frac{2}{5}$
2 $1\frac{1}{3}$
3 $5\frac{1}{4}$
4 $6\frac{2}{3}$
5 $1\frac{1}{3}$
6 $6\frac{2}{3}$
7 $3\frac{3}{5}$
8 $1\frac{2}{7}$
9 $7\frac{1}{2}$
10 $\frac{3}{5}$
11 10
12 $\frac{7}{5}$
13 $5\frac{2}{5}$
14 $7\frac{1}{2}$
15 $2\frac{1}{4}$
16 $\frac{3}{5}$
17 $3\frac{1}{3}$
18 $7\frac{1}{5}$
19 $16\frac{2}{3}$
20 $3\frac{3}{4}$
21 $3\frac{3}{4}$
22 $3\frac{3}{5}$
23 $22.5
24 18 cm
25 98 cm
26 $10\frac{2}{3}$ cm
27 $10\frac{1}{2}$ cm
28 27 cm
29 30 cm
30 12 m

Exercise 6f   page 106

1 −8
2 15
3 −24
4 −3
5 2
6 28
7 −4
8 $\frac{1}{3}$
9 3
10 −2x+17

11 17x−20
12 −15x−30
13 25−14x
14 15x−20
15 −9x+6
16 2
17 25x−46
18 5x−17
19 1
20 3

21 2
22 13
23 $\frac{13}{11}$
24 $\frac{13}{16}$
25 10
26 7
27 −4
28 $\frac{7}{16}$

Exercise 6g   page 108

1 −1
2 −12
3 5
4 33
5 50

6 19
7 16
8 2
9 105
10 $3\frac{1}{3}$

11 $\frac{5}{24}$
12 31
13 −3

14 a 48   **b** −18   **c** 6   **d** 5
15 a 4   **b** 20   **c** 8   **d** −12
16 a 52   **b** 20   **c** 96   **d** −4
17 a 5   **b** 3   **c** 38   **d** −24
18 a $1\frac{1}{4}$ **b** $4\frac{7}{8}$ **c** $12\frac{5}{6}$ **d** $\frac{5}{24}$
19 a 15   **b** −1.1   **c** −15.9   **d** 0.62

Exercise 6h   page 109

1 formula
2 equation
3 equation
4 expression

5 formula
6 expression
7 expression
8 equation

9 equation
10 formula
11 formula
12 expression

Exercise 6i   page 110

1 $2\frac{1}{2}$
2 $3\frac{1}{3}$
3 6x−24

4 12
5 $\frac{1}{3}$
6 5x−8

7 $P = 4l + f + g$
8 7

Exercise 6j   page 110

1 $\frac{1}{3}$
2 2y−4x+23
3 $-\frac{7}{20}$
4 10

5 $P = 6a$
6 3
7 $9\frac{1}{2}$

CHAPTER 7

Exercise 7a   page 112

1 a $\frac{1}{7}$ **b** $\frac{2}{7}$ **c** $\frac{3}{7}$ **d** $\frac{6}{7}$
2 a $\frac{1}{6}$ **b** $\frac{1}{2}$ **c** $\frac{1}{2}$ **d** 0
3 a $\frac{1}{13}$ **b** $\frac{1}{4}$ **c** $\frac{12}{13}$
4 a $\frac{2}{5}$ **b** $\frac{2}{5}$ **c** $\frac{1}{2}$
5 a $\frac{4}{9}$ **b** $\frac{5}{9}$ **c** 0   **d** 1

Exercise 7c   page 114

1 a $\frac{1}{6}$ **b** $\frac{1}{9}$ **c** $\frac{1}{6}$
2 a $\frac{4}{25}$ **b** $\frac{16}{25}$
3 $\frac{1}{6}$
4 a $\frac{1}{16}$ **b** $\frac{1}{8}$ **c** $\frac{3}{16}$ **d** $\frac{5}{8}$
5 $\frac{1}{4}$
6 a $\frac{5}{36}$ **b** $\frac{1}{18}$ **c** $\frac{1}{18}$ **d** $\frac{19}{36}$
7 $\frac{1}{5}$
8 a $\frac{3}{10}$ **b** $\frac{3}{20}$
9 a $\frac{1}{4}$ **b** $\frac{1}{16}$ **c** $\frac{3}{4}$ **d** $\frac{1}{4}$

REVIEW TEST 1   page 120

1 D   **4** A   **7** D   **10** D
2 B   **5** D   **8** C   **11** D
3 D   **6** C   **9** C   **12** C
13 A
14 $a = 80°$, $b = 75°$, $c = 25°$, $d = 75°$, $e = 105°$, $f = 100°$
15 19.5
16 a $x = 50°$, $y = 130°$, $z = 70°$
b BC = AC = 6 cm and angle ABC = 60° because it is an equilateral triangle.
17 a $\frac{1}{3}$ **b** $\frac{8}{15}$ **c** $\frac{8}{15}$
18 a

		red			
		1	2	3	4
blue	1	(1, 1)	(1, 2)	(1, 3)	(1, 4)
	2	(2, 1)	(2, 2)	(2, 3)	(2, 4)
	3	(3, 1)	(3, 2)	(3, 3)	(3, 4)
	4	(4, 1)	(4, 2)	(4, 3)	(4, 4)

b 16   **c** $\frac{3}{16}$

CHAPTER 8

Exercise 8a   page 123

1 150%
2 125%
3 120%
4 160%
5 175%
6 135%
7 148%
8 400%
9 275%
10 112.5%
11 157%
12 115%
13 $\frac{130}{100}$

14 $\frac{180}{100}$
15 $\frac{165}{100}$
16 $\frac{230}{100}$
17 50%
18 75%
19 30%
20 15%
21 65%
22 58%
23 96%
24 34%
25 $37\frac{1}{2}$%
26 $66\frac{2}{3}$%

27 47%
28 90%
29 $\frac{60}{100}$
30 $\frac{25}{100}$
31 $\frac{66}{100}$
32 $\frac{88}{100}$
33 140
34 370
35 493
36 748
37 2768
38 849.3
39 104

40 185
41 319
42 2415
43 70
44 170
45 189
46 652.5
47 2448
48 3312
49 62
50 91
51 26
52 155

Exercise 8b   page 124

1 63.25 kg
2 $345.50
3 84
4 180 cm
15 a $3600   **b** $7650
16 63
17 94.3 kg
18 a $16 320 000   **b** $13 872 000
19 $5 832 000
20 6.75 km
21 a $112   **b** 616 litres   **c** $1008 less

5 33
6 $299 000
7 $4200
8 $3680

9 $8400
10 $10 500
11 $750 000
12 198 kg

13 414
14 $33 000

Exercise 8c   page 127

1 $42\frac{1}{2}\%$
2 2.17 m
3 125%
4 $\frac{145}{100}$
5 a 98 cm   **b** 960 sheep
6 58%
7 0.82
8 a 94.5   **b** 8.8 miles
9 a $87.30   **b** $442.8

CHAPTER 9

Exercise 9a   page 129

1 $6261, $3739 change   **2** $6748, $3252 change
3 $5661, $4339 change   **4** $5045, $4955 change
5 $5642, $4358 change
In questions 6 to 12 the answers in order are:
6 $2880, $360, $1560, $4800
7 $520, $540, $440, $1500
8 $350, $632, $660, $530, $2172
9 $950, $1350, $1040, $2130, $5470
10 $1450, $1335, $300, $270, $245, $699, $4299
11 $11 946   **12** $2 460 000

Exercise 9b   page 131

1 $64 000   **3** $44 100   **5** $40 200   **7** $40 250
2 $36 800   **4** $54 400   **6** $60 000   **8** $32 160
9 a 1125   **b** 500   **c** 625   **d** $48 125

10

	a	bi	bii	c
Ms Arnold	186	100	86	$21 970
Mr Beynon	158	80	78	$18 910
Miss Capstick	194	100	94	$22 130
Mr Davis	225	100	125	$27 625
Mr Edmunds	191	100	91	$22 695

d Thursday

Exercise 9c   page 133

1 a $4000   **b** $16 000
2 a $130 000   **b i** $390 000   **ii** $7500
3 M. Davis $1640, $6560
P. Evans $3075 $9225
G. Brown $4412.50, $13 237.50
A. Khan $5455, $16 365
4 The values in order are:
Mrs Peacock $400, $1000, $ 400, $1800, $18 200
Mr Walters $520, $1300, $520, $2340, $23 660
Ms Morgan $640, $1600, $640, $2880, $ 29 120
Mr Davis $760, $1900, $760, $3420, $34 580
Mrs Evans $960, $2400, $960, $4320, $43 680
Ms Bennett $2400, $6000, $2400, $10 800, $109 200
5 NIS: $1400, NHT; $3500, Education tax: $1400, total of
these deductions: $11 300, remaining income; $58 700,
income tax: $14 675, net income: $44 025
6 a $210   **b** $1410
7 $47 000   **8** $846   **9** $1 703 750
10 $33 350   **11** $5640, $5856
12 a $52 875   **b** no
c he increased the previous GCT price by $2\frac{1}{2}\%$, £54 000

Exercise 9d   page 136

1 $2000   **13** $49 434   **25** $26 369   **37** $8278
2 $2400   **14** $111 360   **26** $24 995   **38** $21 972
3 $2400   **15** $13 032   **27** $990   **39** $30 893
4 $5200   **16** $10 164   **28** $1904   **40** $51 889
5 $7700   **17** $36 456   **29** $34 731   **41** $1052
6 $4000   **18** $5076   **30** $48 788   **42** $1358
7 $8000   **19** $42 120   **31** $5250   **43** $7397
8 $14 400   **20** $116 676   **32** $11 089   **44** $2650
9 $21 600   **21** $12 636   **33** $20 176   **45** $2390
10 $46 200   **22** $3026   **34** $36 800   **46** $484
11 $12 250   **23** $7819   **35** $4861
12 $6000   **24** $42 297   **36** $4125

Exercise 9e   page 137

1 $525   **5** $173 840   **9** $97 047   **13** $29 858
2 $48 720   **6** $158 110   **10** $28 086   **14** $100 440
3 $962   **7** $70 110   **11** $43 890   **15** $113 082
4 $79 002   **8** $88 971   **12** $122 322   **16** $52 828

Exercise 9f   page 138

1 $50 000   **10** 7%   **19** 4 years
2 $60 000   **11** 10%   **20** 3 years
3 $35 000   **12** 16%   **21** $30 000
4 $42 000   **13** 8%   **22** $58 000
5 $84 000   **14** 12%   **23** $43 000
6 10%   **15** 9%   **24** $72 000
7 15%   **16** 2 years   **25** $100 000
8 12%   **17** 5 years
9 8%   **18** 3 years

Exercise 9g   page 140

	Principal in $s	Rate%	Time	Simple interest $s	Amount in $s
1	23 000	10	2 years	4600	27 600
2	18 000	16	8 years	23 040	41 040
3	95 000	9	4 years	34 200	129 200
4	76 000	18	$5\frac{1}{2}$ years	75 240	151 240
5	63 700	15	6 years	57 330	121 030
6	42 400	14	$\frac{3}{4}$ years	4452	46 852
7	82 800	$7\frac{1}{2}$	5 months	2587.50	85 387.50
8	55 500	$12\frac{1}{2}$	144 days	2737	58 237

9 $143 200 before interest is added
10 $94 300   **13** $2 034 375
11 5 years   **14** 125 days
12 7 years   **15** 183 755

Exercise 9h   page 141

1 $4200   **6** $20 654   **11** $4 840 000
2 $7632   **7** $402   **12** $627 200
3 $10 388   **8** $14 098   **13** $7604
4 $19 177   **9** $16 064   **14** $109 350
5 $14 399   **10** $28 490   **15** $1 280 000

Exercise 9i   page 143

1 $6550   **5** $16 877   **9** $9020
2 $8432   **6** $10 075.20   **10** $34 565
3 $10 580   **7** $15 232.50
4 $13 399   **8** $11 640

Exercise 9j   page 144

1 3	**15** 16	**23** 4 hours
2 $\frac{1}{10}$	**16** 1	**24** $\frac{1}{2}$ hour
3 $1\frac{1}{2}$	**17** 12	**25** 10 hours
4 1.2	**18** 3	**26** $2\frac{7}{9}$ hours
5 0.06	**19** 1.8	**27** $3
6 0.02	**20** 0.144	**28** $12
7 8	**21** 0.056	**29** $3024
8 2	**22** 0.84	**30** 1\frac{1}{2}$

Exercise 9k   page 146

1 Mr George $2350, Miss Newton $3330, Mr Khan $2287.50, Mrs Wilton $1905.80, Mr Barnes $4335.50

2 Mr George $235, Miss Newton $333, Mr Khan $228.75, Mrs Wilton $190.58, Mr Barnes $433.55

3 Mrs Wan $5850, Mr Davis $7752, Mr Deats $4023, Miss Beale $9305.40

4 Mrs Wan $6435, Mr Davis $8527.20, Mr Deats $4425.30, Miss Beale $10 235.94.

5

current	previous		usage	rate(c)	charge($)
9421	9175	Energy first	100	600	600
		Energy next	146	1150	1679
		Cost charge			110
		sub-total			2389
		Fuel & IPP charge	246	($)18	4428
				Total	6817

6

current	previous		usage	rate(c)	charge($)
8432	8156	Energy first	100	650	650
		Energy next	176	1250	2200
		Cost charge			130
		sub-total			2980
		Fuel & IPP charge	276	($)17	4692
				Total	7672

7

current	previous		usage	rate(c)	charge($)
11 041	10 762	Energy first	120	590	708
		Energy next	159	1300	2067
		Cost charge			140
		sub-total			2915
		Fuel & IPP charge	279	($)19	5301
				Total	8216

Exercise 9l   page 149

1 $496.30   **2** $248.15   **3** $1219.88   **4** $2115.95
5 a $1372.60   **b** $1523.59
6 a $2469.72   **b** $2741.39
7 a $4233.90   **b** 4699.63
8 a $1854.12

CHAPTER 10

Exercise 10a   page 152

1 a, b and c

2   **5**

3   **6**

4   **7**

Exercise 10b   page 153

1 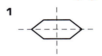   **7**

2   **8**

3 None   **9**

4   **10**

5   **11**

6   **12** 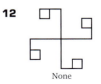 None

13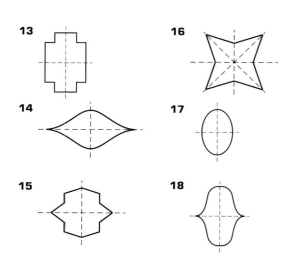

14

15

16

17

18

Exercise 10c   page 155

1

2

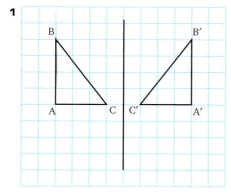

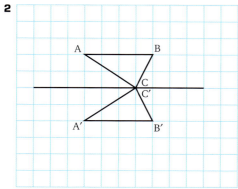

3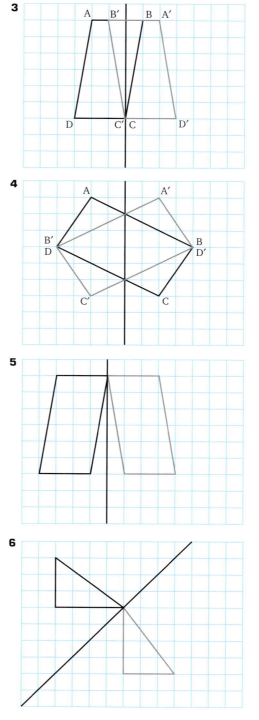

4

5

6

Exercise 10d   page 157

1 a and c
2 Translation e and b, reflection a and c, neither d
3 a yes; the shape and size is the same and it has not been turned or reflected
b yes; the shape and size is the same and it has not been turned or reflected.

Exercise 10e   page 158

1 6 units to the right, 3 units up

2 a 5 units to the left and 4 units up
  b 5 units to the right and 4 units down
3 a 4 units down
  b 6 units left
  c 5 units right and 5 up
  d no change

4

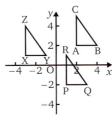

  a 1 unit left and 4 down
  b 1 unit right and 4 up
  c 4 units left, 3 up
  d no change

5

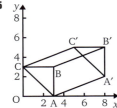

Yes, 5 units right and 2 up, parallelogram – the opposite sides are parallel. AA'C'C, B B'C'C.

6 a

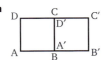

  b

7

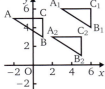

  a 4 units right and 2 down
  b 4 units left and 2 up
  c 1 unit right and 3 up

Exercise 10f   page 159

1 a $\frac{1}{4}$ **b** $\frac{1}{2}$ **c** $\frac{1}{3}$

2 a, **b** and **c**

Exercise 10g   page 160

1 a 4   **b** 2   **c** 3
2 a 6   **b** 3   **c** 2   **d** none

3

6

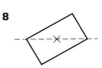

4

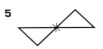

7

5

8

9 90°, 120°, 180°, 90°, 120°, 180°

Exercise 10h   page 161

1 rotational   **4** line   **7** both
2 rotational   **5** both   **8** both
3 line   **6** both   **9** rotational

Exercise 10i   page 163

1 90° clockwise
2 90° clockwise
3 180° either way
4 90° clockwise
5 origin, 180°
6 (1, 0), 90° anticlockwise
7 (1, 0), 180°
8 (2, 0), 180°
9 (2, 1) 90° clockwise
10 (3, 1) 180°

11

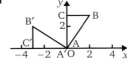

12

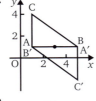

13

14

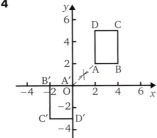

15

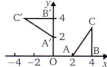

16

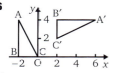

17

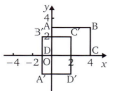

18

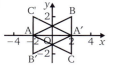

19

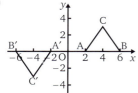

a semicircle **b** OC = OC′, OB = OB′

20 Because for 180° the direction of rotation does not matter.

21

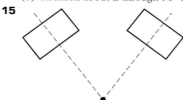

Exercise 1Cj page 167

1 c (0, 4) **e** 90° clockwise
2 c (−2, −2) **e** 90° clockwise
3 c (−1, 3) **e** 90° anticlockwise

Exercise 10k page 168

1 90° anticlockwise **2** 90° clockwise

Exercise 10l page 169

Simple models may again prove useful.
1 Translation given by 2 units left and 2 up
2 Reflection in $x = 0$
3 Reflection in $x = \frac{1}{2}$
4 Translation given by 2 units left
5 Reflection in $y = -x$
6 Rotation through 90° anticlockwise about $(-\frac{1}{2}, -\frac{1}{2})$
7 Rotation through 90° anticlockwise about (0, 1)
8 Rotation through 180° about (0, 2)
9 Rotation through 180° about $\left(\frac{5}{2}, \frac{3}{2}\right)$
10 Reflection in $y = x + 1$
11 Reflection in BC, rotation about B through 90° clockwise
12 Reflection in y-axis, rotation about O through 180°, translation parallel to x axis (8 units to the right).

13 (1) Reflection in OB
    (2) Translation parallel to AB
    (3) Rotation about B through 120° clockwise
    (4) Rotation about O through 120° clockwise
14 (1) Reflection in BE
    (2) Translation parallel to AB
    (3) Rotation about B through 90° clockwise
    (4) Rotation about the midpoint of BE, through 180°
    (5) Rotation about E through 90° anticlockwise

15

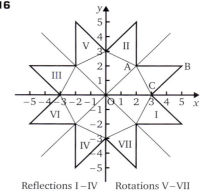

centre of the turning circle

16

Reflections I–IV   Rotations V–VII

17 a Reflection in the line $y = -x$
   b Yes

CHAPTER 11

Exercise 11b page 177

1 12 cm **5** 2 km
2 10 m **6** 9.2 cm
3 30 mm **7** approx 3.14
4 7 cm **8** approx 3.14

Exercise 11c page 178

1 14.5 m **13** 8.80 m
2 28.9 cm **14** 220 mm
3 18.2 cm **15** 35.2 cm
4 333 mm **16** 970 mm
5 54.7 m **17** 88 cm
6 1570 mm **18** 24 m
7 226 cm **19** 1300 mm
8 30.2 m **20** 220 cm
9 11.3 m **21** 1600 mm
10 0.0880 km **22** 2000 cm
11 44.0 cm **23** 29 m
12 176 mm

Exercise 11d page 180

1 10.3 cm **6** 33.6 cm
2 10.7 cm **7** 94.3 cm
3 18.3 cm **8** 62.8 mm
4 20.5 cm **9** 20.6 cm
5 27.9 cm **10** 45.1 cm

Exercise 11e   page 181

1 78.5 mm
2 62.8 mm, 88.0 mm
6 175.9 cm, 200
7 12.6 cm
8 94.2 cm

3 4.4
4 194.2 cm
5 175.9 cm

9 62.8 m
10 6.3 secs, 9.6 revolutions
11 3141.6 cm
12 12.6 m
13 70.7
14 94.2 m

Exercise 11f   page 184

1 7.0 cm
2 19.3 mm
3 87.5 m
4 43.8 cm
5 73.5 mm
6 132.3 cm
7 5.8 mm
8 62.2 m
9 92.6 cm
10 13.9 m
11 16.6 m

12 59.8 m
13 31.8 cm
14 20.1 m
15 4.9 cm
16 9.5 cm each
17 3.8 cm, 45.8 cm
18 37.7 cm
19 4.8 cm
20 9.5 cm
21 9.5 cm, 29.1 cm

Exercise 11g   page 186

1 50.3 cm^2
2 201.1 cm^2
3 78.5 m^2
4 78.5 mm^2
5 38.5 cm^2
6 11309.7 cm^2
7 45.4 m^2
8 9.62 km^2
9 20106.2 m^2
10 25.1 cm^2

11 51.3 m^2
12 58.9 cm^2
13 117.8 mm^2
14 451.2 mm^2
15 373.8 cm^2
16 457.1 cm^2
17 714.2 m^2
18 942.5 cm^2
19 3536.5 cm^2
20 193.1 cm^2

Exercise 11h   page 188

1

707 cm^2

4

26.2 cm^2

2

236 cm^2

3 491 mm^2

5 No
6 21.5 cm^2
7 8,110 cm^2
8 11700 cm^2
9 2

Exercise 11i   page 190

1 17.6 mm
2 9.55 m
3 37.7 cm

4 26.4 m^2
5 491 cm^2
6 28.6 mm

7 7.95 cm^2

Exercise 11j   page 191

1 62.8 m
2 452 cm^2
3 57.3 cm

4 50.2 m^2
5 89.2 mm
6 39.2 cm

7 87.5 cm^2

Exercise 11k   page 191

1 12.6 km^2
2 308 mm

3 14 m
4 154 cm^2

5 32.2 cm^2
6 18.1 m^2

CHAPTER 12

Exercise 12a   page 195

1 No, angles not equal
2 Yes
3 No, sides not equal
4 No, $\begin{cases} \text{sides not equal} \\ \text{angles not equal} \end{cases}$
5 No, $\begin{cases} \text{sides not equal} \\ \text{angles not equal} \end{cases}$
6 No, $\begin{cases} \text{sides not equal} \\ \text{angles not equal} \end{cases}$
7 Yes
8 No, not bounded by straight lines

Exercise 12b   Page 196

1 180°   **2** 360°
3 a $p = 100°$, $r = 135°$, $x = 55°$, $q = 125°$
  b 360°
4 a $w = 120°$, $x = 60°$, $y = 120°$, $z = 60°$
  b 360°
5 a 180°   **b** 540°   **c** 180°   **d** 360°
6 360°
7 a equilateral   **b** 60°   **c** 120°
  d 60°   **e** 360°

Exercise 12c   Page 198

1 60°
2 90°
3 50°
4 50°
5 60°

6 90°
7 95°
8 55°
9 30°
10 125°

11 $x = 50°$
12 $x = 30°$
13 $x = 24°$
14 a 5
   b 8

Exercise 12d   Page 200

1 36°
2 45°
3 30°

4 60°
5 24°
6 20°

7 40°
8 22.5°
9 18°

Exercise 12e   Page 201

1 720°
2 540°
3 1440°

4 360°
5 900°
6 1800°

7 2880°
8 1260°
9 2340°

Exercise 12f   Page 202

1 a 3240°   **b** 2520°   **c** 1620°
2 80°   **6** 85°   **10** 135°
3 120°   **7** 110°   **11** 144°
4 110°   **8** 108°   **12** 150°
5 105°   **9** 120°   **13** 162°
14 a 18   **b** 24
15 a 12   **b** 20
16 a yes, 12   **d** yes, 6
   b yes, 9   **e** no
   c no   **f** yes, 4
17 a yes, 4   **d** yes, 72
   b yes, 6   **e** yes, 36
   c no   **f** yes, 8

Exercise 12g   Page 204

1 54°   **6** 50°   **11** 72°
2 45°   **7** 80°   **12** 45°
3 150° **8** 135° **13** 60°
4 72°   **9** 100° **14** 36°
5 60°   **10** 60°
15 a 36°   **b** 36°
16 a 128.6°   **b** 25.7°
17 77.1°
18 a 22.5°   **b** 22.5°
19 22.5°
20 45°

Exercise 12h   Page 207

1 a The interior angles (135°) do not divide exactly into 360°
  b A square
2 a No
  b A regular ten-sided polygon
4 Square, equilateral triangle

CHAPTER 13

Exercise 13a   page 210

1 The second number is one more than the first.
2 The second number is two more than the first.
3 The second number is the square of the first.
4 {(Maths, Spanish), (Science, Spanish), (Science, Maths)}
5 {(4, □), (5, ◇), (3, ▲)
6 {(Tnd, Jma), (Bds, Jma), (St Lucia, Jma), (Bds, Tnd),
   (St Lucia, Tnd), (St Lucia, Bds)}
7 25, 8
8 7, 2, 17

Exercise 13b   page 212

1 {1, 2, 5, 10}, {2, 3, 6, 11}
2 a {a, b}, {b, c}   **b** {□, △, ▱}, {□, △}
3 {4, 16, 36}
4 a {Dwayne, Fred), (Scott, Fred), (Scott, Dwayne)},
  b {Dwayne, Scott}, {Fred, Dwayne}
5 a {10°, 150°, 45°, 175°}, {acute, obtuse}
  b {t, s}, {t, u, w}

Exercise 13c   page 213

1 a

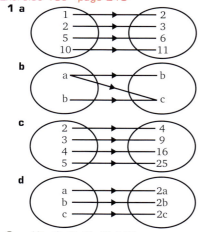

2 a {(1, 4), (3, 8), (5, 14)}
  b {(1, 2), (2, 2), (3, 2), (4, 5)}
  c {(a, b), (a, c), (b, c)}
  d {(a, b), (a, d), (b, c)}

3

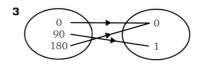

Exercise 13d   page 215

1 a 1:1   **b** $n:n$ **c** 1:1   **d** 1:1
2 a 1:1   **b** $n:1$ **c** $n:n$ **d** $1:n$
3 $n:1$

Exercise 13e   page 216

1 a

x	1	2	3	4
y	2	4	6	8

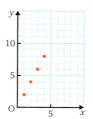

b

x	2	4	6	9
y	1	2	3	4.5

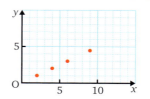

c

x	10	7	5	0
y	0	3	5	10

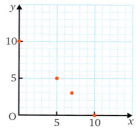

d

x	0	1	2	3	4
y	0	4	6	8	6

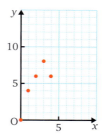

Answers

2

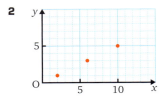

3 a

x	1	4	7	10	13
y	12	10	7	6	4

b {(1, 12), (4, 10), (7, 7), (10, 6), (13, 4)}   **c** 1 : 1

4 a

x	0	5	5	10
y	10	5	15	10

b

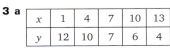

c $n:n$

5 a

x	0	2	4	6	10
y	1	2	3	4	6

b $\frac{1}{2}(x\text{-coordinate} + 1)$ **c** 5

d $3\frac{1}{2}$, 9, $\frac{1}{2}a + 1$

6 a

x	0	2	4	7	8
y	7	6	5	3.5	3

b $7 - \frac{1}{2}(x\text{-coordinate})$ **c** 4.5

d $6\frac{1}{2}$, 1, $7 - \frac{1}{2}a$

7 a

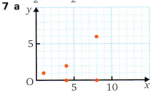

b 1 : n; one value of x maps to different values of y

Exercise 13f   page 219

1 Missing values: 3, 9
2 Missing values: 3, 11
3 Missing values: 8, 4, 2
4 Missing values: 10, 8, 0
5 Missing values: 2, 5, 17
6 Missing values: 1, 7, 31
7 Missing values: 2, 4, 8
8 a Missing values: 0, 6, 12
  b {(1, 2, 3, 4)}, {(0, 2, 6, 12)}
  c

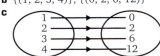

  d 1 : 1
9 a Missing values: 5, $6\frac{1}{2}$
  b $\left\{\left(1, 1\frac{1}{2}, 2\right)\right\}, \left\{\left(3\frac{1}{2}, 5, 6\frac{1}{2}\right)\right\}$
  c

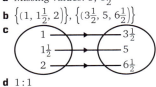

  d 1 : 1

10 a Missing values: 2, 2
  b {(1, 2, 3, 4)}, {(0, 2)}
  c

  d $n : 1$
  e

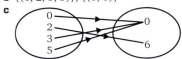

11 a Missing values: 0, 6, 0
  b {(0, 2, 3, 5)}, {(0, 6)}
  c

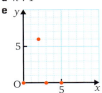

  d $n : 1$
  e

12 a

x	2	5	8	11	14
y	2	3	4	5	6

  b $\frac{1}{3}$ of x-coordinate plus $1\frac{1}{3}$
  c $\{(x, y)\}$ where $y = \frac{1}{3}x + 1\frac{1}{3}$ for $x = 2, 5, 8, 11, 14$

CHAPTER 14

Exercise 14a   page 223

1 yes   **5** yes   **9** no
2 no   **6** yes   **10** no
3 yes   **7** yes   **11** A and D
4 no   **8** no

Exercise 14b   page 224

1 a yes
  b AC = 4.1 cm, CB = 3.2 cm, A'C' = 8.2 cm, C'B' = 6.4 cm
  c each is 2
  d all are equal to 2
2 a yes
  b AC = 8.6 cm, CB = 7.7 cm, A'C' = 5.7 cm, C'B' = 5.1 cm
  c each is 0.67 or $\frac{2}{3}$
  d all equal to 0.67
3 a yes
  b AC = 7.9 cm, CB = 6.4 cm, A'C' = 3.9 cm, C'B' = 3.2 cm
  c each is 0.5 or $\frac{1}{2}$
  d all equal to 0.5
4 a yes
  b AC = 10.1 cm, CB = 6.6 cm, A'C' = 7.6 cm, C'B' = 4.9 cm
  c each is 0.75 or $\frac{3}{4}$
  d all equal 0.75
5 a yes
  b AC = 6.1 cm, CB = 9.2 cm, A'C' = 9.2 cm, C'B' = 13.8 cm
  c each is 1.5 or $\frac{3}{2}$
  d all equal 1.5

6 80°, 52°, yes
7 72°, 72°, yes
8 70°, 70°, yes
9 93°, 52°, no

Exercise 14c   page 226
1 yes, $\frac{AB}{PQ}=\frac{BC}{QR}=\frac{AC}{PR}$ **5** yes, $\frac{AB}{PQ}=\frac{BC}{QR}=\frac{AC}{PR}$

2 yes, $\frac{AB}{PR}=\frac{BC}{RQ}=\frac{AC}{PQ}$ **6** yes, $\frac{AB}{RP}=\frac{BC}{PQ}=\frac{AC}{RQ}$

3 no   **7** yes, $\frac{AB}{RQ}=\frac{BC}{QP}=\frac{AC}{RP}$

4 yes, $\frac{AC}{QP}=\frac{CB}{PR}=\frac{AB}{QR}$ **8** no

Exercise 14d   page 227
1 yes, 2.5 cm   **4** yes, 6.3 cm   **7** $8\frac{1}{3}$ cm

2 yes, 7.2 cm   **5** 7.5 cm   **8** $4\frac{1}{2}$ cm

3 no   **6** 7.5 cm   **9 b** 4 cm

10 b CD = 9 cm, DE = 10.5 cm
11 b 5.0 cm
12 b DE = 18 cm, AE = 13.5 cm, CE = 4.5 cm

Exercise 14e   page 231
1 8 cm   **4** 30 cm
2 6 cm   **5** 24 cm
3 10 cm

Exercise 14f   page 232
1 yes, $\widehat{P}$ **4** yes, $\widehat{P}$
2 yes, $\widehat{Q}$ **5** no
3 no   **6** yes, $\widehat{P}$
7 yes, $\widehat{B}=\widehat{D}$, $\widehat{C}=\widehat{E}$, yes, corresponding angles equal

Exercise 14g   page 233
1 yes, CB = 3.6 cm   **5** yes, AC = $10\frac{2}{3}$ cm
2 no   **6** 5.1 cm
3 yes, RQ = 35 cm   **7** 3 cm
4 yes, RQ = 7.2 cm

Exercise 14h   page 235
1 yes, 4 cm   **5** yes, 34°
2 yes, 2.4 cm   **6** yes, 32°
3 yes, 83°   **7** yes, $3\frac{1}{2}$ cm
4 no   **8** yes, 18 cm
9 b AC = 3.15 cm, CE = 1.05 cm
10 b 143 cm   **13** 19.2 m
11 c yes   **14** 60 cm
12 10 m

REVIEW TEST 2   page 238
1 C   **3** B   **5** C   **7** B   **9** B
2 D   **4** B   **6** B   **8** B   **10** C
11 a 154 cm²   **b** 66 cm   **c** 7 cm
12 $3587.50
13 $39 420
14 a $6312 50   **b** $18 687.50
15 They are, 3.6 cm
16 a (2, −5), (3, −8), (7, −1)
b (−2, 5), (−3, 8), (−7, 1)
c (5, 2), (8, 3), (1, 7)
17 a a translation 6 units to the left and 3 units down
b rotation clockwise through 90° about the origin

CHAPTER 15

Exercise 15a   page 241
1 48 cm²   **17** 33 cm²
2 1.56 m²   **18** 75 cm²
3 80 cm²   **19** 70 cm²
4 3.2 cm²   **20** 24.4 cm²
5 100 cm²   **21** 82.5 cm²
6 399 cm²   **22** 30 cm²
7 2.4 cm, 12 cm, 25 cm   **23** 96 cm²
8 24 cm²   **24** 21 cm²
9 14.4 cm²   **25** 8.32 cm²
10 40 cm²   **26** 10 sq. units
11 32.4 m²   **27** 12 sq. units
12 22.2 cm²   **28** 10 sq. units
13 45 cm²   **29** 15 sq. units
14 44 cm²   **30** 10 sq. units
15 64 cm²   **31** $7\frac{1}{2}$ sq. units
16 540 cm²

Exercise 15b   page 244
1 8 cm   **5** 3 cm   **9** 0.4 cm
2 6 cm   **6** 36 cm   **10** 6 cm
3 6 cm   **7** 3 cm   **11** 8 cm
4 2 cm   **8** $2\frac{2}{3}$ cm   **12** 4 cm

Exercise 15c   page 245
1 78 cm²   **6** 75 cm²
2 22.5 cm²   **7** 18 cm²
3 20 cm²   **8** 68 cm²
4 54 cm²   **9** 38.5 cm²
5 60 cm²   **10** 48 cm²

Exercise 15d   page 248
1 48 cm³   **8** 160 m³   **15** 8 m³
2 1600 mm³   **9** 12 cm³   **16** $\frac{1}{8}$ cm³
3 5400 mm³   **10** 7.2 cm³   **17** 15.625 cm³
4 16 mm³   **11** 4.32 m³   **18** 27 km³
5 31.72 m³   **12** 0.756 m³   **19** 512 km³
6 10.5 cm³   **13** 64 cm³   **20** $3\frac{3}{8}$ km³
7 24 m³   **14** 125 cm³   **21** 39.304 m³

Exercise 15e   page 249
1 8   **2** 6   **3** 8   **4** 12

Exercise 15f   page 250
1 a m³   **b** mm³   **c** cm³
2 8000 mm³   **8** 3 000 000 cm³
3 14 000 mm³   **9** 2 500 000 cm³
4 6200 mm³   **10** 420 000 cm³
5 430 mm³   **11** 6300 cm³
6 92 000 000 mm³   **12** 0.022 cm³
7 40 mm³   **13** 0.731 cm³

Exercise 15g   page 251
1 2500 cm³   **8** 4 litres
2 1760 cm³   **9** 24 litres
3 540 cm³   **10** 0.6 litres
4 7.5 cm³   **11** 5000 litres
5 35 000 cm³   **12** 12 000 litres
6 28 cm³   **13** 4600 litres
7 7 litres   **14** 67 litres

Answers

Exercise 15h   page 252

1 160 cm³	**5** 403.2 mm³	**9** 36 000 cm³
2 35 cm³	**6** 480 mm³	**10** 168 cm³
3 2.25 m³	**7** 1.12 m³	**11** 1.89 cm³
4 1500 mm³	**8** 1.944 m³	**12** 230 400 cm³

Exercise 15i   page 255

1 720 cm³	**11** 315 cm³
2 2160 cm³	**12** 450 cm³
3 1120 cm³	**13** 690 cm³
4 720 cm³	**14** 624 cm³
5 1242 cm³	**15** 864 cm³
6 128 cm³	**16** 720 cm³
7 660 cm³	**17** 5.184 m³
8 192 cm³	**18** 1344 cm³
9 2400 cm³	**19** 624 m³
10 2880 cm³	

Exercise 15j   page 258

1 126 cm³	**11** 322 cm³
2 113 cm³	**12** 407 cm³
3 314 cm³	**13** 330 cm³
4 59.4 cm³	**14** 651 cm³
5 3.14 cm³	**15** 70 800 cm³
6 15.1 m³	**16** 30 cm³
7 37.7 cm³	**17** 235 mm³
8 50.9 cm³	**18** 825 cm³
9 4520 cm³	**19** 1 596 844 cm³
10 1390 cm³	**20** 44.0 cm³

Exercise 15k   page 259

1 1005 cm³	**3** 35 cm³	**5** 628 cm³
2 402 cm³	**4** 204 cm³	**6** 2160 cm³

CHAPTER 16

Exercise 16a   page 263

1 9	**10** 100	**19** 0.09
2 25	**11** 0.09	**20** 64
3 81	**12** 4 000 000	**21** 1600
4 900	**13** 0.000 016	**22** 1 000 000
5 0.16	**14** 1	**23** 4900
6 2500	**15** 0.0009	**24** 0.0009
7 90 000	**16** 900	**25** 8100
8 0.0004	**17** 10 000	**26** 0.0064
9 250 000	**18** 16	**27** 40 000

Exercise 16b   page 264

1 60.84	**11** 146.41	**21** 14.2884
2 1444	**12** 8.6436	**22** 0.0576
3 6272.64	**13** 1.0404	**23** 0.005 184
4 0.1681	**14** 184.96	**24** 196
5 0.0256	**15** 289	**25** 19 600
6 0.001 024	**16** 58 081	**26** 0.0196
7 2323.24	**17** 0.678 976	**27 c** 4.8, 3.2,
8 127.69	**18** 0.772 641	9.6, 7.3
9 2631.69	**19** 0.001 296	**28 c** 30, 70,
10 96.04	**20** 5241.76	164, 185

Exercise 16c   page 265

1 5.76 cm²	**4** 1.1236 m²	**7** 0.003 844 m²
2 92.2 m²	**5** 295.84 cm²	**8** 102 400 km²
3 1049.76 cm²	**6** 2704 mm²	**9** 0.0961 cm²

Exercise 16d   page 266

1 3	**6** 6	**11** 0.9	**16** 20	**21** 0.02
2 5	**7** 7	**12** 0.8	**17** 50	**22** 500
3 2	**8** 8	**13** 70	**18** 100	**23** 2000
4 9	**9** 1	**14** 700	**19** 0.3	**24** 0.004
5 10	**10** 90	**15** 0.2	**20** 0.4	

Exercise 16e   page 267

1 4. _ _ _ _ _	**6** 3. _ _ _ _ _	**11** 0.4 _ _ _ _ _
2 3. _ _ _ _ _	**7** 9. _ _ _ _ _	**12** 9. _ _ _ _ _
3 6. _ _ _ _ _	**8** 4. _ _ _ _ _	**13** 3. _ _ _ _ _
4 6. _ _ _ _ _	**9** 2. _ _ _ _ _	**14** 0.7 _ _ _ _ _
5 1. _ _ _ _ _	**10** 0.2 _ _ _ _ _	**15** 2. _ _ _ _ _

Exercise 16f   page 267

1 30	**8** 200	**15** 2000
2 200	**9** 60	**16** 60
3 20	**10** 100	**17** 20
4 80	**11** 600	**18** 3
5 20	**12** 10	**19** 1
6 100	**13** 20	**20** 6
7 50	**14** 200	

Exercise 16g   page 268

1 6.2	**8** 2.4	**15** 101.5
2 4.4	**9** 25.5	**16** 641.9
3 20.7	**10** 8.1	**17** 27.0
4 65.0	**11** 3.3	**18** 85.3
5 5.7	**12** 7.6	**19** 7.8
6 3.1	**13** 4.9	**20** 2698.1
7 8.2	**14** 4.4	

Exercise 16h   page 269

1 0.205	**8** 0.527	**15** 0.208
2 0.648	**9** 0.167	**16** 0.098
3 0.118	**10** 0.053	**17** 0.912
4 0.748	**11** 0.548	**18** 0.566
5 0.012	**12** 0.416	**19** 0.228
6 0.707	**13** 0.447	**20** 0.866
7 0.775	**14** 0.831	**21** 0.009

Exercise 16i   page 269

1 9.22 cm	**5** 0.24 m	**9** 0.09 km
2 10.95 cm	**6** 3.89 cm	**10** 7.68 cm
3 22.36 m	**7** 27.37 mm	**11** 15.52 m
4 5.66 m	**8** 290.34 km	**12** 7.81 cm

CHAPTER 17

Exercise 17a   page 272

1 a 90 km	**b** 2 hours	**c** 45 km
2 a 30 km	**b** 3 hours	**c** 10 km
3 a 107 km	**b** 3.2 hours	**c** 33.4 km
4 a 50 miles	**b** 2 hours	**c** 25 miles

Exercise 17b   page 274

The scales in some of these answers have been halved.

1

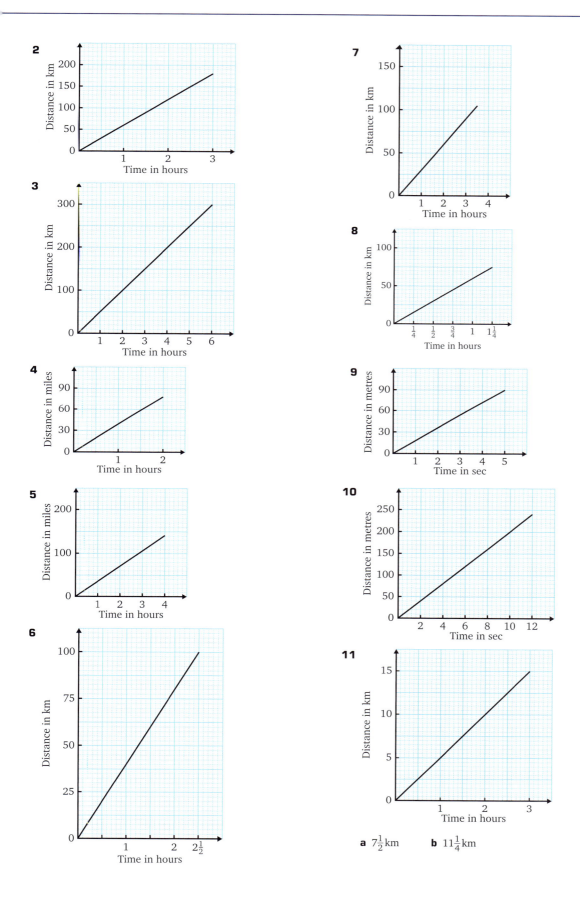

2

3

4

5

6

7

8

9

10

11

a $7\frac{1}{2}$ km   **b** $11\frac{1}{4}$ km

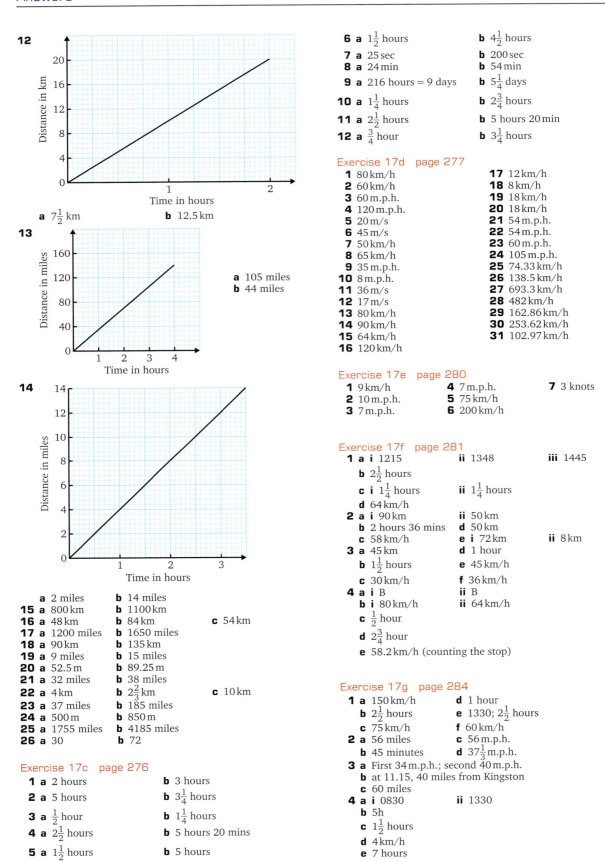

12

a $7\frac{1}{2}$ km   b 12.5 km

13

a 105 miles
b 44 miles

14

a 2 miles   b 14 miles
15 a 800 km   **b** 1100 km
16 a 48 km   **b** 84 km   **c** 54 km
17 a 1200 miles   **b** 1650 miles
18 a 90 km   **b** 135 km
19 a 9 miles   **b** 15 miles
20 a 52.5 m   **b** 89.25 m
21 a 32 miles   **b** 38 miles
22 a 4 km   **b** $2\frac{2}{3}$ km   **c** 10 km
23 a 37 miles   **b** 185 miles
24 a 500 m   **b** 850 m
25 a 1755 miles   **b** 4185 miles
26 a 30   **b** 72

Exercise 17c page 276

1 a 2 hours   **b** 3 hours
2 a 5 hours   **b** $3\frac{1}{4}$ hours
3 a $\frac{1}{2}$ hour   **b** $1\frac{1}{4}$ hours
4 a $2\frac{1}{2}$ hours   **b** 5 hours 20 mins
5 a $1\frac{1}{2}$ hours   **b** 5 hours

6 a $1\frac{1}{2}$ hours   **b** $4\frac{1}{2}$ hours
7 a 25 sec   **b** 200 sec
8 a 24 min   **b** 54 min
9 a 216 hours = 9 days **b** $5\frac{1}{4}$ days
10 a $1\frac{1}{4}$ hours   **b** $2\frac{3}{4}$ hours
11 a $2\frac{1}{2}$ hours   **b** 5 hours 20 min
12 a $\frac{3}{4}$ hour   **b** $3\frac{1}{4}$ hours

Exercise 17d page 277

1 80 km/h   **17** 12 km/h
2 60 km/h   **18** 8 km/h
3 60 m.p.h.   **19** 18 km/h
4 120 m.p.h.   **20** 18 km/h
5 20 m/s   **21** 54 m.p.h.
6 45 m/s   **22** 54 m.p.h.
7 50 km/h   **23** 60 m.p.h.
8 65 km/h   **24** 105 m.p.h.
9 35 m.p.h.   **25** 74.33 km/h
10 8 m.p.h.   **26** 138.5 km/h
11 36 m/s   **27** 693.3 km/h
12 17 m/s   **28** 482 km/h
13 80 km/h   **29** 162.86 km/h
14 90 km/h   **30** 253.62 km/h
15 64 km/h   **31** 102.97 km/h
16 120 km/h

Exercise 17e page 280

1 9 km/h   **4** 7 m.p.h.   **7** 3 knots
2 10 m.p.h. **5** 75 km/h
3 7 m.p.h.   **6** 200 km/h

Exercise 17f page 281

1 a i 1215   **ii** 1348   **iii** 1445
  b $2\frac{1}{2}$ hours
  c i $1\frac{1}{4}$ hours **ii** $1\frac{1}{4}$ hours
  d 64 km/h
2 a i 90 km   **ii** 50 km
  b 2 hours 36 mins **d** 50 km
  c 58 km/h   **e i** 72 km   **ii** 8 km
3 a 45 km   **d** 1 hour
  b $1\frac{1}{2}$ hours   **e** 45 km/h
  c 30 km/h   **f** 36 km/h
4 a i B   **ii** B
  b i 80 km/h   **ii** 64 km/h
  c $\frac{1}{2}$ hour
  d $2\frac{3}{4}$ hour
  e 58.2 km/h (counting the stop)

Exercise 17g page 284

1 a 150 km/h   **d** 1 hour
  b $2\frac{1}{2}$ hours   **e** 1330; $2\frac{1}{2}$ hours
  c 75 km/h   **f** 60 km/h
2 a 56 miles   **c** 56 m.p.h.
  b 45 minutes **d** $37\frac{1}{3}$ m.p.h.
3 a First 34 m.p.h.; second 40 m.p.h.
  b at 11.15, 40 miles from Kingston
  c 60 miles
4 a i 0830   **ii** 1330
  b 5h
  c $1\frac{1}{2}$ hours
  d 4 km/h
  e 7 hours

5 a 80 km/h, 1430
  b 100 km/h, 1354
  c at 1∡10, 153 miles from A
  d 44 miles
6 a Betty, Chris, Audrey
  b 10 km/h
  c 15 km/h
  d 20 km/h
  e at 2.30 p.m., after 25 km
  f Audrey 10 km, Betty 9 km, Chris 15 km
  g $2\frac{1}{2}$ km
7 a at 3.23 p.m., $9\frac{1}{2}$ miles from Jane's home
  b 3.8 miles
8 a at 3.14 p.m., 60 miles from A
  b 6 miles from B
  c 26 miles from A

Exercise 17h　page 288

1 a 20 km　　　**b** $2\frac{1}{2}$ hours　　**c** 8 km/h

2 a 28 km　　　**b** 1 hr 34 mins

3 a 14 hours　　**b** 57 hours
4 80 km/h
5 420 m.p.h.
6 5 km/h
7 a 15 km　　　　**d** 10 km/h
  b $1\frac{1}{2}$ hours　　**e** 45 km/h
  c 10 min

Exercise 17i　page 290

1 a 175 km　　　　**c** it stopped
  b $\frac{3}{4}$ hours　　　**d** 240 km/h
2

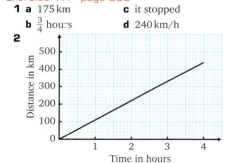

Distance in km vs Time in hours

3 a 900 m　　　　**b** 1575 m, 54 km/h

4 a 3 hours　　　**b** $1\frac{3}{4}$ hours

5 200 km/h by 5.6 m/s or 20 km/h
6 12 km/h
7 48 m.p.h.

CHAPTER 18

Exercise 18a　page 293

1 a 36 °C　　**b** 78 °C　　　**c** 77 °F　　**d** 176 °F
2 a £112　　**b** £67　　　**c** $174　　**d** $109
3 a 54%, 77%　　　　　**b** $32\frac{1}{2}$, 52
4 a $4375　　**b** $8400　　**c** $11 725
  d $11 429　　**e** $25 143
5 a 34 mpg　**b** 22 km/l　　**c** 64 mpg
  d 8 km/l (to nearest unit)
6 a 39 m/s　**b** 166 km/h　　**c** 65 km/h
  d 49 m/s (to nearest unit)
7 a 9.5 cm　**b** 5.8 cm　　　**c** 6.5 cm　　**d** 9.2 cm

Exercise 18b　page 297

1 a 1290 g　　　　　　**b** 7 mm
2 a i $8\frac{3}{4}$ s　　　　**ii** $15\frac{1}{2}$ s
  b i 136 km/h　　　**ii** 191 km/h
3 a i 290 g　　　　　**ii** 930 g
  b i 65 days　　　　**ii** 150 days
  c 240 g
  d 20 g
4 a 84 m/s when $t = 4.55$
  b i 81 m/s　　　　**ii** 61.5 m/s
  c 2.25 s and 6.6 s
5 a 19 knots, £16.40
  b 14.5 knots and 24.2 knots
  c i £17.57　　　　**ii** £17.04
6 a i 386 g　　　　　**ii** 1340 g
  b i 3.82 cm　　　　**ii** 5.5 cm
7 a 28.8 °C, 27.7 °C
  b 1120 a.m., 7.40 p.m.
8 a 1731　　　**b** November, December
9 a i 1.7 cm　　　**ii** 10 cm
  b i 1.3 cm　　　**ii** 8.6 cm

CHAPTER 19

Exercise 19a　page 302

1 a 2　　　**b** 3　　　　**c** 7　　　**d** 12
2 a −1　　**b** −6　　　　**c** −8　　**d** −20
3 a $-3\frac{1}{2}$　**b** $4\frac{1}{2}$　　　**c** −6.1　**d** 8.3
4 a −7　　**b** 2　　　　**c** $-5\frac{1}{2}$　**d** 4.2
5 a 10　　**b** −8　　　　**c** 7　　**d** −5.2
6 a −1　　**b** 3　　　　**c** −2　　**d** $\frac{4}{3}$
7 a 12　　**b** −24　　　**c** 1　　**d** −16.4
8 a −2　　**b** 4　　　　**c** $-\frac{3}{2}$　**d** $\frac{3}{4}$
9 $a = -5, b = 3, c = -4$
10 $a = -2, b = 8, c = 18$
11 $y = 3x$　　　　　　**13** $y = -\frac{1}{3}x$
12 $y = -2x$　　　　　**14** $y = \frac{2}{3}x$
15 (−2, −4), (6, 12)
16 (−2, 6), (1, −3), (8, −24)
17 a above (2, 2), (−2, 1), (−4.2, −2)
  b below (3, 0)

Exercise 19b　page 305

1–6

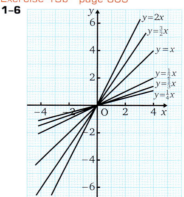

7–12

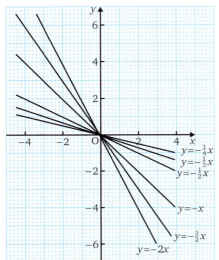

Exercise 19c   page 306
1 a 2   **b** 2   **c** 2
2 a −4   **b** −4   **c** −4
3 a 3   **b** 3   **c** 3
4 a −4   **b** −4   **c** −4
5 2.5
6 −0.5
7 a +   **c** +   **e** −
  b −   **d** −   **f** −

Exercise 19d   page 308
1 $y = 5x$

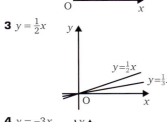

2 $y = 5x$

3 $y = \frac{1}{2}x$

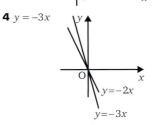

4 $y = -3x$

5 $y = 10x$

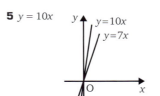

6 $y = -\frac{1}{2}x$

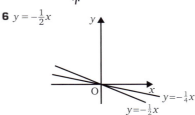

7 $y = -6x$

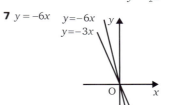

8 $y = 0.75x$

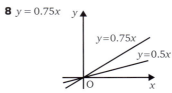

9 acute   **13** acute   **17** obtuse
10 obtuse   **14** acute   **18** obtuse
11 obtuse   **15** acute   **19** obtuse
12 acute   **16** acute   **20** obtuse
21 approximately $\frac{1}{3}$, 1, $-\frac{2}{3}$, 0

Exercise 19e   page 310
1 gradient   3, y-intercept   1,   **a** −5   **b** 7
2 gradient   −3, y-intercept   4,   **a** 7   **b** −5
3 gradient $\frac{1}{2}$, y-intercept   4,   **a** 3   **b** 4
4 gradient   1, y-intercept   −3,   **a** 7   **b** −2
5 gradient $\frac{3}{4}$, y-intercept   3,   **a** 4   **b** 2
6 gradient   2, y-intercept   −2
7 gradient   −2, y-intercept   4
8 gradient   3, y-intercept   −4
9 gradient $\frac{1}{2}$, y-intercept   3
10 gradient $-\frac{3}{2}$, y-intercept   3
11 gradient 2, y-intercept   5
12 gradient −2, y-intercept   −7
13 gradient −3, y-intercept   +2

Exercise 19f   page 312
1 $m = 4, c = 7$ **5** $m = 7, c = 6$
2 $m = \frac{1}{2}, c = -4$ **6** $m = \frac{2}{5}, c = -3$
3 $m = 3, c = -2$ **7** $m = \frac{3}{4}, c = 7$
4 $m = -4, c = 5$ **8** $m = -3, c = 4$

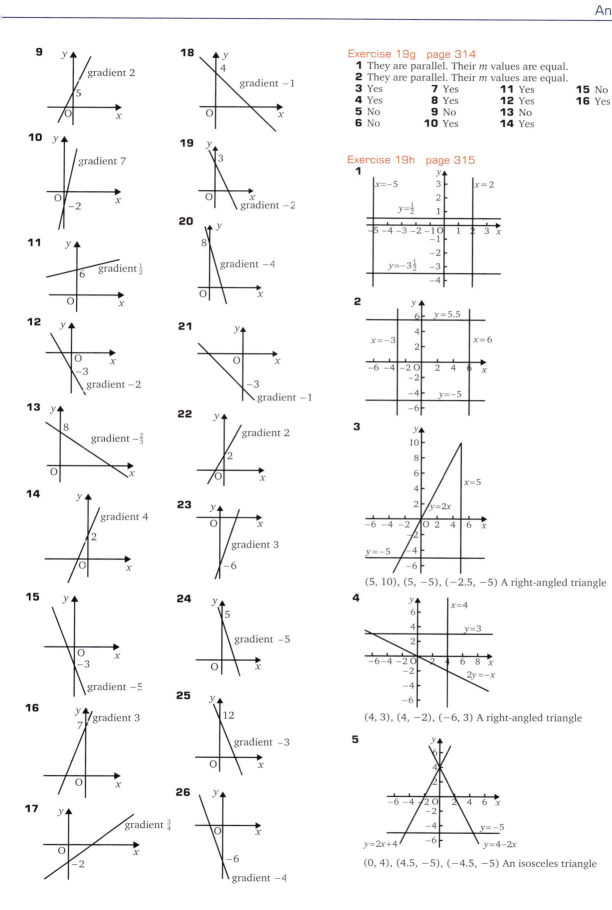

9 gradient 2, 5

10 gradient 7, −2

11 gradient $\frac{1}{2}$, 6

12 gradient −2, −3

13 gradient −$\frac{2}{3}$, 8

14 gradient 4, 2

15 gradient −5, −3

16 gradient 3, 7

17 gradient $\frac{3}{4}$, −2

18 gradient −1, 4

19 gradient −2, 3

20 gradient −4, 8

21 gradient −1, −3

22 gradient 2, 2

23 gradient 3, −6

24 gradient −5, 5

25 gradient −3, 12

26 gradient −4, −6

Exercise 19g   page 314

1 They are parallel. Their m values are equal.
2 They are parallel. Their m values are equal.

3 Yes	**7** Yes	**11** Yes	**15** No
4 Yes	**8** Yes	**12** Yes	**16** Yes
5 No	**9** No	**13** No	
6 No	**10** Yes	**14** Yes	

Exercise 19h   page 315

1

2

3

$(5, 10), (5, -5), (-2.5, -5)$ A right-angled triangle

4

$(4, 3), (4, -2), (-6, 3)$ A right-angled triangle

5

$(0, 4), (4.5, -5), (-4.5, -5)$ An isosceles triangle

407

Exercise 19i   page 317

1 $x = 1.5, y = 4.5$
2 $x = 1\frac{1}{3}, y = 3\frac{2}{3}$
3 $x = 1\frac{1}{3}, y = 5\frac{1}{2}$
4 $x = -\frac{1}{2}, y = 1\frac{1}{2}$
5 $x = -\frac{1}{2}, y = 2$
6 $x = 1\frac{1}{2}, y = 3\frac{1}{2}$
7 $x = 2\frac{2}{5}, y = \frac{9}{10}$
8 $x = -\frac{2}{5}, y = 1\frac{3}{5}$
9 $x = 2.4, y = 1.2$
10 $x = \frac{1}{3}, y = 1\frac{2}{3}$

Exercise 19j   page 317

1–4 Each pair of equations gives parallel lines. Parallel lines do not intersect, so there are no solutions.

Exercise 19k   page 318

1 $x + y = 8, 2x - y = 1$
2 $2x - y = 5, x + 2y = 10$
3 $2x + 2y = 12, x = 2y$
4 $x + y = 18, x = 2y$
5 $8 = 2m + c, 2 = -m + c$
6 $x = 2y, x + y = 90$

Exercise 19l   page 319

1 a 2   **b** -4 **c** $\frac{2}{3}$
2 a $= -4$,   **b** $= \frac{1}{3}$,   **c** $= -1.5$
3 a $+$ **b** $-$ **c** $+$
4

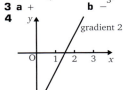
gradient 2

5 a obtuse   **c** obtuse
  b acute   **d** obtuse
6 $(-1, -6)$

Exercise 19m   page 320

1 a 10   **b** 15   **c** $\frac{5}{2}$
2 $a = -5, b = 3, c = -4$
3 a $+$ **b** $-$ **c** $-$
4 a gradients 4, y-intercept -7
  b gradient $\frac{5}{2}$, y-intercept 1
  c gradient 3, y-intercept 2
  d gradient$-\frac{1}{3}$, y-intercept-4
5 a Yes   **b** No
6

$(-3, 4)$, $(8, 4)$, $\left(-3, -1\frac{1}{2}\right)$

Exercise 19n   page 320

1 a 11   **b** -10 **c** -31
2 $a = -4, b = 11, c = 5$
3

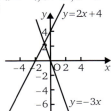

4

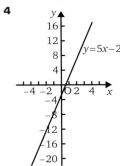

gradient 5
y intercept -2

5 a $y = 2x - 4$
  b $2y = x + 10$
  c $y = -4x - 3$
6

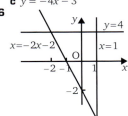

$(1, 4)$, $(1, -4)$, $(-3, 4)$
7 They all pass through the point $(0, 4)$.
8 They all pass through the point $(-1, -6)$.

CHAPTER 20

Exercise 20a   page 323

1
7

2
4

3
-2

4
0

5
-2

6
$\frac{1}{2}$

7
5

8
0

9
1.5

10 a 2, 3, 4, 6, 7   **b** 2, 5, 7, 8, 9
   c 2, 3, 7, 9   **d** 2, 3, 4, 6, 7
   e 2, 3, 4, 7, 9
12 a $5 > 3$; yes   **b** $1 > -1$; yes
   c $-2 > -4$; yes   **d** $7 > 5$; yes
13 a $0 > -1$; yes   **b** $-4 > -5$; yes
   c $-7 > -8$; yes   **d** $2 > 1$; yes
14 a $1 < 6$; yes   **b** $-3 < 2$; yes
   c $-6 < -1$; yes   **d** $3 < 8$; yes

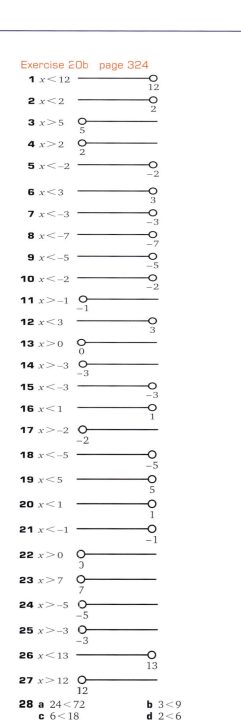

Exercise 20b   page 324

1 $x < 12$
2 $x < 2$
3 $x > 5$
4 $x > 2$
5 $x < -2$
6 $x < 3$
7 $x < -3$
8 $x < -7$
9 $x < -5$
10 $x < -2$
11 $x > -1$
12 $x < 3$
13 $x > 0$
14 $x > -3$
15 $x < -3$
16 $x < 1$
17 $x > -2$
18 $x < -5$
19 $x < 5$
20 $x < 1$
21 $x < -1$
22 $x > 0$
23 $x > 7$
24 $x > -5$
25 $x > -3$
26 $x < 13$
27 $x > 12$

28 **a** $24 < 72$ **b** $3 < 9$
   c $6 < 18$ **d** $2 < 6$
   e $-24 < -72$ **f** $-4 < -12$
   a Yes   **b** Yes   **c** Yes
   d Yes   **e** No   **f** No
29 **a** $72 > -24$ **b** $9 > -3$
   c $18 > -6$ **d** $6 > -2$
   e $-72 > 24$ **f** $-12 > 4$
   a Yes   **b** Yes   **c** Yes
   d Yes   **e** No   **f** No
30 **a** $-36 < -12$ **b** $-4\frac{1}{2} < -1\frac{1}{2}$
   c $-9 < -3$ **d** $-3 < -1$
   e $36 < 12$ **f** $6 < 2$
31 **a** Yes   **b** Yes   **c** Yes
   d Yes   **e** No   **f** No

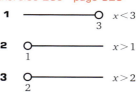

32 only when you are multiplying by a positive number.

Exercise 20c   page 325

1   $x < 3$
2   $x > 1$
3   $x > 2$
4   $x < 1$
5   $x < \frac{1}{2}$
6   $x > 1\frac{1}{3}$
7   $x < 2\frac{1}{4}$
8   $x > 1\frac{1}{2}$
9   $x \leqslant 1$
10 $x \leqslant 4$
11 $x \geqslant -2$
12 $x \geqslant 1$
13 $x < -1$
14 $x \leqslant 2$
15 $x > 1$
16 $x \geqslant 1\frac{1}{3}$
17 $x \geqslant 0$
18 $x \leqslant 1$
19 $x < 1$
20 $x < -3$

21 **a** $x > 3$ **b** $2 \leqslant x \leqslant 3$
   c No values of x
22 **a** $0 < x < 1$ **b** $x \leqslant 0$
   c No values of x
23 **a** $-2 < x \leqslant 4$ **b** No values of x
   c $x < -2$
24 **a** $-3 < x < -1$ **b** $x < -3$
   c No values of x
25 $x < 12$; $x > -1$; $-1 < x < 12$
26 $x \leqslant -1$; $x \geqslant 3$; no values of x
27 $x \leqslant 7$; $x \geqslant -2$; $-2 \leqslant x \leqslant 7$

28 $x > 1$; $x < 2$; $1 < x < 2$
29 $x > 2$; $x < 3$; $2 < x < 3$
30 $x < 2$; $x > -1$; $-1 < x < 2$
31 $x \geqslant -1$; $x < 2$; $-1 \leqslant x < 2$
32 $x > \frac{1}{2}$; $x \leqslant 3$; $\frac{1}{2} < x \leqslant 3$

33 $2 < x < 5$　　**38** $-4 < x < 2$
34 $-3 \leqslant x \leqslant 2$　　**39** $x < -3$
35 $x < -2$　　**40** $x < -1$
36 $0 < x < 2$　　**41** $1\frac{4}{5} < x < 3$
37 $x \geqslant 1$　　**42** $\frac{1}{2} < x < 1$

Exercise 20d　page 329

1
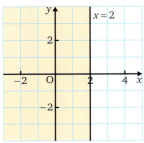

2
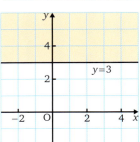

3
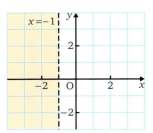

4
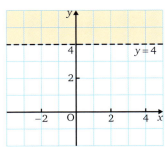

5
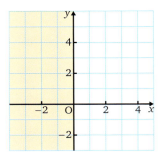

6
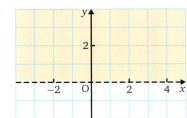

7
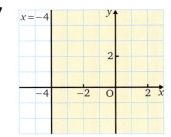

8
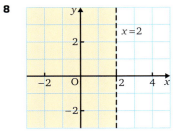

9
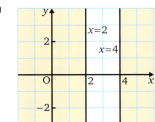

10
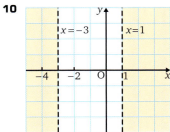

11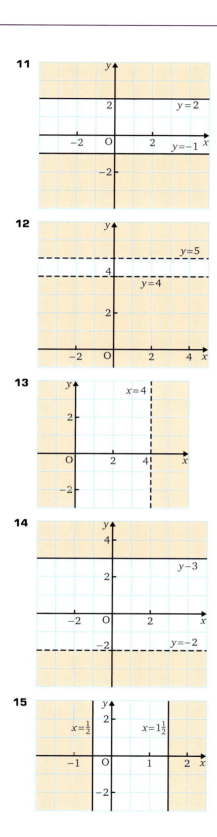

12

13

14

15

16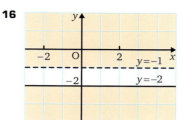

17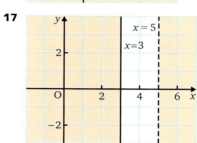

18 9 : No   10 : No   11 : No
19 $x \leqslant 2$　　　　**21** $x < -1$　　　**23** $-1 \leqslant x < 2$
20 $y < 3$　　　　**22** $-2 \leqslant y \leqslant 2$　　**24** $-\frac{1}{2} < y < 2\frac{1}{2}$
25 19 : Yes   20 : Yes   21 : No   22 : Yes   23 : No   24 : No
26 $-3 \leqslant x \leqslant 1$
27 $-4 \leqslant y \leqslant -1$
28 $2 \leqslant y < 3$
29 $3 \leqslant x \leqslant 6$
30 26 : Yes   27 : No   28 : Yes   29 : No

Exercise 20e   page 332

1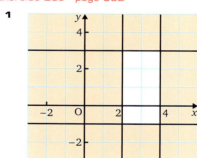

2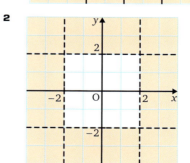

411

3

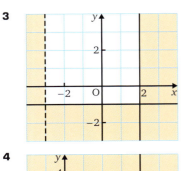

4

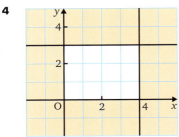

5

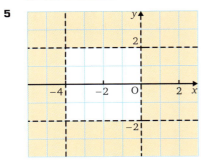

6

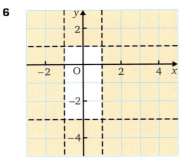

7

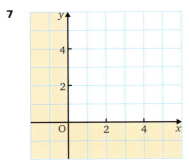

8

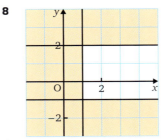

9 $-2 \leqslant x \leqslant 3, -1 \leqslant y \leqslant 2$
10 $-2 < x \leqslant 2, -2 \leqslant y \leqslant 1$
11 9: Yes   10: No
12 $-2 \leqslant x \leqslant 1, y > -1$
13 $x < 0, y > 0$

CHAPTER 21

Exercise 21a   page 335

1 a i 11 **ii** 15 **b i** 12 **ii** 16 **c** 2 **d i** 4 **ii** 1 **e** 163
2 a i 30 cm **ii** 55 cm **b i** 10 cm **ii** 40 cm **c** 2nd
3 a i $22 500 000 **ii** $20 000 000
   b i $27 500 000 **ii** $17 500 000
   c i 4th **ii** 4th in 2006
   d Greatest in 4th quarter then decreases.
   e Concerned, the trend in sales is down.
4 a Calcutta **b i** 32.5 cm **ii** 1 cm   **c i** 23 cm **ii** 5 cm
   d i 4 cm **ii** 14.5 cm
   e Calcutta between November and April
5 a i 9°C **ii** 11°C   **b** 8°C   **c i** January **ii** May   **d** 3
6 b Yes, profits are increasing
7 b e.g. Falling unemployment for males, recent increase in
   unemployment for females
   c any reasonable suggestion
8 b New York. Greater range of temperatures
   c e.g. June to October, any reasonable suggestion

Exercise 21b   page 340

1 8
2 7
3 15
4 29
5 16
6 28
7 3
8 40
9 6.2
10 3.5
11 50
12 0.62
13 63
14 96
15 16.5
16 63
17 74

18 1.35
19 0.875
20 5.8
21 2 mm
22 86 kg, 81 kg
23 1837 miles
24 2583 km
25 72 mm
26 131.6 hours
27 134
28 $110
29 92
30 9
31 61, 21
32 233, 193
33 106, 238

34 11.8 hours, 5.6 hours
35 68; reduces it to 67
36 158 cm; increases it to 159 cm
37 63 610, 12 722, 8294
38 136.4 kg

39 160.6 cm
40 55.6 kg
41 26
42 285 cm
43 2652

Exercise 21c   page 345

1 12
2 9
3 1.8
4 56
5 5.9
6 26.4
7 1
8 8
9 155 cm
10 31, 3
11 36, 6

Exercise 21d   page 347

1 5 **5** 3.2 **9** 1.885
2 42 **6** 12 **10** 15
3 17 **7** 98
4 16 **8** 36

Exercise 21e   page 348

1 a 23 **b** 21 **c** 21
2 a 71 **b** 66, 67 **c** 69
3 a 45 **b** 43 **c** 45
4 a 43 **b** 13 **c** 32
5 a 28 **b** 27 **c** 27
6 77, 72, 73
7 a 157 cm **b** 157 cm **c** 157 cm
8 a 54 **b** 52 **c** 52
9 83, 84, 83.5
10 a 0 **b** 0 **c** 1.5
11 a 3 **b** 3.5 **c** 3.77

Exercise 21f   page 350

1 2.00 (to 3 s.f.)
2 3.8
3 a 4.45 **b** 5
4 a 3.64 **b** 6
5 a 1.57 **b** 1

Exercise 21g   page 351

1 2
2 1
3 3.5
4 a 3.5 **b** 3 **c** 3.48
5 a 2 **b** 1 **c** 2.2

CHAPTER 22

Exercise 22a   page 355

1 a {teachers in my school}
  b {books I have read}
3 a odd numbers up to 9
  b the days of the week from Monday to Friday
4 a {European countries}, France
  b {multiples of 10}, 60
5 John ∈ {boys' names}
6 English ∈ {school subjects}
7 June ∉ {days of the week}
8 Monday ∉ {domestic furniture}
9 false **10** true **11** true **12** true

Exercise 22b   page 356

1 infinite **5** 5 **9** 11 **13** no
2 infinite **6** 8 **10** no **14** yes
3 finite **7** 6 **11** yes **15** yes
4 infinite **8** 21 **12** yes **16** no

Exercise 22c   page 357

1 cutlery
2 whole numbers less than 50
3 whole numbers less than 25
4 $n(A) = 8$, $n(B) = 6$
  $B = \{3, 6, 9, 12, 15, 18\}$
5 $A = \{1, 2, 3, 4, 6, 12\}$ $B = \{2, 3, 5, 7, 11, 13\}$
  $C = \{6, 12\}$
6 $n(A) = 4$, $n(B) = 2$, $n(C) = 5$

Exercise 22d   page 358

1 {Audrey, Janet}, {Audrey, Jill}, {Jill, Janet}
2 $A = \{2, 4, 6, 8, 10, 12, 14\}$,

$B = \{2, 3, 5, 7, 11, 13\}$,
$C = \{3, 6, 9, 12, 15\}$
yes, 3
3 $B = \{6, 12, 18\}$, $C = \{2\}$,
  $D = \{13, 14, 15, 16, 17, 18, 19, 20\}$

Exercise 22e   page 360

1 Your own answers.
2 my friends who do not like coming to my school
3 my friends who like coming to my school
4 all pupils at my school except those who are not my friends and like coming to school
5 My friends who like coming to school and the pupils who are not my friends who do not like coming to school.

Exercise 22f   page 361

1
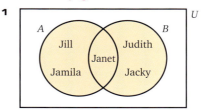

$A \cup B = \{\text{Janet, Jill, Jamila, Judith, Jacky}\}$

2
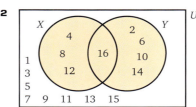

$X \cup Y = \{2, 4, 6, 8, 10, 12, 14, 16\}$

3
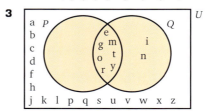

$P \cup Q = \{\text{e, g, i, m, n, o, t, r, y}\}$

4 a
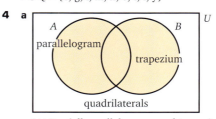

$A \cup B = \{\text{all parallelograms and trapeziums}\}$

b
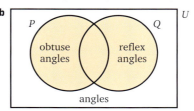

$P \cup Q = \{\text{angles that are either obtuse or reflex}\}$

5

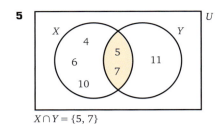

$X \cap Y = \{5, 7\}$

6

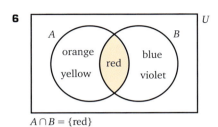

$A \cap B = \{red\}$

7

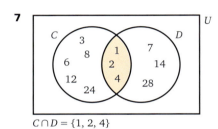

$C \cap D = \{1, 2, 4\}$

8

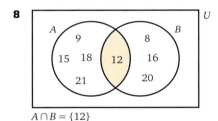

$A \cap B = \{12\}$

Exercise 22g   page 362

1 a Lenny, Sylvia
  b Adam, Richard
  c Jack, Scott, Lee
2 a David, Joe, Tariq, Paul
  b Tariq, Paul
  c Claude, Alan, Clive
3 a Emma, Majid, Clive, Sean, Ann
  b Emma, Majid, Clive
  c Sean, Ann

4

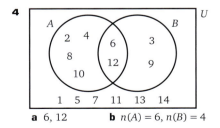

  a 6, 12   **b** $n(A) = 6, n(B) = 4$

5

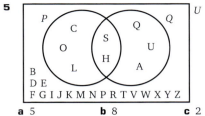

  a 5   **b** 8   **c** 2

6 a 26, 6, 6
  b

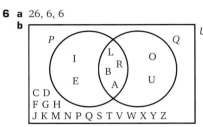

  i 4   **ii** 8 letters in both Liberal and Labour;
                   letters in either Liberal or Labour
                   (or both)

7 a 11, 5, 6
  b

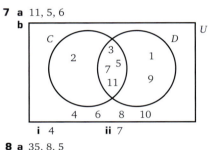

  i 4   **ii** 7

8 a 35, 8, 5
  b i 2   **ii** 11
9 10

CHAPTER 23

Exercise 23a   page 367

1 213_5 **2** 2014_5 **3** 41240_5 **4** 30201_5

	5^3	5^2	5	unit
5			3	1
6			4	2
7		4	1	0
8		2	3	1
9			3	4
10			1	0
11		2	0	4
12		4	0	0

13 16_{10} **19** 17_{10} **25** 13_5 **31** 12_5
14 14_{10} **20** 10_{10} **26** 23_5 **32** 41_5
15 20_{10} **21** 4_{10} **27** 20_5 **33** 110_5
16 36_{10} **22** 100_{10} **28** 124_5 **34** 1003_5
17 54_{10} **23** 70_{10} **29** 133_5 **35** 312_5
18 23_{10} **24** 75_{10} **30** 1100_5 **36** 400_5

Exercise 23b   page 370

1 43_5 **7** 1113_5 **13** 31_5 **19** 214_5
2 104_5 **8** 1024_5 **14** 14_5 **20** 40_5
3 104_5 **9** 410_5 **15** 10_5 **21** 102_5
4 112_5 **10** 1003_5 **16** 4_5 **22** 34_5
5 304_5 **11** 1114_5 **17** 101_5 **23** 143_5
6 344_5 **12** 1201_5 **18** 103_5 **24** 44_5

25 22_5 **32** 233_5 **39** 332_5
26 31_5 **33** 13_5 **40** 13_5
27 44_5 **34** 234_5 **41** 114_5
28 132_5 **35** 1013_5 **42** 134_5
29 $204_5 - 121_5$ **36** 204_5 **43** 222_5
30 231_5 **37** 1322_5 **44** 1012_5
31 242_5 **38** 1013_5
45 a and **c** base 5, **b** and **d** base 10.

Exercise 23c   page 372

1 1010   **3** 110001   **5** 1011   **7** 10111
2 101010   **4** 11100   **6** 10100   **8** 111001

9

	$64\,(2^6)$	$32(2^5)$	$16\,(2^4)$	$8(2^3)$	$4\,(2^2)$	2	units
a				1	1	0	1
b		1	0	0	0	0	1
c		1	1	0	1	0	1
d		1	0	1	0	1	1
e	1	0	1	0	1	0	1

10 a 23   **b** 5   **c** 15   **d** 57   **e** 119
11 a 11   **b** 27   **c** 14   **d** 60
    e 49   **f** 87   **g** 44   **h** 94
12 a 110   **b** 1101   **c** 111000   **d** 110101
13 a 1100   **b** 10011   **c** 11001
    d 100110   **e** 111100
14 a 10111   **b** 11011   **c** 100010   **d** 100110
    e 1011011

Exercise 23d   page 374

1 a 1100   **b** 10001   **c** 11101
2 a 10100   **b** 101111   **c** 1000100
3 a 1011   **b** 10011   **c** 10101
4 a 100011   **b** 100111   **c** 101110
5 a 10110101   **b** 1111000
6 a 101   **b** 1001   **c** 111
7 a 1000   **b** 1011   **c** 1110
8 a 10001   **b** 1110   **c** 1111
9 a 11001   **b** 100111   **c** 1001101
10 a 100011   **b** 10111101   **c** 100101001
11 a 100   **b** 10101   **c** 111111
12 11, 101, 111, 1001, 1011. Last digit is 1
13 10, 100, 110, 1000, 1010, 1100. Last digit is 0
14 b, **c** and **e**
15 a divisible by 4   **b** divisible by 8
16 a 1111   **b** 1000
17 1000, 1100, 11100
18 1000, 10000, 1100100
19 10100, 101000, 1010000
20 11001, 110010, 1100100
21 a 33_5 **b** 140_5 **c** 22_5 **d** 104_5
22 a 10001 **b** 1001   **c** 10100 **d** 1101
    e 1001110   **f** 1000101
23 111110_2

Exercise 23e   page 376

1 a 1111   **b** 11100   **c** 100010   **d** 110101
2 $300 = 2^8 + 2^5 + 2^3 + 2^2$, $300_{10} = 100101100_2$
3 5 i.e. 0, 1, 2, 3, 4.
4 a 14_5 **b** 31_5 **c** 134_5 **d** 1013_5
5 a 1100000   **b** 11010   **c** 101000101
6 a 1204_5 **b** 3_5 **c** 2401_5
7 a 1000000   **b** 11111
8 a 4420_5 **b** 444_5
9 0
10 a 4. $5^3 = 1000_5$ **b** 125   **c** 1111101_2

REVIEW TEST 3   page 378

1 A   **6** C
2 A   **7** D
3 B   **8** D
4 D   **9** B
5 B
10 a i $2\,000\,000\,\text{cm}^3$ **ii** $0.425\,\text{cm}^3$ **iii** $4500\,\text{cm}^3$
    b $105\,000\,\text{cm}^3$
11 $49.82\,\text{cm}^2$ **12** $480\,\text{cm}^3$
13 a i 11.6964   **ii** 0.4   **iii** 2.43
    b i 53   **ii** 7   **iii** 6.17 (2 d.p.)
14 missing values are 60, 80, 160, 200 **a** 140 cm   **b** 4.5 seconds
15 4 km/h
16 a i A = {prime numbers less than 18}
    ii B = {multiples of 3 less than 18}
    b i 11   **ii** 1
17 a 114_5 **b** 421_5 **c** 24_5 **d** 124_5 **e** 100011_2
    f 10101_2 **g** 11000100_2
18 a i 89   **ii** 22   **b i** 1134_5 **ii** 10101001

REVIEW TEST 4   page 381

1 C   **13** D   **25** C
2 A   **14** D   **26** B
3 D   **15** A   **27** B
4 D   **16** D   **28** D
5 B   **17** C   **29** B
6 B   **18** D   **30** A
7 B   **19** A   **31** C
8 D   **20** B   **32** D
9 B   **21** B   **33** D
10 A   **22** B   **34** D
11 D   **23** D   **35** B
12 B   **24** D

Index